International Review of Science

# Physical Chemistry
# Series Two

Consultant Editor
**A. D. Buckingham, F.R.S.**

# Publisher's Note

The International Review of Science is an important venture in scientific publishing, presented by Butterworths. The basic concept of the Review is to provide regular authoritative reviews of entire disciplines. Chemistry was taken first as the problems of literature survey are probably more acute in this subject than in any other. Biochemistry and Physiology followed naturally. As a matter of policy, the authorship of the Review of Science is international and distinguished, the subject coverage is extensive, systematic and critical.

The Review has been conceived within a carefully organised editorial framework. The overall plan was drawn up and the volume editors appointed by seven consultant editors. In turn, each volume editor planned the coverage of his field and appointed authors to write on subjects which were within the area of their own research experience. No geographical restriction was imposed. Hence the 500 or so contributions to the Review of Science come from many countries of the world and provide an authoritative account of progress. The publication of Physical Chemistry Series One was completed in 1973 with thirteen text volumes and one index volume; in accordance with the stated policy of issuing regular reviews to keep the series up to date, volumes of Series Two will be published between the middle of 1975 and early 1976; Series Two of Organic Chemistry will be published at the same time, while Inorganic Chemistry Series Two was published during the first half of 1975. Volume titles are the same as in Series One but the articles themselves either cover recent advances in the same subject or deal with a different aspect of the main theme of the volume. In Series Two an index is incorporated in each volume and there is no separate index volume.

Butterworth & Co. (Publishers) Ltd.

## PHYSICAL CHEMISTRY SERIES TWO

Consultant Editor
A. D. Buckingham, F.R.S.,
*Department of Chemistry University of Cambridge*

*Volume titles and Editors*

1 **THEORETICAL CHEMISTRY**
Professor A. D. Buckingham, F.R.S., *University of Cambridge* and Professor C. A. Coulson, F.R.S., *University of Oxford*

2 **MOLECULAR STRUCTURE AND PROPERTIES**
Professor A. D. Buckingham, F.R.S., *University of Cambridge*

3 **SPECTROSCOPY**
Dr. D. A. Ramsay, F.R.S.C., *National Research Council of Canada*

4 **MAGNETIC RESONANCE**
Professor C. A. McDowell, F.R.S.C., *University of British Columbia*

5 **MASS SPECTROMETRY**
Professor A. Maccoll, *University College, University of London*

6 **ELECTROCHEMISTRY**
Professor J. O'M Bockris, *The Flinders University of S. Australia*

7 **SURFACE CHEMISTRY AND COLLOIDS**
Professor M. Kerker, *Clarkson College of Technology, New York*

8 **MACROMOLECULAR SCIENCE**
Professor C. E. H. Bawn, C.B.E., F.R.S., *formerly of the University of Liverpool*

9 **CHEMICAL KINETICS**
Professor D. R. Herschbach *Harvard University*

10 **THERMOCHEMISTRY AND THERMO-DYNAMICS**
Dr. H. A. Skinner, *University of Manchester*

11 **CHEMICAL CRYSTALLOGRAPHY**
Professor J. Monteath Robertson, C.B.E., F.R.S., *formerly of the University of Glasgow*

12 **ANALYTICAL CHEMISTRY —PART 1**
Professor T. S. West, *Imperial College, University of London*

13 **ANALYTICAL CHEMISTRY —PART 2**
Professor T. S. West, *Imperial College, University of London*

---

## INORGANIC CHEMISTRY SERIES TWO

Consultant Editor
H. J. Emeléus, C.B.E., F.R.S.
*Department of Chemistry University of Cambridge*

*Volume titles and Editors*

1 **MAIN GROUP ELEMENTS—HYDROGEN AND GROUPS I–III**
Professor M. F. Lappert, *University of Sussex*

2 **MAIN GROUP ELEMENTS—GROUPS IV AND V**
Dr. D. B. Sowerby, *University of Nottingham*

3 **MAIN GROUP ELEMENTS—GROUPS VI AND VII**
Professor V. Gutmann, *Technical University of Vienna*

4 **ORGANOMETALLIC DERIVATIVES OF THE MAIN GROUP ELEMENTS**
Professor B. J. Aylett, *Westfield College, University of London*

5 **TRANSITION METALS—PART 1**
Professor D. W. A. Sharp, *University of Glasgow*

6 **TRANSITION METALS—PART 2**
Dr. M. J. Mays, *University of Cambridge*

7 **LANTHANIDES AND ACTINIDES**
Professor K. W. Bagnall, *University of Manchester*

8 **RADIOCHEMISTRY**
Dr. A. G. Maddock, *University of Cambridge*

9 **REACTION MECHANISMS IN INORGANIC CHEMISTRY**
Professor M. L. Tobe, *University College, University of London*

10 **SOLID STATE CHEMISTRY**
Dr. L. E. J. Roberts, *Atomic Energy Research Establishment, Harwell*

---

## ORGANIC CHEMISTRY SERIES TWO

Consultant Editor
D. H. Hey, F.R.S., *formerly of the Department of Chemistry, King's College, University of London*

*Volume titles and Editors*

1 **STRUCTURE DETERMINATION IN ORGANIC CHEMISTRY**
Professor L. M. Jackman, *Pennsylvania State University*

2 **ALIPHATIC COMPOUNDS**
Professor N. B. Chapman, *Hull University*

3 **AROMATIC COMPOUNDS**
Professor H. Zollinger, *Eidgenossische Technische Hochschule, Zurich*

4 **HETEROCYCLIC COMPOUNDS**
Dr. K. Schofield, *University of Exeter*

5 **ALICYCLIC COMPOUNDS**
Professor D. Ginsburg, *Technion-Israel Institute of Technology, Haifa*

6 **AMINO ACIDS, PEPTIDES AND RELATED COMPOUNDS**
Professor N. H. Rydon, *University of Exeter*

7 **CARBOHYDRATES**
Professor G. O. Aspinall, *York University, Ontario*

8 **STEROIDS**
Dr. W. F. Johns, *G. D. Searle & Co., Chicago*

9 **ALKALOIDS**
Professor K. Wiesner, F.R.S., *University of New Brunswick*

10 **FREE RADICAL REACTIONS**
Professor W. A. Waters, F.R.S., *formerly of the University of Oxford*

**BIOCHEMISTRY SERIES ONE**

Consultant Editors
H. L. Kornberg, F.R.S.
*Department of Biochemistry University of Leicester* and
D. C. Phillips, F.R.S., *Department of Zoology, University of Oxford*

*Volume titles and Editors*

1 **CHEMISTRY OF MACRO-MOLECULES**
Professor H. Gutfreund, *University of Bristol*

2 **BIOCHEMISTRY OF CELL WALLS AND MEMBRANES**
Dr. C. F. Fox, *University of California, Los Angeles*

3 **ENERGY TRANSDUCING MECHANISMS**
Professor E. Racker, *Cornell University, New York*

4 **BIOCHEMISTRY OF LIPIDS**
Professor T. W. Goodwin, F.R.S., *University of Liverpool*

5 **BIOCHEMISTRY OF CARBO-HYDRATES**
Professor W. J. Whelan, *University of Miami*

6 **BIOCHEMISTRY OF NUCLEIC ACIDS**
Professor K. Burton, F.R.S., *University of Newcastle upon Tyne*

7 **SYNTHESIS OF AMINO ACIDS AND PROTEINS**
Professor H. R. V. Arnstein, *King's College, University of London*

8 **BIOCHEMISTRY OF HORMONES**
Professor H. V. Rickenberg, *National Jewish Hospital & Research Center, Colorado*

9 **BIOCHEMISTRY OF CELL DIFFER-ENTIATION**
Professor J. Paul, *The Beatson Institute for Cancer Research, Glasgow*

10 **DEFENCE AND RECOGNITION**
Professor R. R. Porter, F.R.S., *University of Oxford*

11 **PLANT BIOCHEMISTRY**
Professor D. H. Northcote, F.R.S., *University of Cambridge*

12 **PHYSIOLOGICAL AND PHARMACO-LOGICAL BIOCHEMISTRY**
Dr. H. K. F. Blaschko, F.R.S., *University of Oxford*

---

**PHYSIOLOGY SERIES ONE**

Consultant Editors
A. C. Guyton,
*Department of Physiology and Biophysics, University of Mississippi Medical Center* and
D. F. Horrobin,
*Department of Physiology, University of Newcastle upon Tyne*

*Volume titles and Editors*

1 **CARDIOVASCULAR PHYSIOLOGY**
Professor A. C. Guyton and Dr. C. E. Jones, *University of Mississippi Medical Center*

2 **RESPIRATORY PHYSIOLOGY**
Professor J. G. Widdicombe, *St. George's Hospital, London*

3 **NEUROPHYSIOLOGY**
Professor C. C. Hunt, *Washington University School of Medicine, St. Louis*

4 **GASTROINTESTINAL PHYSIOLOGY**
Professor E. D. Jacobson and Dr. L. L. Shanbour, *University of Texas Medical School*

5 **ENDOCRINE PHYSIOLOGY**
Professor S. M. McCann, *University of Texas*

6 **KIDNEY AND URINARY TRACT PHYSIOLOGY**
Professor K. Thurau, *University of Munich*

7 **ENVIRONMENTAL PHYSIOLOGY**
Professor D. Robertshaw, *University of Nairobi*

8 **REPRODUCTIVE PHYSIOLOGY**
Professor R. O. Greep, *Harvard Medical School*

International Review of Science

# Physical Chemistry
# Series Two

## Volume 11

## Chemical Crystallography

Edited by **J. M. Robertson, C.B.E., F.R.S.**
formerly of the University of Glasgow

**Butterworths** · London and Boston

THE BUTTERWORTH GROUP

ENGLAND
Butterworth & Co (Publishers) Ltd
London: 88 Kingsway, WC2B 6AB

AUSTRALIA
Butterworths Pty Ltd
Sydney: 586 Pacific Highway 2067
Melbourne: 343 Little Collins Street, 3000
Brisbane: 240 Queen Street, 4000

CANADA
Butterworth & Co (Canada) Ltd
Toronto: 2265 Midland Avenue, Scarborough, Ontario M1P 4S1

NEW ZEALAND
Butterworths of New Zealand Ltd
Wellington: 26–28 Waring Taylor Street, 1

SOUTH AFRICA
Butterworth & Co (South Africa) (Pty) Ltd
Durban: 152–154 Gale Street

U.S.A.
Butterworth (Publishers) Inc
Reading: 161 Ash Street, Mass., 01867

Library of Congress Cataloging in Publication Data

Robertson, John Monteath.
Chemical crystallography.

(Physical chemistry, series two; v. 11) (International review of science)
Includes bibliographies and index.
1. Crystallography. I. Title. II. Series.
III. Series: International review of science.
QD450.2.P59 vol. 11 [QD951] 541′.3′08s [548′.3]
ISBN 0 408 70610 4 75–4592

First Published 1975 and © 1975



Typeset, printed and bound in Great Britain by
REDWOOD BURN LIMITED
Trowbridge & Esher

# Consultant Editor's Note

The International Review of Science was conceived as a comprehensive, critical and continuing survey of progress in research. The difficult problem of keeping up with advances on a reasonably broad front makes the idea of the Review especially appealing, and I was grateful to be given the opportunity of helping to plan it.

Physical Chemistry Series One was published in 1972/1973. Its success was assured by the very great distinction of its editors and authors. Our need for critical reviews at a high level has not diminished. In the rather difficult times being experienced in most parts of the world, research workers should seek to broaden the range of their expertise; it is hoped that this Series will be of use in this connection.

Like its forerunner, Series Two consists of thirteen volumes covering Physical and Theoretical Chemistry, Chemical Crystallography and Analytical Chemistry. Each volume has been edited by a distinguished chemist; in several cases the person responsible for Series One has acted again as editor. The editors have assembled a strong team of authors, each of whom has assessed and interpreted recent progress in a specialised field in terms of his own experience. I believe that their efforts have again produced useful and timely articles which will help us in our efforts to keep abreast of progress in research.

It is my pleasure to thank all those who have collaborated in this venture —the volume editors, the authors and the publishers.

Cambridge

A. D. Buckingham

# Preface

The field of chemical crystallography is wide and continues to expand rapidly. Several important and comprehensive reviews of recent work have been published within the past year. The first volume in a series of the Chemical Society Specialist Periodical Reports on *Molecular Structure by Diffraction Methods*, compiled by G. A. Sim, L. E. Sutton and others, is available and covers the literature from January 1971 to March 1972. This work runs to over 800 pages and has detailed references and diagrams.

The invaluable classified bibliography of organic and organometallic crystal structures prepared by the Crystallographic Data Centre, Cambridge, continues, and the fifth volume, covering the literature published during 1972–1973 has appeared. Also from the Data Centre in Cambridge, the first volume of an important work on *Molecular Structure and Dimensions* is now available. This gives stereoscopic drawings and all the numerical data for organic and organometallic structures and covers the years 1960–1965. The volume runs to 600 large pages and costs £24.50. A computerised file at the Crystallographic Data Centre contains all this information and about 1500 new entries are added annually. Further numeric volumes for the post-1965 period are planned.

In view of these publications it would be unnecessary as well as impossible to attempt anything like a complete review of chemical crystallography in the present small volume. Instead we have confined our review to a number of special topics. In the field of natural products, a condensed review of alkaloid structures is provided. Another short chapter covers the astonishing molecular rearrangements and transformations of a single sesquiterpene, caryophyllene, while Speakman deals with further aspects of hydrogen bonding. Organo-transition metal chemistry continues to be a subject of very rapid development, stimulated by the synthesis of sandwich compounds and the availability of instrumentation which allows their rapid characterisation. This subject is reviewed in an important chapter by Mason and Mingos. The determination of structure, however, is not the only and perhaps not even the most important aspect of modern crystallography. Studies of electron density are of central importance in our understanding of the nature of molecules and solids. As J. R. Platt has said, a theory of chemistry and the chemical bond is primarily a theory of electron density. It was clear from the very beginning that x-ray crystallography was potentially capable of completely defining the distribution of electron density in molecules and solids. However, for significant results the requirements of accuracy in intensity measurements are extremely stringent and low-angle scattering data must be as complete as possible. Recent advances in experimental techniques have now made some detailed studies of electron distribution feasible, and the review of this subject by Coppens is timely.

Another subject which is substantially non-structural, in the sense that it is not directly concerned with finding the positions of the atoms within a molecule, is covered in the chapter by Dunitz and Burgi on non-bonded interactions. The calculations involved in the study of such interactions are extremely formidable even when all kinds of approximations are made, and this chapter can perhaps be best described as a guide to empirical force field calculations. Applications to small and medium ring cycloalkanes and cyclo-olefins are also described, as well as a number of other topics.

Glasgow J. M. Robertson

# Contents

# 1
# Aspects of Hydrogen Bonding

**J. C. SPEAKMAN**
University of Glasgow



## 1.1 INTRODUCTION

The article on hydrogen bonding in the first volume of this series (hereinafter referred to as HB. 1) included a general elementary review of the subject with special reference to the study of O—H···O bonds by crystallographic methods. There followed a report on some of the results then recently published, coverage being to the end of 1971 approximately. In this second volume, we shall continue to be selective, concentrating on topics likely to be of more general interest from the literature to November, 1973. Many of the researches are of importance in other fields of chemistry, though they are considered here because they involve hydrogen bonding.

Mention should be made of a new general text: *Hydrogen Bonding* by Vinogradov and Linnell[1]. A supplement to Pimentel and McClellan's authoritative book (1960) has also been published[2] as well as a recent review of the theory of the hydrogen bond[3].

As before, we shall give interatomic distances in terms of the ångstrom: 1 Å = 10 pm. Estimated standard deviations of cited results are shown in parentheses; they should be accepted with the usual qualifications.

## 1.2 POLYMORPHS OF ICE

Definitive papers have now been published describing neutron diffraction (ND) work on the two forms of heavy ice, $D_2O$ (II)[4] and $D_2O$ (IX)[5]. The former is completely ordered with respect to the deuterons. (Whalley has stated[6] that were this structure disordered, its higher entropy might render it stable in conditions that could occur terrestrially. As ice (II) is denser than water, disastrous consequences might then ensue.)

Ice (IX) has now had its structure more precisely determined than that of any other form of ice (see HB. 1, Figure 1.9). It is nearly completely ordered but has perhaps 4% of its deuterons in alternative sites. In general, there is little evidence in any form of ice for notable deviation of the water molecule from its gas-phase geometry, though this is not true for water in crystalline hydrates where experimental studies have been easier and much more numerous (see Section 1.3). In $D_2O$ (IX), however, there is probably a significant lengthening of the hydrogen bonded O—D bonds. After correction for libration (see HB. 1, Section 1.6.4), O—D bond lengths average at 0.982(3) Å compared with 0.970 Å in the gas. The validity of the lengthening is supported by the O—D stretching frequency which is 2454 $cm^{-1}$ in $D_2O$ (IX) as against 2727 $cm^{-1}$ in gaseous HOD. On the other hand, and more surprisingly, the D—O—D angles are not significantly below 104.5°, the gas-phase value, despite the circumstance that the acceptor angles, $O^1 \cdots D—O—D \cdots O^2$, are 98° and 101° for the respective molecules in $D_2O$ (IX).

## 1.3 THE WATER MOLECULE IN SALT AND OTHER CRYSTALLINE HYDRATES

In salt hydrates, the water molecule engages in one, or more usually both, of two bonding roles: it directs one — sometimes both — of its lone pairs of electrons towards the (usually metallic) cation; or it donates its protons to form hydrogen bonds with the anion, either directly if this is a monatomic ion, or to the oxygen atoms of oxy anions. It may also be hydrogen-bonded to other water molecules. The possibilities have been classified in more detail[7].

To understand such bonding in hydrates, we need to know the positions of the hydrogen atoms, preferably by ND but alternatively by less direct inference[8]. Ferraris and Franchini-Angela[9] have made a detailed statistical survey of some 90 water molecules whose structures have been determined accurately by ND in about 40 different crystalline hydrates. Amongst other features, the geometry of the hydrogen-bonded molecule (1) has been

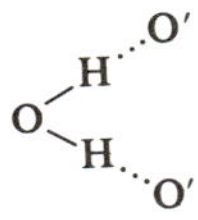

(1)

examined. On an average, the water molecule in a hydrate has the dimensions, O—H = 0.96, O···O′ = 2.81, H···O′ = 1.88 Å, H—O—H = 108°, O′···(HOH)···O′ = 108°. The molecule itself is not much different from the isolated molecule.

However, the average dimensions have internal standard deviations larger than those of the individual values. The scatter is significant. The interpretation is that the water of hydration is acting as a 'flexible strain-absorber'. It helps to stabilise the interactions between charged ions by relieving any imbalance of bond strengths, in the sense of Pauling's electrostatic valency principles.

One of the correlations published in this paper is a histogram of the frequencies of occurrence of various O—H···O angles. As had been found previously, 180° is not the commonest angle: there is a maximum for the range 165–170° which has nearly twice as many examples as has the range 175–180°. This has been taken by various commentators as a possible indication of a preference for O—H···O bonds that are slightly bent. It has already been suggested (see, e.g. HB. 1, Section 1.3) that this deduction may be invalid. In a bond with O—H···O = 180°, the acceptor atom must lie somewhere on the O—H line produced beyond H. For an angle of (say) 165°, the acceptor has a wider choice: it is merely required to lie somewhere on the surface of the cone generated by rotating H···O about O—H. There is more phase-space in the latter situation and this must be taken into account in assessing the meaning of the histogram[10].

The hydrogen bonds occurring in hydrates are fairly weak bonds. For such, the effectiveness of the bonding will not be sensitive to small deviations of O—H···O from strict linearity. For stronger bonds, with O···O less than 2.5 Å, linearity is probably more important; but certainly some very short OHO bonds are not exactly linear.

## 1.4 THE HYDRATED 'HYDROGEN ION'

For over half a century it has been realised that the 'hydrogen ion' of aqueous-solution chemistry must be hydrated, and the formula $H_3O^+$ has been written, when necessary, to betoken the hydration. This particular entity has been characterised in many crystals. As long ago as 1924, Volmer noticed the similarity between the x-ray powder patterns of ammonium perchlorate and of the hydrate of perchloric acid: as the former was best represented as $NH_4^+ClO_4^-$, so $OH_3^+ClO_4^-$ was preferable to $HClO_4{\cdot}H_2O$ for the latter. This suggestion has subsequently been confirmed by more direct experiments. From recent x-ray analyses, examples of crystals containing the 'oxonium ion' are $[H_3O^+]_2[O_3S{\cdot}CH_2{\cdot}SO_3^{2-}]$ [11] and $[H_3O^+][CF_3{\cdot}SO_3^-]$ [12].

In early x-ray work, the positions of the hydrogen atoms were merely inferred, and even now they are not accurately found. So there is a lack of firm information about the exact geometry of $H_3O^+$ in crystals. A recent account of a ND study of the monohydrate of toluene-*p*-sulphonic acid[13] therefore supplies valuable information. In this compound, the cation has no built-in crystal symmetry; the four atoms are all in general positions. All the same, its structure does not deviate significantly from the ideal of $3m$ ($C_{3v}$) symmetry. The O—H distances are 1.008, 1.011 and 1.013(8) Å, and the H—O—H angles are 109.2, 110.7 and 111.2(5)°. These results imply that the oxygen atom is 0.322 Å out of the plane of the three hydrogen atoms.

The finding agrees satisfactorily with one's expectations. However, it is fair to point out — as the authors do — that each hydrogen atom takes part in a strong hydrogen bond (O···O = 2.528 Å, average), and that the acceptor oxygen atoms, belonging to three different sulphonate groups, lie in positions of approximate trigonal symmetry with respect to the oxonium ion. No doubt, in a less favourably symmetrical environment the $H_3O^+$ ion would be more distorted[12].

More hydrated species of the proton are well documented in the crystallographic literature. We instanced $H_5O_2^+$ in HB. 1, Section 1.4.4, and a recent example was found in x-ray work on the tetrahydrate of sulphuric acid, where '$H_2SO_4{\cdot}4H_2O$' turn out to be $[H_5O_2^+]_2[SO_4^{2-}]$ [14].

Almlöf, using x-rays[15], has found a complex hydrogen-bonded system in $HClO_4{\cdot}3.5H_2O$. In this crystal, two independent $H_9O_4^+$ units can be recognised, and they share the same water molecule. The entity (2) thus produced

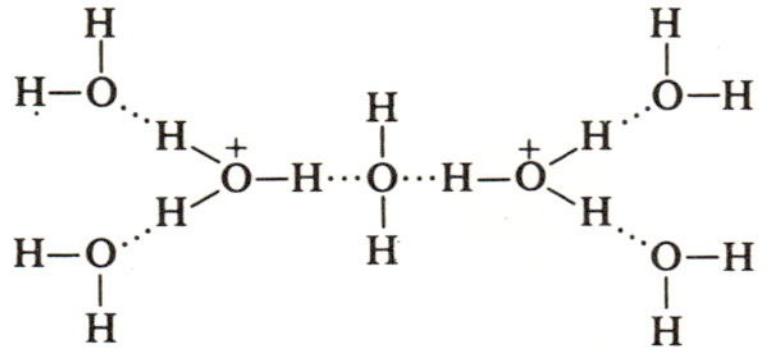


(2)

could be regarded as $H_{16}O_7^{2+}$, though it does not possess the crystal symmetry suggested by this simplified diagram. The hydrogen bonds to the central water molecule have O···O = 2.68 and 2.61 Å; the other four bonds are even stronger with O···O = 2.48–2.59 Å.

As is well known (e.g. see HB. 1, Section 1.1.1), acidity in the proton donor and basicity in the acceptor favour hydrogen bonding up to a point, but if they become too strong an ionised structure results. This is illustrated by ND work on the molecular compounds between acetic acid and a range of other acids. With sulphuric acid the crystal is ionic, being made up of $MeC(OH)_2^+$ and $HSO_4^-$ [16]. With phosphoric acid, the dimeric hydrogen-bonded molecule

(3) is formed[17]. The intercarboxyl bonds [O—H···O = 2.685(3) Å] are significantly longer than those in a normal carboxylic acid dimer (~2.64 Å).

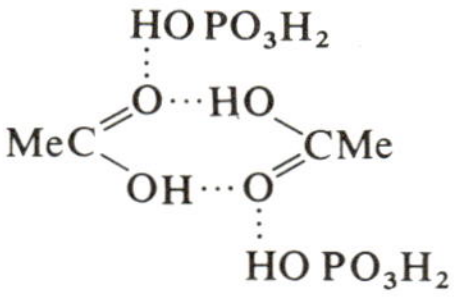

(3)

Such weakening is attributable to the acceptance by each carbonyl oxygen atom of a second hydrogen atom from a phosphoric acid molecule.

## 1.5 SOME ACID SALTS

### 1.5.1 The bifluoride anion

For many years the $FHF^-$ ion has been regarded as the firmest example — possibly the only sound example — of a symmetrical hydrogen bond. Structurally, it has been studied by particularly careful ND work on the sodium and potassium salts ($MHF_2$, where M = Na, K; see HB. 1, Section 1.6.2). According to a preliminary report[18], Williams and Schneemeyer have discovered by ND an unsymmetrical $FHF^-$ ion in *p*-toluidinium bifluoride ($M = C_7H_7NH_3^+$). In this crystal the anion has the geometry F···F = 2.260(4), F—H = 1.025(6), F···H = 1.235(6) Å; F—H···F = 178°. The distortion here is similar to, but more marked than, that in the intramolecular O—H···O bond reported by Schlemper *et al.*[19] (see HB. 1, Section 1.4.3), and similarly it may be attributed to the asymmetry of the crystal environment. Indeed, in the toluidinium bifluoride the asymmetry can be quantified simply. At each end the F(2)—H···F(1) anion accepts two hydrogen bonds from different —$NH_3^+$ groups; the N—H···F bonds to F(1) are stronger than those to F(2), the H···F distances being, respectively, 1.608, 1.675 and 1.777, 2.518 Å†. The stronger external bonding renders the corresponding internal (half-)bond the weaker, and the longer.

The obvious interpretation is that an isolated bifluoride ion would have its proton moving in a symmetrical, single-minimum* (but flattened) potential well. At a crystallographically symmetrical site, the bond retains its symmetry; at one that lacks symmetry, the point of minimum energy is readily displaced towards one of the fluorine atoms (see HB. 1, Figure 1.5).

Is the symmetry of $HF_2^-$ in the simple alkali bifluorides merely a statistical effect, due to two opposite but equally probable orientations of $F—H···F^-$, itself unsymmetrical and with a geometry like that specified above? This seems improbable. It is true that the ND measurements, considered alone, on $NaHF_2$ and $KHF_2$ can equally well be reconciled with a single-minimum potential, or with 'half-protons' vibrating less vigorously about two minima, provided these minima are *close together*. But a minimum 0.10 Å from the

* Or *effectively* single, and flattened, at the bottom of the well, relative to a parabolic shape.
† The last H···F distance is really too long for significant bonding.

mid-point, as found in the toluidinium bifluoride, is rather beyond the limits allowed by the ND measurements, for $NaHF_2$ in particular. There were, too, other reasons for supposing the anions in these latter salts to be genuinely symmetrical[20].

### 1.5.2 Symmetrical, or nearly symmetrical, OHO bonds

In HB. 1, Sections 1.4.1 and 1.4.2, the type-A acid salts of carboxylic acids were cited as a class of crystalline compounds in which very short, and effectively symmetrical O···H···O bonds have been found. Five more well-authenticated examples are now known to the reviewer. X-ray analyses have been reported on two type-A acid salts ($MHX_2$, where M is a univalent cation) of monocarboxylic acids (HX)[21, 22] and on two type-$A_2$ acid salts (MHY) of the symmetrically dicarboxylic acids ($H_2Y$)[23, 24]. There has also been reported a ND analysis[25] of RbHY, where $H_2Y$ is oxydiacetic acid $O(CH_2CO_2H)_2$. Some particulars of the hydrogen bonding in these acid salts are listed in Tables 1.1 and 1.2, which are to supplement Tables HB. 1,

**Table 1.1 Intercarboxylic hydrogen bonding in type-A acid salts, $MHX_2$.**

| HX | M | *Symmetry of bond* | O···O /Å | C—O··H /degrees | *Ref.* |
|---|---|---|---|---|---|
| Dichloroacetic | $NH_4$ | $\bar{1}$ | 2.432(5) | 114.8 | 21 |
| *m*-Chlorobenzoic | K | $\bar{1}$ | 2.437(6) | 111.6 | 22 |

**Table 1.2 Intercarboxylic hydrogen bonding in type-$A_2$ acid salts, MHY**

| $H_2Y$ | M | *Symmetry of bond* | O···O /Å | C—O··H /degrees | *Ref.* |
|---|---|---|---|---|---|
| Oxalic | $N_2H_5$ | $\bar{1}$ | 2.457(1) | 114.0 | 23 |
| Acetylene-dicarboxylic | K (*a* form) | 2 | 2.445(3) | 116.2 | 24 |
| Oxydiacetic | Rb | 2 | 2.450(3) | 116.1 | 25 |

1.3 and HB. 1, 1.4, respectively. Hydrazinium hydrogen oxalate has already appeared in Table HB. 1, 1.4, but the recent analysis[23] is more precise and to a large extent supersedes the earlier analyses.

The weighted averages for the O···O distances (including those from HB.1) are 2.443(2) Å for the monobasic series and 2.454(1) Å for the dibasic. This method of averaging and the resulting standard deviations of the means are based on the unrealistic assumption that all the hydrogen bonds in these different crystals have the same length*, whereas undoubtedly there is a

* The average for the dibasic series of 2.454 Å may have been unduly biased by the heavy weight given to the value for the acid oxalate with its very low standard deviation.

significant scatter within each series. Nevertheless, these acid salts constitute a notable set of crystals with very short hydrogen bonds, all lying across elements of symmetry.

The C—O···H angles found in the 17 most precise analyses all lie in the range 110.2–116.2°, with a mean ~115°. An angle near to this value seems to be necessary for optimal bonding between carboxyl groups.

Interest attaches to structures which just fail to achieve strict crystal symmetry with respect to the OHO bond. Three examples have a close relationship to the structures listed in Tables 1.1 and 1.2.

Thomas has studied the deuteriated acid salt $[N_2D_5^+][DC_2O_4^-]$ [23]. Apart from the hydrogen bond, this structure is closely similar to that of $[N_2H_5^+]$ $[HC_2O_4^-]$, but it belongs to the space group $P2_1$ which lacks the centres of symmetry included in $P2_1/m$, the space group adopted by the ordinary acid salt. The ODO bond therefore does not lie across a centre of symmetry. Location of the deuterium atom with x-ray data runs into difficulty with the KKM effect (see Section 1.5.3), but the atom seems to be more associated with one oxygen than with the other. The O—D···O distance is 2.466(2) Å. The increase of 0.009 Å should be significant, but there is evidence[26] that deuteriation may possibly change the character of hydrogen bonds that are short, but not 'very short'.

There is a rubidium hydrogen acetylenedicarboxylate which is isomorphous with the potassium salt listed in Table 1.2. This monoclinic *α* form (of RbHY) has not been studied in any structural detail. However, there has been an x-ray study of a triclinic *β* form of RbHY [27], with a comparison of the structures of *α*-KHY and *β*-RbHY. Chemically they are identical, and most of the geometric parameters are virtually identical. In each structure there are infinite chains of $HY^-$ anions linked by OHO bonds, but whereas in the *α* form these bonds lie across twofold axes of the crystal, there is no symmetry constraint on the OHO bonds of the *β* form. The O···O distances differ: *α* form, 2.446(3); *β* form, 2.464(8) Å. Unfortunately, the difference is of dubious significance especially as different cations are involved. (The precision in the analysis of the rubidium salt was impaired by high absorption errors.)

Grenthe and co-workers have also reported[28] x-ray analyses of the isostructural (isomorphous?) monoclinic acid salts of oxydiacetic acid, NaHY and KHY. (They are not isomorphous with RbHY which belongs to the tetragonal system*.) Again, there are infinite chains of $HY^-$ units, but in (Na/K)HY no crystal symmetry is involved in the chains. Though not symmetrical, the hydrogen bonds are short with O···O = 2.462(3) in NaHY and 2.480(2) Å in KHY, but slightly longer than the symmetrical bond in RbHY [2.450(3) Å]. Surprisingly, in view of the difficulty of locating hydrogen atoms with x-rays in such situations (see Section 1.5.3 and HB. 1, Section 1.2.3), the acidic hydrogen atoms were found in both NaHY and KHY at reasonable positions, 1.01–1.05 Å from one oxygen atom, and with C—O—H ≈ 115° in each.

* This is an uncommon situation. Whilst K, Rb and $NH_4$ salts are often isomorphous, it is very rare to find isomorphism between Na and K salts of organic acids, the sizes of the cations differing too much. $NaHX_2$ and $KHX_2$, where HX = acetic acid, crystallise in different systems and with totally different structures[29].

### 1.5.3 The Kroon–Kanters–McAdam (KKM) effect

This phenomenon is normally associated with short OHO bonds lying across a point of crystal symmetry (see HB. 1, Section 1.7.3.). A 'difference' electron-density synthesis, based on a precise x-ray study, may reveal a double peak which, at its face value, implies that the acidic hydrogen atom is disordered between alternative sites up to 0.5 Å on either side of the bond centre. One of the most striking examples, illustrated in HB.1, Figure 1.8 and reproduced in Figure 1.1(a), was discovered in an accurate study of potassium hydrogen *meso*-tartrate at $-160\,°C$[30]. The 'half-hydrogens' in this map are 0.37 Å away from the centre.

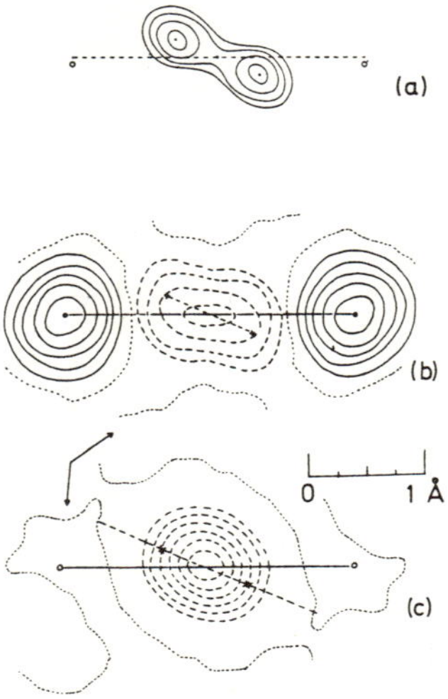


**Figure 1.1** A short O···H···O bond in potassium hydrogen *meso*-tartrate. (a) Electron-density 'difference' synthesis, showing the KKM effect; (b) neutron-scattering density in the same region; (c) corresponding ND 'difference' synthesis, the scattering due to the oxygen atoms having been removed. [In (b) and (c), negative levels are shown by broken contour lines]

This crystal structure has now been examined by ND at room temperature[31]. Though the number of ND reflections measurable was much smaller, the shortness of the relevant hydrogen bond is confirmed; O···O = 2.474(13) Å as against 2.483(2) Å by x-rays. Figure 1.1(b) shows the neutron-scattering density in the same region as 1.1(a), the negative contours round the proton being represented by broken lines. The 'peak' corresponding to the proton is indeed elongated, but not sufficiently to indicate statistical half-protons ~0.74 Å apart. Figure 1.1(c) is a neutron 'difference' map, all nuclei having been taken out except the acidic proton. Though it is still elongated, the residual peak is even less compatible with 1.1(a) than is 1.1(b). Were the proton sites really 0.74 Å apart, they should be easily resolved in this ND map. Figures 1.1(b) and 1.1(c) could be explained by either of two models: (1) the proton has its point of minimum energy at the middle, but vibrates very anistropically with its greatest amplitude in a direction decidedly away from the direct O···O line, or (2) it occupies one or other of equivalent, alternative sites, perhaps as far as 0.15 Å (but not more) from the middle, with a reduced amplitude of vibration. In the case of this particular bond in potassium hydrogen *meso*-tartrate, the second explanation seems more likely. One reason is that, were the proton at the mid-point, the C—O··H angle would be very unfavourable at 131°.

It is relevant to point out that this hydrogen bond connects two extremely unorthodox carboxyl groups. For one thing, they break the almost universal rule that unequal C—O distances in the same carboxyl are matched by an inverse inequality in the C—C—O angles*. Typically, as in salicyclic acid dimer (see Section 1.6), C—O(1) = 1.308, C—O(2) = 1.237 Å; C—C—O(1) = 116, C—C—O(2) = 123°. In the *meso*-tartrate, corresponding dimensions are 1.297(1), 1.231(2) Å; 121.2, 117.3°. In view of these anomalies, the KKM effect in this bond might be exceptional; we notice also that the same crystal includes a second symmetrical OHO bond between more conventional carboxyl groups, and that the KKM effect at this second bond is rather less spectacular though easily detectable. However, there are numerous other examples[33].

In this context, great interest attaches to the paper[23] by Thomas describing x-ray work on hydrazinium hydrogen oxalate, as well as its deuteriate. The former is a type-$A_2$ crystal, some details of which were given in Table 1.2, and the OHO bond is short and crystallographically symmetrical. Its carboxylate groups are equivalent and of unimpeachable orthodoxy: C—O(H) = 1.279(1), C—O = 1.224(1) Å; C—C—O(H) = 113.4, C—C—O = 120.3°. Further, the x-ray analysis is of high precision. At the end of the refinement, the 'difference' electron-density distribution was calculated over the region of the hydrogen bond. The outcome is represented by the three sections, from the three-dimensional distribution, shown in Figures 1.2(a), 1.2(b) and 1.2(c). This is very similar to the effect discovered by Kroon *et al.*[34] and reproduced in Figure 1.1(a), but here the setting is without abnormal

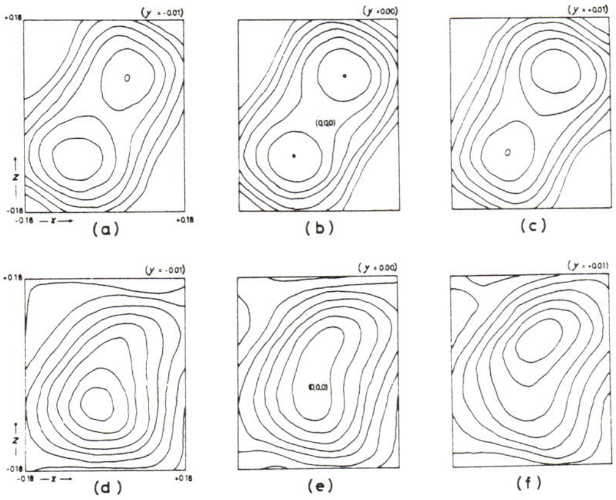


**Figure 1.2** Electron-density 'difference' distribution near the short hydrogen bonds in (a), (b) and (c) hydrazinium hydrogen oxalate, and (d), (e) and (f) its deuteriate. In each case the density has been sampled at three levels: above, at, and below, the mid-point of the bond. [Reproduced, by permission, from (1973). *Acta Crystallogr.*, **B29**, 1769]

* The hydrogen maleate anion also breaks this rule. The severe strains arising from ring closure, by a very short intramolecular O···H···O bond, account for the anomaly[32].

features. It constitutes the best authenticated example so far of the KKM effect.

Some details of the situation are picked out in Figure 1.3, which represents a projection down the *b* axis. The origin is taken at the mid-point of the OHO bond and only half the actual bond is shown. Besides the relevant oxygen atom (O), P marks the position of peak density in Figure 1.2(a) and 1.2(c), and L marks the position found for the hydrogen atoms from a least-squares refinement based on a disordered model with two equivalent 'half-hydrogens' on either side of the centre of symmetry at the origin. P is 0.84 Å from the oxygen atom and L is 0.76(5) Å. Neither is an acceptable position for the proton and neither seems a very plausible position for the local electron-density maximum at, or very near to, the nucleus of the hydrogen atom.

The following interpretation of results such as those depicted in Figure 1.2(b) seems tenable. As is well known (see, e.g., HB. 1, Section 1.2.3), x-ray analysis which is otherwise precise always locates a hydrogen atom too close to the carbon, nitrogen or oxygen atom to which it is covalently bound. This is illustrated in Figure 1.4(a)*, which represents the ($F_X - F_N$) 'difference' map in the region of a long hydrogen bond. ($F_X$ is the x-ray structure factor observed; $F_N$ that calculated, for x-rays, from the neutron-diffraction parameters of all the atoms.) The peak is ~0.7 Å from the right-hand atom, which is too close to be the true position of the hydrogen atom. Such errors are explained by the *asphericity effect*[35], though in an exaggerated form in this case since the bonding hydrogen atom has been included in calculating $F_N$. For strong practical reasons, x-ray analysis must always be organised on the assumption that the atoms are spherical masses of electron density. Such an assumption is not strictly valid (see Coppens, Chapter 2) and, although it works remarkably well in many cases, it fails particularly badly for a hydrogen atom bonded to an electronegative neighbour. For optimal matching of the actual, aspherical electron density round the hydrogen nucleus, the spherical-model atom has to be displaced nearer to the heavier atom. Now in regions including a symmetry element of the crystal, symmetry is built into any sort of structure refinement including a Fourier synthesis. The asphericity effect, forced to act both ways by the overall symmetry, will tend to pull apart any mass of electron density at the bond centre. It may produce the artefact illustrated in Figure 1.2(a). In Figure 1.4(b), we have forced symmetry on 1.4(a) and the result is very similar to the KKM effect.

Small errors in scaling our observed structure factors may have a large effect on the amount of electron density to be disposed of at our centre of symmetry, or other special position. So the exact type of effect produced is sensitive to scaling errors (see Coppens, Chapter 2).

Without any doubt, the three-dimensional map revealed in Figures 1.2(a), 1.2(b) and 1.2(c) is loaded with information about the electron-density

* Figure 1.4(a) is based on a photograph kindly supplied by Dr. J. O. Thomas of Uppsala. It actually covers an N—H···O bond, and it is an ($F_X - F_N$) map, the positions of the atomic nuclei having been found by ND. A similar map for an O—H···O bond would have been more appropriate for our present purposes but Thomas' map has been used because it is based on very precise work. So far as the relevant details are concerned, the 'difference' map at a rather long O—H···O bond would be very similar. Figure 1.4(b) is a symmetrised map, obtained by averaging 1.4(a) with its own inversion about the mid-point; negative contours have been omitted for clarity in this case.

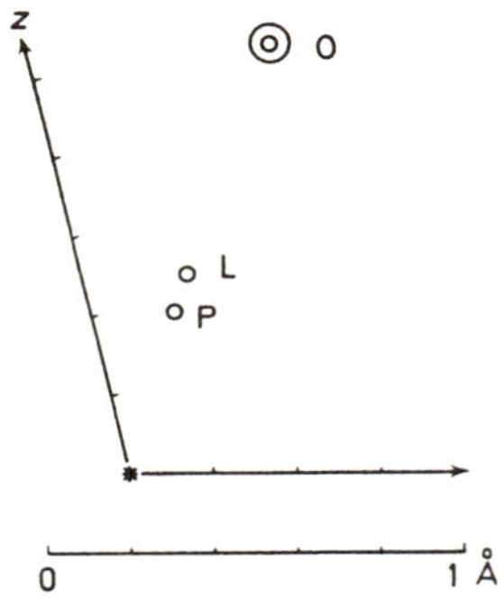


**Figure 1.3** Some details of the hydrogen bonding in hydrazinium hydrogen oxalate; relative positions found for the oxygen atom (O), and for the electron density associated with the hydrogen atom (L) as found by least-squares analysis and (P) from the peak featured in Figure 1.2(b)

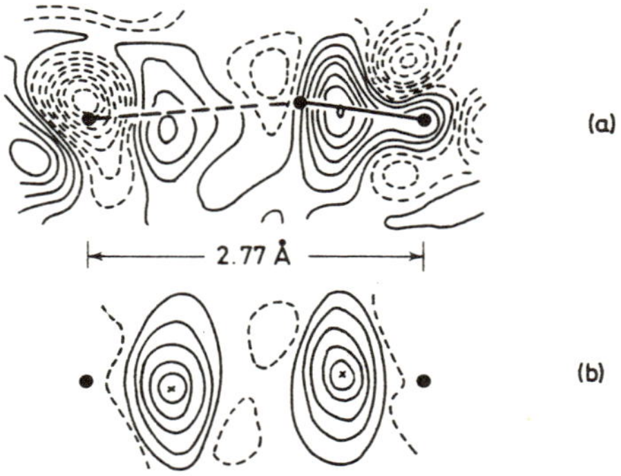


**Figure 1.4** (a) The electron-density 'difference' distribution at a rather long hydrogen bond. (Zero and negative levels are shown by broken contour lines.) (b) Symmetrised version of (a). [The positions found by ND for the nuclei are marked by small filled circles; negative levels are omitted in (b).] [Based on work by Thomas —details in the footnote to p. 10]

distribution in a short 'symmetrical' hydrogen bond. The information is disguised, however, and the disguise is hard to penetrate. It should not be taken as a direct indication of the position, or positions, of the hydrogen atom.

Figures 1.2(d), 1.2(e) and 1.2(f) were derived in a similar way from the x-ray work on the deuteriated compound. Though the space group was now the unsymmetrical $P2_1$ the model of disordered 'half-deuterium' atoms was used as the basis for this 'difference' synthesis. Unlike the distribution explored in the upper half of Figure 1.2, there is now no constraint requiring strict centrosymmetry. Nevertheless, because all the heavier atoms are very close to the symmetrically related pairs of positions required by $P2_1/m$, there is a bias towards symmetry in the difference distribution as can be seen. The ODO bond in $[N_2D_5^+][DC_2O_4^-]$ appears to be unsymmetrical. Whether it is statically so (as we can symbolise by writing the formula O—D····O), or disordered with different proportions of the arrangements O—D····O and O····D—O, is uncertain. Certainly, when we ask what the lower half of the diagram tells us about the real position of the deuterium atom, we may well conclude that the disguise is even more impenetrable.

### 1.5.4 Neutron-diffraction studies of two 'super-acid' salts

The acid salts of the *dibasic* acids considered in the previous sections are normal in the sense that their existence is envisaged in elementary chemical theory: $H_2Y \rightarrow MHY \rightarrow M_2Y$. Acid salts of monobasic acids, on the other hand, are anomalous; HX should yield a salt MX, but not $MHX_2$, which in the elementary context can be regarded as a 'super-acid' salt of HX. Similarly, super-acid salts of dibasic acids are very common; the tetroxalates, $MH_3(C_2O_4)_2{\cdot}2H_2O$, are a long familiar example. Two super-acid salts from the dicarboxylic acid series have recently been examined by ND. They are both represented by the formula $KH_3Y_2$, $H_2Y$ being malonic acid in one case[36] and succinic in the other[37].

Discrete $H_2Y$ molecules and $HY^-$ ions can be characterised in each salt, so that they are stoichiometrically identical, i.e. $K^+HY^-{\cdot}H_2Y$, but their crystal structures are wholly different. They are two more examples illustrating the almost infinite variety of possible hydrogen-bonding schemes[38] and how the details of a particular structure depend on a whole range of intermolecular, and intramolecular, forces, whose final effect is well beyond our present powers of prediction.

As in most acid salts of dicarboxylic acids, there are infinite chains of hydrogen-bonded units, such as those schematically illustrated in HB. 1, Section 1.4.2, structures (4)* and (5), though here without any participation of crystal symmetry in the chains. In the dimalonate[36], the chains consist of alternate $H_2Y$ molecules and $HY^-$ ions, and the situation may be stylised as in (4). The two O—H···O bonds donated by each HYH are non-equivalent. Their geometries (O···O, O—H and O—H···O) may be expressed as 2.543(8), 1.050(12) Å, 172° and 2.554(7), 1.018(12) Å, 163°, respectively.

* Some of the symmetry symbols were accidentally lost during the final printing of structure (4).

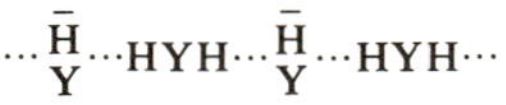

(4)

The proton at one end of $HY^-$ is donated to the carboxylate residue at the other and the six atoms of the ring thus completed are nearly co-planar. The intramolecular O—H···O bond has the dimensions 2.513(9), 1.023(14) Å, 155°.

In the disuccinate[37], the main chain consists only of $HY^-$ ions linked by a single type of bond [2.579(9), 1.009(12) Å; 178°]. In a manner suggested by (5), the $H_2Y$ molecules hang as festoons from the main chain. The two bonds

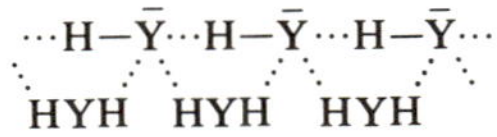

(5)

attaching $H_2Y$ to the chain have the dimensions 2.573(11), 0.991(12) Å, 163° and 2.630(10), 0.996(15) Å, 172°

All the unsymmetrical hydrogen bonds along these two types of chain are significantly longer than the symmetrical bonds along the chains in the simple (KHY) acid salts of malonic and succinic acids (see HB. 1, Table 1.4).

## 1.6 SALICYCLIC ACID AND α-RESORCINOL

Recent ND work throws light on the intramolecular hydrogen bond in salicyclic acid[39]. The dimer (6), in which the asterisk marks the position of a

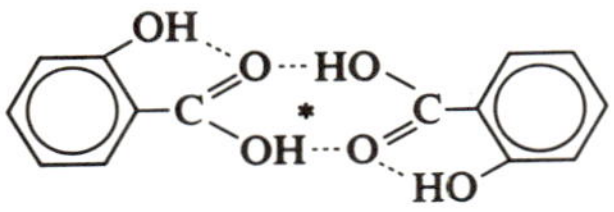

(6)

crystallographic centre of symmetry, was one of the first systems in which hydrogen atoms engaged in hydrogen bonding were located from electron-density 'difference' maps (Cochran, 1953). The detailed geometry of the intra-molecular bond is pictured in Figure 1.5, which is based on the ND study. Were we to ignore the proton, whose exact position was hitherto uncertain, the O(2)···O(3) distance corresponds to a rather strong hydrogen bond; for instance, it would be judged to be much stronger than the bond in ice and about as strong as the intermolecular O—H···O bonds which hold the dimer together. However, we then see that the C—O(3)···O(2) and C—O(2)··· O(3) angles, each less than 90°, are very unfavourable. Location of the proton, as shown in the diagram, now enables us to examine the consequences of this misfit in detail; the O(3)—H distance and the C—O(3)—H angle differ hardly at all from those in the molecule of gaseous phenol, while the O(3)—

H···O(2) angle is far below the optimum (180°) for hydrogen bonding. On the other hand, H···O(2) is much less than the sum (2.4 Å; see the footnote to p. 19) of the van der Waals' radii of hydrogen and oxygen. The general impression is of a strong repulsion between H and O(2) only just compensated by electrostatic attraction. The proton must lie somewhere on the circle it generates when O—H is rotated about C—O. It finds a position of minimum energy at the place shown in the diagram, only about 0.04 Å from the mean plane of the other five atoms of the chelate ring. There is, perhaps, no very positive act of bonding.

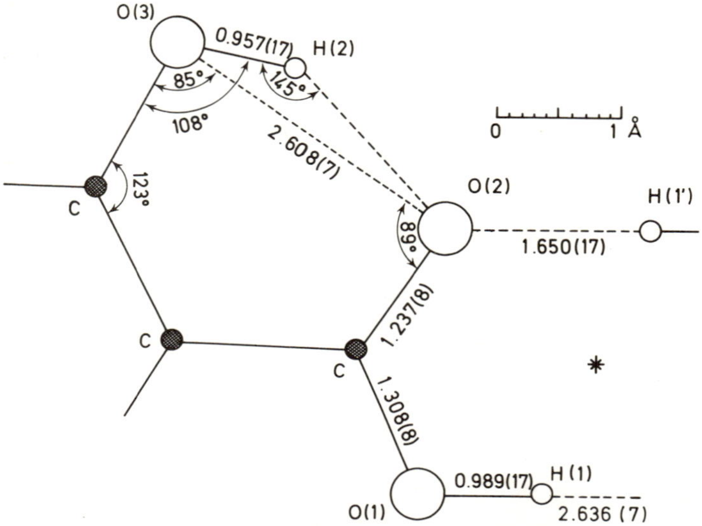


**Figure 1.5** Some geometric details of the hydrogen bonding in salicyclic acid dimer as found by ND. (Distances in Å, with standard deviations in parentheses; the asterisk marks the position of a centre of symmetry)

Nakamoto, *et al.* drew attention[40] to a close inverse correlation between the overall length of an O—H···O bond and the decrease ($\Delta\nu$) in the O—H stretching frequency below that of an unperturbed hydroxyl group (~3700 $cm^{-1}$). In the spectrum of solid salicyclic acid, there are two frequencies due to O—H stretchings at 3225 and 2564 $cm^{-1}$. There are the two hydrogen bonds with O···O equal to 2.636 and 2.608 Å. The allocation must be anomalous. The lower frequency has to be associated with the longer intercarboxyl bond since there is a band at 2630 $cm^{-1}$ in the spectrum of the methyl ether of salicyclic acid, while the higher frequency has to be associated with the shorter bond since methyl salicylate has a band at 3220 $cm^{-1}$. In a nearly straight bond, a $\Delta\nu$ value of 475 $cm^{-1}$ would correspond to an O···O distance of 2.8 Å. It indicates a weak bond, which is undoubtedly true in this case, despite an O···O distance of only ~2.6 Å.

The same workers[39] have reported a precise ND analysis of $\alpha$-resorcinol (*m*-dihydroxybenzene), one of the earliest hydrogen-bonded structures to be studied by x-rays (Robertson, 1936). There are two rather weak intermolecular hydrogen bonds with the geometries O···O = 2.717(5), O—H = 0.989(5) Å, O—H···O = 175.9°, and 2.731(5), 0.991(7) Å, 165.7°.

## 1.7 HYDROGEN BONDING IN HYDRATED COPPER ACETATE

X-Ray work by van Niekerk and Schoening (1953) showed that copper(II) acetate monohydrate contained the dimeric molecule $Cu_2(C_2H_3O_2)_4$, this affording one of the first examples of a direct metal–metal bond. This crystal has now been studied by precise ND analysis[41]. [The Cu—Cu distance is now refined to 2.6143(17) Å.]

The water molecule fits into the structure in a way typical of salt hydrates (see Section 1.3). The hydrogen-bonding scheme involves eight-membered rings like that illustrated in Figure 1.6. Formally, such a ring closely resembles the ring closed by the pair of hydrogen bonds in a carboxylic acid dimer [see structure (6) and Figure 1.5]. In copper acetate there are, in fact, two independent types of eight-membered ring. One type lies about the centre of symmetry, marked by an asterisk in the diagram. For this ring, the resemblance to (6) is increased by the circumstance that all eight atoms are nearly co-planar. This is a necessary condition in doubly hydrogen-bonded carboxyls.

However the similarity, though pleasing to the eye, must not be taken too seriously. The dimensions given in the diagram show that the hydrogen bonding in this case is very weak; O—H···O is nearly 0.3 Å longer than in typical carboxylic acids, and the condition of co-planarity is not nearly so stringent. The rings of the second type are chemically identical with that shown in Figure 1.6 and most of the dimensions are about the same. But

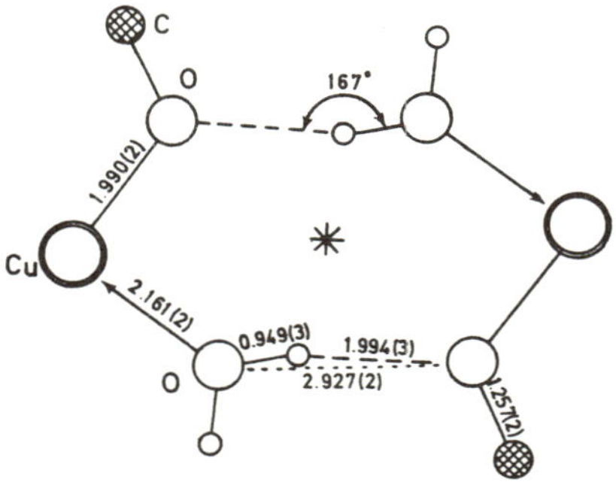


**Figure 1.6** Part of the hydrogen-bonding pattern in hydrated copper acetate as found by ND. (Distances in Å)

instead of a centre, a crystallographic twofold axis passes through the middle of the ring and the eight atoms are far from co-planar. Four atoms constituting each half of the ring do indeed lie in the same plane, but the ring is buckled through ~62° about a line joining the oxygen atoms of the two water molecules. Evidently a ring of the sort shown in the diagram is compliant in adapting itself to its environment.

## 1.8 PRECISION NEUTRON-DIFFRACTION ANALYSES OF AMINO ACIDS

### 1.8.1 General

The late Walter Hamilton initiated at the Brookhaven National Laboratory a project for the 'precision neutron-diffraction structure determination of protein and nucleic acid components'. This is enabling the protons to be located with standard deviations normally in the range 0.001–0.003 Å, and in most cases it has also improved the precision with which the heavier atoms had been found by x-rays. Some of the advantages of ND were summarised in HB. 1, Section 1.2.2. One advantage, not mentioned, is that the vibrational behaviour of atoms is more directly and more accurately studied by ND. Neutrons are scattered by the atomic nuclei which are virtually points so far as thermal neutrons are concerned. X-Rays, on the other hand, are scattered by the total electron density around the individual atoms in the crystal. When the density around a given atom is found to be non-spherical or distorted in any other way from the ideal shape of the atom, it is hard to differentiate distortion due to bonding from that due to the various modes of vibration which the atom may be undergoing (see Coppens, Chapter 2). For this reason, amongst others, ND is inherently a more accurate method for studying crystal structure in all its aspects — always provided we restrict the word 'structure' to the description of the positions and motions of the atomic centres only.

Twelve papers have been published from Brookhaven at the time of writing. They deal with (Part I) L-alanine[42], (II) diglycylglycine hydrochloride hydrate[43], (III) $\alpha$-glycine[44], (IV) orthorhombic L-histidine[45], (V) L-arginine dihydrate[46], (VI) L-asparagine hydrate[47], (VII) L-lysine hydrochloride dihydrate[48], (VIII) $\beta$-L-glutamic acid[49], (IX) 4-hydroxy-L-proline[50], (X) L-tyrosine and its hydrochloride[51], (XI) L-serine hydrate and DL-serine[52], and (XII) the base-pair 9-methyladenine–1-methylthymine[53]. In addition, from Bombay, there are papers describing ND work on L-glutamic acid hydrochloride[54] and the aminosulphonic acid L-cysteic acid monohydrate[55].

These papers are a mine of precise information of many kinds. In this short review we shall confine ourselves to matters directly, or indirectly, connected with hydrogen bonding. In nearly every case, all hydrogen atoms attached to oxygen or nitrogen take part in hydrogen bonding, in accord with Donohue's rule: 'A structure (is most stable) in which all possible hydrogen bonds are formed. . . . It is exceedingly rare to find an "unused" hydrogen atom'[56]. Dihydrated L-asparagine involves an exception. The protons of its $\alpha$-amino group are unused although the group accepts a proton from an $>$NH group of another molecule.

An elementary structural problem with amino acids concerns the particular tautomeric form adopted by the molecule. In most cases the zwitterionic structure (7) was inferred long ago from indirect evidence, subsequently confirmed and amplified by traditional x-ray analysis. Consideration of structures (8) and (9), for instance, suggests that the carboxyl group should be ionised when the C—O distances are equal, or nearly so. Such a problem is directly settled by successful ND work which locates all protons. There have been no surprises. In all but one of the acids so far examined by ND the neutral molecule has the form (7). The exception is argenine, in the

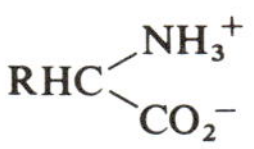

(7)

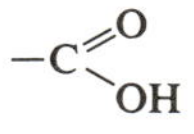

(8)

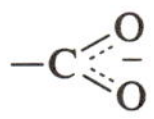

(9)

dihydrate of which the positive charge is carried by the guanine residue as in (10). (In the uncharged state, guanine is a much stronger base than $—NH_2$.)

When any amino acid is examined in the form of its hydrochloride, the extra proton is normally accommodated at the carboxyl group as in (11).

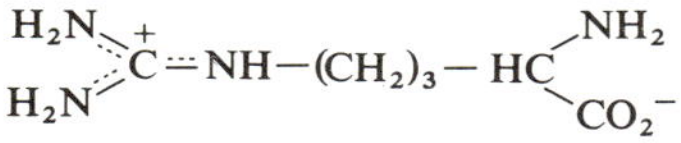

(10)

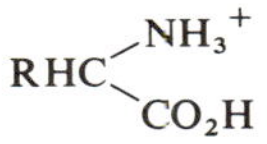

(11)

This happens in the hydrochlorides of tyrosine and of glutamic acid. Lysine hydrochloride (in its dihydrate), however, has the structure (12); the second

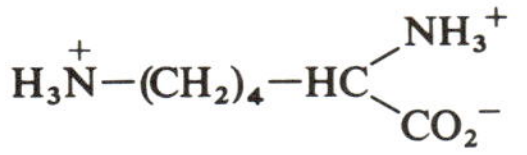

(12)

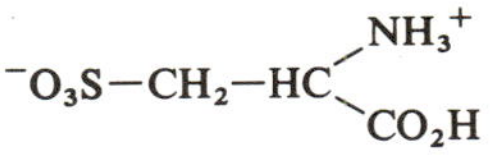

(13)

amino group, in the side-chain of the neutral molecule, is still a much stronger base than the carboxylate residue. In cysteic acid, on the other hand, carboxylate is more basic than sulphonate so that the structure is represented by (13).

### 1.8.2 Torsional oscillation of $—CH_3$ and $—NH_3^+$ groups

The results of precise ND study of a fairly simple molecule usually make it possible to undertake a detailed vibrational analysis. The vibration observed for a particular atom is the resultant of various orders of movement: (a) rigid-body translational vibration of the whole molecule; (b) rigid body torsional oscillation (libration) of the whole molecule; and (c) internal vibrations — notably torsional and 'wagging' vibrations of parts of the molecule with respect to the whole. In the simple amino acids, at any rate, the heavier (non-hydrogen) atoms usually vibrate in a manner fairly consistent with a rigid-body motion; (c) contributes little. But the hydrogen atoms of groups such as $—CH_3$ or $—NH_3^+$ often have extra vibrational motion which can be satisfactorily accounted for as due to libration about the $C—CH_3$ or

$C—NH_3^+$ bond. The results often fit a model with a triple-minimum potential-energy function, with minima 120° apart.

In alanine[42], $CH_3CH(NH_3^+)CO_2^-$, for example, the vibrations of the methyl hydrogen atoms relative to the rigid skeleton of the molecule can be analysed as due to a libration about the C—C bond of r.m.s. amplitude 9.5°. If we assume that the libration is harmonic, with a potential-energy function of the form $V(\alpha) = \frac{1}{2}V_0(1 - \cos 3\alpha)$, where $\alpha$ is the angular displacement, the energy barrier, $V_0$, restricting internal rotation of the methyl group can be calculated as 5.6 kcal mol$^{-1}$. This is a reasonable value, comparable with estimates for methyl groups in other systems.

In contrast, the r.m.s. amplitude for the $C—NH_3^+$ group is much smaller, 5.9°, though the relevant masses and dimensions are similar. On the same assumptions this corresponds to an energy barrier of ~20 kcal mol$^{-1}$. This severe restriction on internal rotation is attributable to the three rather strong hydrogen bonds which exist in the crystal. The difference between $—CH_3$ and $—NH_3^+$ is a striking example of the effect of hydrogen bonding.

In a similar situation in asparagine[47], the torsional vibration of the $—NH_3^+$ group is appreciably greater than in alanine and the estimated barrier height smaller, ~7 kcal mol$^{-1}$. The three hydrogen bonds around $—NH_3^+$ in asparagine are slightly longer and therefore may reasonably be supposed to have less effect on internal libration.

### 1.8.3 The bifurcated hydrogen bond

One of the first examples of possible bifurcation was discovered by Albrecht and Corey[57]. The situation they found in $\alpha$-glycine is sketched in Figure 1.7 where the dimensions given, however, are the results of the recent ND work[44]. The early x-ray work (1939) was of astonishing accuracy for its period and yielded distances of 2.93 and 3.05 Å between a nitrogen atom of the $—NH_3^+$ group and two oxygen atoms of neighbouring molecules. Considered individually, these distances warrant the supposition that each corresponds to an N—H···O bond. Although the relevant hydrogen atom could not be located in 1939, its position was inferred, from general arguments, to be very near to that actually found by ND; the N—H distance was assessed at 0.99 Å.

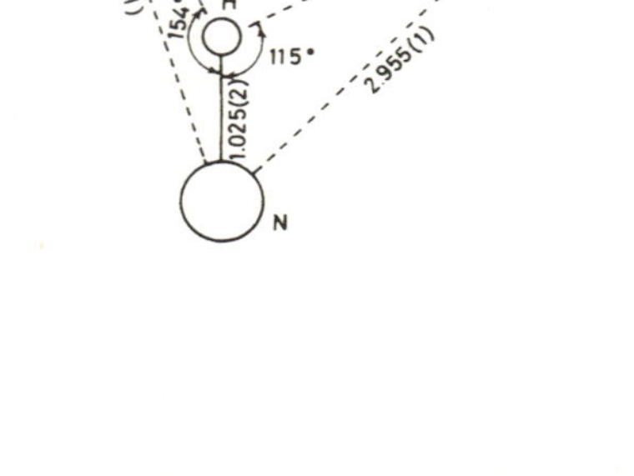


**Figure 1.7** The supposed bifurcated hydrogen bond in $\alpha$-glycine as surveyed by ND. (Distances in Å)

Thus the N—H bond was directed towards a point between the two oxygen atoms. A bifurcated bond was assumed, not unreasonably.

The ND work confirms Albrecht and Corey's N···O distances. The precise location of the proton, however, shows that whilst one H···O distance [2.121(2) Å] is indicative of hydrogen bonding, the other [2.365(2) Å] is near enough to the sum of the van der Waals' radii of hydrogen* and oxygen (1.0 + 1.4 = 2.4 Å) to rule out any substantial bonding. The *longer* N···O distance corresponds to the more respectable hydrogen bond. Its angle (N—H···O = 154°) is much more favourable than 115° at the other side.

Cases of supposed bifurcated (even trifurcated) hydrogen bonds are reported from time to time. It now seems wiser to regard them with some scepticism. An argument such as that advanced in Section 1.6 may apply. The hydrogen atom attached to (say) nitrogen or oxygen has to be somewhere in space. When, occasionally, it can do no better than find a position of minimum potential energy more or less equidistant from two (or even three) oxygen atoms, this may not imply any positive act of bonding to both oxygens (or to all three). Repulsive forces are also relevant. What might have become one good hydrogen bond may well have been thwarted, and flawed, by the abortive intrusion of another proton acceptor.

## References


1. Vinogradov, S. N. and Linnell, R. H. (1971). *Hydrogen Bonding* (New York: Van Nostrand Reinhold)
2. Pimentel, G. C. and McClellan, A. L. (1971). *Ann. Rev. Phys. Chem.*, **22,** 356
3. Kollman, P. A. and Allen, L. C. (1972). *Chem. Rev.*, **72,** 283
4. Kamb, B., Hamilton, W. C., La Placa, S. J. and Prakash, A. (1971). *J. Chem. Phys.*, **55,** 1934
5. La Placa, S. J., Hamilton, W. C., Kamb, B. and Prakash, A. (1973). *J. Chem. Phys.*, **58,** 567
6. Whalley, E. (1967). *Ann. Rev. Phys. Chem.*, **18,** 222
7. Chidambaram, R., Sequeira, A. and Sikka, S. K. (1964). *J. Chem. Phys.*, **41,** 3616; see also Hamilton, W. C. and Ibers, J. A. (1968). *Hydrogen Bonding in Solids* (New York: Benjamin)
8. Baur, W. H. (1972). *Acta Crystallogr.*, **B28,** 1456
9. Ferraris, G. and Franchini-Angela, M. (1972). *Acta Crystallogr.*, **B28,** 3572
10. Kroon, J. and Kanters, J. A. (1973). Personal communication
11. Mootz, D. and Wunderlich, H. (1970). *Acta Crystallogr.*, **B26,** 1820
12. Spencer, J. B. and Lundgren, J.-O. (1973). *Acta Crystallogr.*, **B29,** 1923
13. Lundgren, J.-O. and Williams, J. M. (1973). *J. Chem. Phys.*, **58,** 788
14. Kjällman, T. and Olovsson, I. (1972). *Acta Crystallogr.*, **B28,** 1692
15. Almlöf, J. (1973). *Chemica Scripta*, **3,** 73
16. Jönsson, P.-G. and Olovsson, I. (1968). *Acta Crystallogr.*, **B24,** 559
17. Jönsson, P.-G. (1972). *Acta Chem. Scand.*, **26,** 1599
18. (1973). Reported in *Chem. Eng. News*, see also (1973). *J. Amer. Chem. Soc.*, **95,** 5780
19. Schlemper, E. O., Hamilton, W. C. and La Placa, S. J. (1971). *J. Chem. Phys.*, **54,** 3990
20. McGaw, B. L. and Ibers, J. A. (1963). *J. Chem. Phys.*, **39,** 2677; Ibers, J. A. (1964). *J. Chem. Phys.*, **41,** 25
21. Ichikawa, M. (1972). *Acta Crystallogr.*, **B28,** 755 (the paraelectric phase of this acid salt; for the structure of the ferroelectric phase, see (1974). *Acta Crystallogr.*, **B30,** 651)


* The van der Waals' radius of hydrogen was taken to be 1.2 Å by Pauling. This quantity is difficult to assess and there is now evidence from several sources that a lower value, 1.0 Å, is to be preferred[8, 58]. This has been used throughout this review.

22. Macdonald, A. L. and Speakman, J. C. (1972). *J. Chem. Soc.* (*Perkin Trans. II*), 1584
23. Thomas, J. O. (1973). *Acta Crystallogr.*, **B29**, 1767
24. Leban, I., Golić, Lj. and Speakman, J. C. (1973). *J. Chem. Soc.* (*Perkin Trans. II*), 703
25. Albertsson, J. and Grenthe, I. (1973). To be published in *Acta Crystallogr.*; information personally supplied by Dr. Grenthe
26. Snyder, R. G. and Ibers, J. A. (1962). *J. Chem. Phys.*, **36,** 1356
27. Blain, J., Speakman, J. C., Stamp, L. A., Golić, Lj. and Leban, I. (1973). *J. Chem. Soc.* (*Perkin Trans. II*), 706
28. Albertsson, J., Grenthe, I. and Herbertsson, H. (1973). *Acta Crystallogr.*, **B29,** 1855
29. Currie, M. (1972). *J. Chem. Soc.* (*Perkins Trans. II*), 832
30. Kroon, J. and Kanters, J. A. (1972). *Acta Crystallogr.*, **B28,** 714
31. Currie, M., Muir, K. W., Speakman, J. C., Kanters, J. A. and Kroon, J. (1974). To be published, perhaps in *J. Chem. Soc.* (*Perkin Trans. II*)
32. Darlow, S. F. (1961). *Acta Crystallogr.*, **14,** 1257
33. Ferraris, G. and Franchini-Angela, M. (1973). *Acta Crystallogr.*, **B29,** 286
34. Kroon, J., Kanters, J. A. and Peerdeman, A. F. (1971). *Nature* (*Phys. Sci.*), **229,** 120
35. Coppens, P. and Coulson, C. A. (1967). *Acta Crystallogr.*, **23,** 718
36. Currie, M. (1972). *J. Chem. Soc.* (*Perkin Trans. II.*), 972
37. Dunlop, R. S. and Speakman, J. C. (1973). *Z. Kristallogr.*, **138,** 100
38. Wells, A. F. (1954). *Acta Crystallogr.*, **7,** 525; (1969). ibid., **B25,** 1711; see also (1962). *The Third Dimension in Chemistry*, (Oxford: Oxford Univ. Press)
39. Bacon, G. E. and Jude, R. J. (1973). *Z. Kristallogr.*, **138,** 19
40. Nakamoto, K., Margoshes, M. and Rundle, R. E. (1955). *J. Amer. Chem. Soc.*, **77,** 6480
41. Brown, G. M. and Chidambaram, R. (1973). *Acta Crystallogr.*, **B29,** 2393
42. Lehmann, M. S., Koetzle, T. F. and Hamilton, W. C. (1972). *J. Amer. Chem. Soc.*, **94,** 2657
43. Koetzle, T. F., Hamilton, W. C. and Parthasarathy, R. (1972). *Acta Crystallogr.*, **B28,** 2083
44. Jönsson, P.-G. and Kvick, A. (1972). *Acta Crystallogr.*, **B28,** 1827
45. Lehmann, M. S., Koetzle, T. F. and Hamilton, W. C. (1972). *Int. J. Peptide Protein Res.*, **4,** 229
46. Lehmann, M. S., Verbist, J. J., Hamilton, W. C. and Koetzle, T. F. (1973). *J. Chem. Soc.* (*Perkin Trans. II*), 133
47. Verbist, J. J., Lehmann, M. S., Koetzle, T. F. and Hamilton, W. C. (1972). *Acta Crystallogr.*, **B28,** 3006
48. Koetzle, T. F., Lehmann, M. S., Verbist, J. J. and Hamilton, W. C. (1972). *Acta Crystallogr.*, **B28,** 3207
49. Lehmann, M. S., Koetzle, T. F. and Hamilton, W. C. (1972). *J. Cryst. Mol. Struct.*, **2,** 225
50. Koetzle, T. F., Lehmann, M. S. and Hamilton, W. C. (1973). *Acta Crystallogr.*, **B29,** 231
51. Frey, M. N., Koetzle, T. F., Lehmann, M. S. and Hamilton, W. C. (1973). *J. Chem. Phys.*, **58,** 2549
52. Frey, M. N., Lehmann, M. S., Koetzle, T. F. and Hamilton, W. C. (1973). *Acta Crystallogr.*, **B29,** 876
53. Frey, M. N., Koetzle, T. F., Lehmann, M. S. and Hamilton, W. C. (1973). *J. Chem. Phys.*, **59,** 915
54. Sequeira, A., Rajagopal, H. and Chidambaram, R. (1972). *Acta Crystallogr.*, **B28,** 2514
55. Ramanadham, M., Sikka, S. K. and Chidambaram, R. (1973). *Acta Crystallogr.*, **B29,** 1167
56. Donohue, J. (1952). *J. Amer. Chem. Soc.*, **56,** 502
57. Albrecht, G. and Corey, R. B. (1939). *J. Amer. Chem. Soc.*, **61,** 1087; see also Marsh, R. E. (1958). *Acta Crystallogr.*, **11,** 654
58. Ermer, O. and Dunitz, J. D. (1971). *Chem. Commun.*, 178

# 2
# Measurement of Electron Densities in Solids by X-Ray Diffraction

**P. COPPENS**
State University of New York, Buffalo

---

## 2.1 INTRODUCTION

The rapid development of x-ray crystallography in the past decades has provided an often routine tool for the elucidation of simple and moderately complex structures, and has resulted in the determination of an increasing array of very large molecules. At the same time, improvements in experimental accuracy, the ability to collect and process large amounts of data, and developments in theoretical chemistry and physics have revived interest in very-high-resolution studies of crystals.

Only six years after the discovery of x-ray diffraction in 1912, Debye and Scherrer[1] discussed the possible determination of 'electron orbits' through diffraction experiments. Further developments were slow because of limited experimental accuracy, and also because the simple solids which could be studied had few or no x-ray reflections in the low-angle region where scattering of valence electrons is concentrated. Black and Taylor[2] realised this limitation when they defined in 1958 that a complex crystallographic structure is one of the conditions for electron-density studies. This somewhat paradoxical conclusion is a result of the reciprocal relationship between a crystal and its diffraction pattern, which leads to a very dense diffraction pattern (small distance between reflections) for a large crystallographic unit cell (complicated structure).

It is therefore not accidental that the interest in this field revived in the 1960s when more complicated structures became accessible and experimental accuracy was increased through a number of technical developments.

Early work by Brill *et al.*[3], by Wölfel[4] *et al.* and by Calder *et al.*[5] was done on inorganic solids such as NaCl, ZnS, LiH and diamond. The present review is mainly, but not exclusively, concerned with the more intricate structures analysed by x-ray diffraction in the past eight years. Surveys of electron diffraction of gas-phase molecules[6,7], of Compton profile studies of the momentum density[8] and of magnetic neutron-scattering measurements of spin densities[9] have appeared elsewhere.

The requirements of accuracy of the experimental observations in charge-density studies are quite severe. Symmetry-equivalent reflections should be measured and should agree to better than 2% in the intensities, systematic effects like multiple reflection, absorption of x-rays in the crystal and thermal diffuse scattering should be considered and eliminated experimentally or corrected for in data reduction or refinement. Furthermore, it is increasingly evident that for the study of molecular crystals, data should be collected at liquid-nitrogen or liquid-helium temperatures. Such considerations are the subject of continuing discussions[10, 11], but will not be included in the present review.

## 2.2 THE ASSUMPTIONS OF X-RAY STRUCTURE REFINEMENT

### 2.2.1 The isolated-atom approximation and its breakdown

The observed amplitude of a reflection against a plane $hkl$, the structure factor $F_{hkl}$(obs), can be compared with a calculated value $F_{hkl}$(calc) once an approximate model for the structure has been obtained. This model is most commonly refined by least-squares minimisation of

$$\varepsilon = \sum_{\substack{\text{all}\\ \text{observations}}} w\Delta^2 \tag{2.1}$$

where $w$ are weights assigned to the individual observations, and the discrepancy $\Delta$ in its simplest form is $F(\text{obs}) - kF(\text{calc})$. The scale factor $k$ is required because x-ray observations are generally on an arbitrary rather than absolute scale. The least-squares variables are defined by the model for the structure, and include the scale factor $k$ and the positional and thermal parameters of each of the $n$ atoms, $x_n$, $y_n$, $z_n$ and $T_n$ respectively, in the expression for $F$(calc):

$$F_{hkl}(\text{calc}) = \sum_{\substack{\text{all}\\ \text{atoms}}} f_n \exp{(2\pi \mathrm{i}\, \boldsymbol{H}\cdot\boldsymbol{r}_n)} T_n \tag{2.2}$$

Here $\boldsymbol{H}$ is the reciprocal lattice vector defined by $(hkl)$ and $\boldsymbol{r}_n = x_n\boldsymbol{a} + y_n\boldsymbol{b} + z_n\boldsymbol{c}$ describes the position of the $n$th atom in terms of the unit-cell edges $\boldsymbol{a}$, $\boldsymbol{b}$ and $\boldsymbol{c}$.

Expression (2.2) implies two important assumptions. The first concerns the form of the temperature factor $T_n$. In the harmonic approximation, $T_n$ depends on the mean-square amplitude of atomic displacement, which may be a scalar $U_n$ if the displacements are isotropic or a second-rank tensor $U_{ij,n}$ in the case of anisotropic harmonic motion. The second assumption is that the atomic scattering $f_n$ can be derived from the wave function of the isolated spherical atoms, which is strictly valid only if the electron distribution in the crystal can be represented by a superposition of spherical free-space atoms. It is this assumption that is challenged in most of the electron-density work. The two assumptions are not independent as a deviation from atomic sphericity can be a result of thermal vibrations and/or of the anisotropy of the chemical bonds. This is one of the reasons why it is important to reduce thermal vibrations by cooling the crystal and to collect independent structural information by an alternative technique, such as neutron diffraction, for which the second assumption is eliminated.

The structure factor applicable to the diffraction of neutrons is identical to that expressed in (2.2) but with the replacement of $f_n$ by the nuclear neutron-scattering length $b_n$. Since the dimension of the atomic nuclei is much smaller than the wavelength of the thermal neutrons to be used for diffraction (by a factor of $10^5$), the neutron-scattering length, unlike the x-ray scattering factor, is not dependent on the Bragg angle $\theta$ (more specifically sin $\theta/\lambda$). Once $b_n$ is determined experimentally, no further assumptions as to the distribution of the atomic scattering matter are required. Thus, errors in x-ray thermal and positional parameters due to the isolated-atom approximation can be investigated by comparing x-ray and neutron-diffraction results on the same crystal structure.

Discrepancies between x-ray and neutron thermal parameters are often as high as 10% for first-row atoms[12], but typically less for heavier atoms where the valence electrons constitute a smaller fraction of the total electron density. The anisotropy of the discrepancies in oxalic acid dihydrate and *sym*-triazine is such that in the x-ray analysis the atoms appear to be smeared out in directions corresponding to the chemical bonds and the lone-pair electrons[12] (Figure 2.1). On the other hand, x-ray parameters describing vibrations

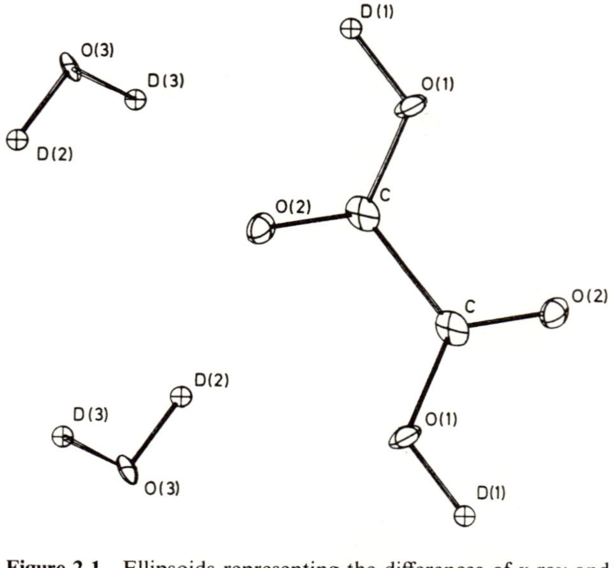


**Figure 2.1** Ellipsoids representing the differences of x-ray and neutron diffraction thermal parameters for $\alpha$-deutero-oxalic acid dideuterate (80% probability ellipsoids). (From Coppens[12], by permission of the International Union of Crystallography)

perpendicular to the molecular plane in aromatic molecules are often too small, suggesting a contraction of electron density into the molecular plane[13,14]. Thermal mean-square amplitudes from x-ray diffraction are therefore not a reliable basis for the study of molecular dynamics in solids, unless the effect of bonding on the molecular electron density is incorporated into the model for the structure.

A similar, but less pronounced, effect is observed for x-ray positional parameters of atoms which are in an asymmetric bonding environment[15,16,20]. Hydrogen atoms with their single (valence) electron are typically displaced by 0.1–0.2 Å into the bond to the adjacent atom in comparison with neutron positions[17, 18, 27], the displacement being larger for OH than for CH [19]. For first-row atoms, the asphericity shift is smaller by a factor of ten or more. Atoms with lone-pair electrons may be displaced away from the bonds, towards the lone-pair region[16,20], while carbon atoms are shifted into triple bonds[21, 22, 175] and towards the centre of the ring in aromatic compounds[23]. In *p*-terphenyl[24], pyrene[25] and phenanthrene[26], the bond length discrepancies are largest for C—C bonds between two CH groups (0.021 and 0.012 Å for pyrene and phenanthrene, respectively) and smaller for the other bonds. This is to be expected if the asphericity shift results from the imbalance between the C—H and C—C overlap densities. Hazel *et al.*[25] have correlated the shifts with the $\pi$-bond order of the C—C bonds. A quantitative calculation of the shifts should account for both effects. Asphericity shifts of the oxygen atoms in a number of molecules are listed in Table 2.1. The origin of the shifts is fairly well understood but their magnitude is dependent on the completeness of the x-ray data set.

To understand the nature of this dependence, it is necessary to separate the scattering of the core electrons from the valence electrons. A typical

**Table 2.1 Examples of x-ray asphericity shifts for oxygen atoms**

| *Compound* | *Direction* | *Magnitude/Å* | *Ref.* |
|---|---|---|---|
| Tetracyanoethylene oxide | 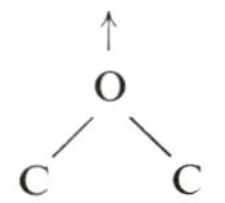 | 0.013(4) | 28 |
| Oxalic acid dihydrate | 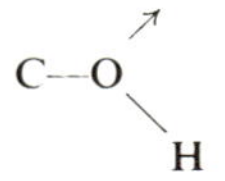 | 0.008(2) | 29, 182 |
| Sucrose | 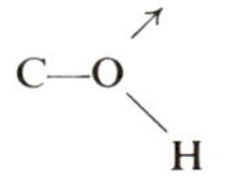 | 0.008(2)<br>0.007(2) | 19 |
| | and 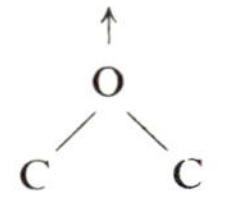 | | |
| Ammonium oxalate monohydrate | 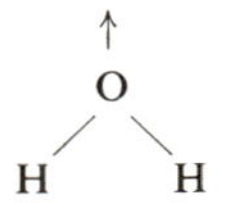 | 0.013(3) | 30 |
| Cyanuric acid | C=O → | 0.005(1) | 31 |

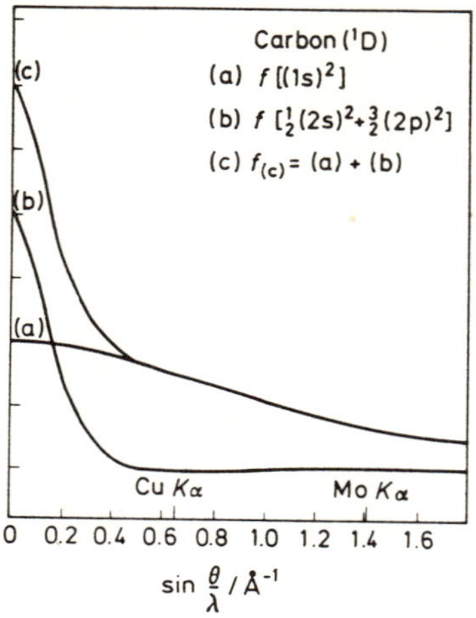


**Figure 2.2** Atomic scattering factor curve for carbon ($^1$D). (From Stewart[32], by courtesy of the American Institute of Physics)

curve for a first-row atom ground state is given in Figure 2.2[32]. Since the valence electrons are relatively diffuse, their scattering function is compact, i.e. the valence scattering of the *isolated* atom is concentrated in the low-order region. This suggests that refinements based on high-order reflections would emphasise the core electrons, which are relatively unaffected by bonding.

### 2.2.2 Modified spherical x-ray refinements

Three different methods, within the spherical approximation, have been applied to reduce the bias in the x-ray parameters:

(i) High-angle (HA) refinement[33, 34, 176] in which all the data below an arbitrary cut-off in sin $\theta/\lambda$ are eliminated. The cut-off is often taken as 0.64 Å$^{-1}$, (the limit accessible with Cu *Ka* radiation) but other values have been used (see Table 2.3).

(ii) A double-atom (DA) refinement[35, 36] in which the core and valence components are treated as entities with different positional but common thermal parameters, and the occupancy of the valence shell is taken as an additional variable in a least-square refinement based on

$$F_{hkl}(\text{calc}) = \sum_{\substack{\text{all}\\ \text{atoms}}} a_n \exp\,(2\pi i\, \boldsymbol{H}\cdot\boldsymbol{r}_{n,\text{core}}) + b_n \exp\,(2\pi i \boldsymbol{H}\cdot\boldsymbol{r}_{n,\text{valence}}) \qquad (2.3)$$

The positional parameters for the core electrons, $x_c$, $y_c$, $z_c$, are derived mainly from the high-order reflections which, according to Figure 2.2, have a small valence contribution, while the separate valence-shell positional parameters allow for the displacement of the valence-shell centroid from the position of the atomic nucleus.

(iii) The sharpened-atom (SA) refinement of Dunitz and Seiler[37], in which the least-squares weights $w$ in equation (2.1) assigned to the x-ray observations are modified to emphasise the high-order data. This method corresponds to fitting spherical atoms to the experimental electron density by emphasising the fit near the nuclear positions at the expense of the fit in the bonding and lone-pair regions.

There is considerable evidence that such refinement modifications eliminate part of the bias in the x-ray parameters. For example, HA refinements of uracil[33], cyanuric acid[31], sucrose[19], hexamethylenetetramine[39] and of cyanoguanidine[38] consistently yield thermal parameters lower than those obtained in a refinement of all the data, in qualitative agreement with the conclusions drawn from the comparison of x-ray and neutron results.

Similar changes are found when the bond lengths and atomic positions are examined. Thus the interatomic distances in the $N_2C$—N group of cyanoguanidine are reported as (high-angle, low-angle, respectively): N≡C (1.167 Å, 1.155 Å) and C—N (1.302 Å, 1.311 Å)[38]. A shift in the carbon atom accounts for the differences, in good agreement with the discrepancies between the x-ray and neutron results on tetracyanoethylene[22]. The sharpened-atom refinement of the crown ether 1,4,7,10,13,16-hexaoxa-cyclo-octadecane and two of its complexes with thiocyanates leads to C—C bonds lengthened by about 0.01 Å from a mean of 1.492 to 1.504 Å, and to C—O bonds shortened by 0.006 Å from an average of 1.419 to 1.413 Å[37]. The DA refinement correctly reproduces the core position for the hydroxylic oxygen atom[36], which was displaced by 0.008 Å in the conventional x-ray refinements of both isotopic species[29].

These results agree well with the asphericity shifts discussed in the previous section, but inconsistencies have appeared in some studies, indicating the necessity of explicit allowance for atomic asphericity. For example, the low-angle refinements of cyanuric acid[31] and ammonium tetraoxalate[39] give better positional parameters for the carbonyl oxygen atoms (as judged by comparison with the neutron results) than the high-angle refinements. Such behaviour cannot be explained with the scattering factors of Figure 2.2. But if the valence electrons in the lone-pair region would scatter at higher angles than indicated by Figure 2.2, the balance between the scattering of the lone pair and the bonding electrons at opposite sides of the oxygen atom would be poorer in the high-order region. Indeed, calculations indicate that the concentration of electrons in the lone-pair region leads to valence scattering at higher angles than indicated by the isolated spherical-atom calculation[40, 41]. The same shortcoming of the model affects the double-atom refinement of cubic tetracyanoethylene, in which the correct C—N bond length is obtained by a shift of the nitrogen towards the lone pair, rather than by the shift of the cyano carbon atom towards the triple bond indicated by the neutron analysis[22]. Non-spherical formalisms for treating the x-ray data, discussed in Section 2.5, eliminate such inconsistencies.

## 2.3 EXPERIMENTAL DIFFERENCE DENSITIES

### 2.3.1 Theory

The deviation from the charge density implicit in the model structure is commonly obtained by a Fourier transform of the differences between the observed and calculated structure factors:

$$\rho(\boldsymbol{r}) = 1/V\sum_{\substack{\text{all}\\ \text{observations}}}(\boldsymbol{F}_{\text{obs}} - \boldsymbol{F}_{\text{calc}}) \exp(-2\pi i \boldsymbol{H}\cdot\boldsymbol{r}) \tag{2.4}$$

where $\boldsymbol{F}_{\text{obs}}$ and $\boldsymbol{F}_{\text{calc}}$ are vectors in the complex plane. When the calculated values are based on the x-ray least-squares parameters, the difference function $\rho(\boldsymbol{r})$ will be relatively featureless because the bias in the positional and the temperature parameters is such as to produce the best fit between $\boldsymbol{F}_{\text{obs}}$ and $\boldsymbol{F}_{\text{calc}}$. More informative maps are obtained with parameters from an independent neutron experiment, from which

$$\boldsymbol{F}_{\text{calc,N}} = \sum_{\text{atoms}} f_{i,\text{X}} \exp(2\pi i\, \boldsymbol{H}\cdot\boldsymbol{r}_{i,\text{N}}) T_{i,\text{N}} \tag{2.5}$$

where the $f_{i,\text{X}}$ is the x-ray scattering factor of the $i$th atom, and $r_{i,\text{N}}$ and $T_{\text{N}}$ represent the neutron parameters[42, 43]. With scattering factors based on the ground states of the isolated atoms, the resulting function $\rho_{\text{X-N}}$ represents the redistribution of the electron density upon molecule formation, time-averaged over the vibrational states in the crystal.

The structure factors $\boldsymbol{F}$ are in general complex, so both magnitude and phase is required for the summation (2.4). As the phase angle is not experimentally accessible, a calculated value must be used, which in centrosymmetric crystals is either 0 or $\pi$ (i.e. $\boldsymbol{F}_{\text{obs}}$ is positive or negative). In the centrosymmetric case, the calculated phase can be used since model errors lead to few sign changes:

$$\rho_{\text{X-N}} = 1/V\sum_{\substack{\text{all}\\ \text{observations}}}\left(\boldsymbol{F}_{\text{obs,X}}\frac{\boldsymbol{F}_{\text{calc,N}}}{|\boldsymbol{F}_{\text{calc,N}}|} - \boldsymbol{F}_{\text{calc,N}}\right) \cos 2\pi \boldsymbol{H}\cdot\boldsymbol{r} \tag{2.6}$$

For acentric crystals, the phase difference between $\boldsymbol{F}_{\text{obs,X}}$ and $\boldsymbol{F}_{\text{calc,N}}$ may be typically 2–3° [19, 39, 43] and its neglect leads to an under-estimate of the magnitude of the difference vector $\Delta\boldsymbol{F} = \boldsymbol{F}_{\text{obs,X}} - \boldsymbol{F}_{\text{calc,N}}$ and a large random error in the phase of $\Delta\boldsymbol{F}$. The average error in the difference density which is dependent on errors in both the magnitude and the phase $\phi_{\text{X-N}}$ of $\Delta\boldsymbol{F}$ is[42, 43]

$$\langle\sigma^2(\rho_{\text{X-N}})\rangle = 2/V^2\sum_{0}^{\infty}\sigma^2\Delta\boldsymbol{F} + \Delta\boldsymbol{F}^2\sigma^2(\phi_{\text{X-N}}) \tag{2.7}$$

The last term represents the phase ambiguity which can give a considerable contribution for acentric crystals. It would seem that the best approximation to $\phi_{\text{X}}$ can be obtained from a refinement of the x-ray data with a generalised model, sufficiently flexible to account for all asymmetric density features.

### 2.3.2 Qualitative discussion of X–N densities

Compounds studied by combined x-ray and neutron diffraction are listed in Table 2.2. Typical difference maps (e.g. Figures 2.3 and 2.4) show excess density in the bond and lone-pair regions and an electron deficiency near the nuclear positions. A quantitative analysis requires the consideration of experimental errors and the effect of averaging over the populated vibrational

**Table 2.2 Combined x-ray and neutron charge-density studies***

| *Compound* | *Centric/acentric–temperature*/K | *Ref.* |
|---|---|---|
| *sym*-Triazine | c–300 | 42 |
| Uronium nitrate | c–300 | 45, 177 |
| Oxalic acid dihydrate | c–300 | 29 |
| Tetracyanoethylene oxide | c–300 | 28 |
| Cubic tetracyanoethylene | c–300 | 22 |
| Ammonium tetraoxalate | c–300 | 39 |
| 1,3,5-Triacetylbenzene | c–300 | 13 |
| α-Glycine | c–300 | 46 |
| α-Glycylglycine | c–300 | 60, 183 |
| Orthorhombic sulphur | c–300 | 47 |
| Cyanuric acid | c–100 | 31 |
| Decaborane | c–113 | 48 |
| Benzenechromium tricarbonyl | c–78 | 49 |
| *p*-Nitropyride *N*-oxide | c–30 | 14 |
| Sucrose | a–300 | 19, 50 |
| Ammonium oxalate monohydrate | a–300 | 30 |
| Hexamethylenetetramine | a–300 | 39, 51 |

*As at January 1974

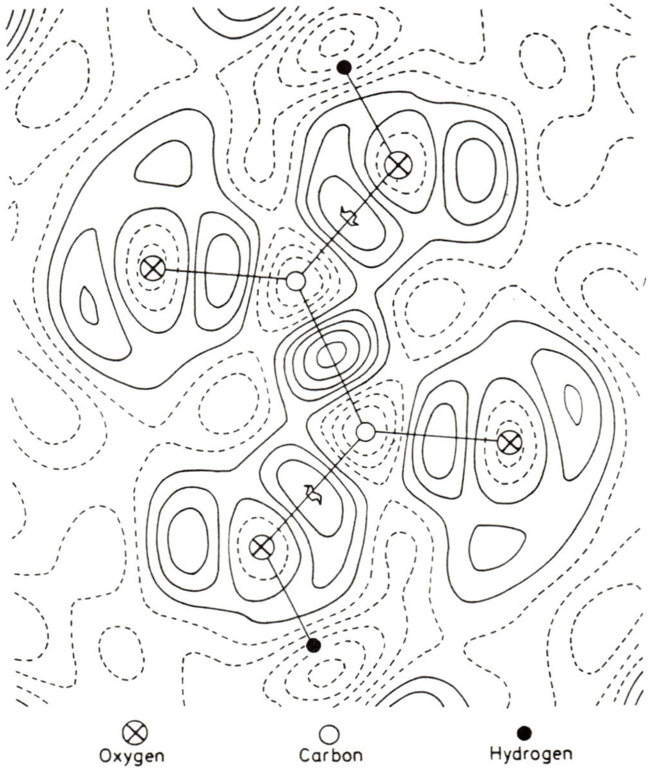


**Figure 2.3** X–N map through the molecular plane of the oxalic acid molecule. Contours at 0.05 e Å$^{-3}$. Zero and negative contours dashed. (From Coppens *et al.*[29], by courtesy of the International Union of Crystallography)

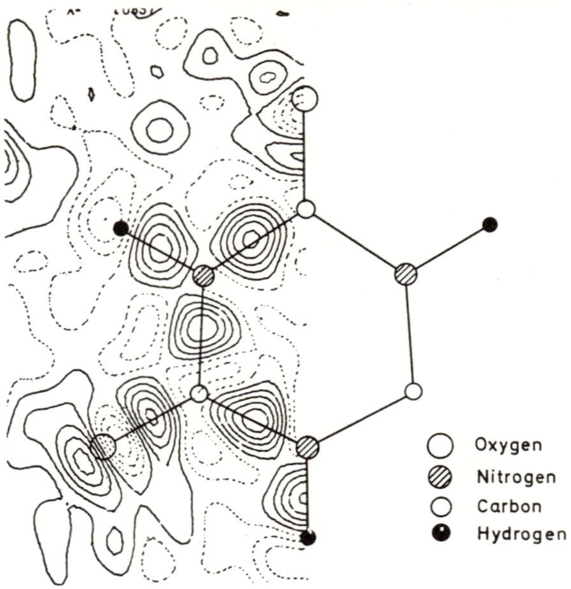


**Figure 2.4** X–N map through the molecular plane of cyanuric acid. Contours at 0.1 e Å$^{-3}$, negative contours dotted. (From Coppens and Vos[31], by courtesy of the International Union of Crystallography)

states (Section 2.4.2), but the qualitative aspects are remarkably consistent in different maps. We will consider the lone-pair and bond features separately:

(a) *Lone-pair peaks*: CO oxygen atoms in conjugated systems show a broad 'half-moon'-shaped maximum extended in the molecular plane (cyanuric acid[31], oxalic acid dihydrate[29], glycine[46], ammonium oxalate[30]) or two separated maxima (the second CO in cyanuric acid, glycine and ammonium oxalate), suggesting two lone pairs with approximate $sp^2$ hybridisation. A single maximum is found on the hydroxyl oxygen in oxalic acid. Very sharp, single peaks also appear along the extension of the C≡N bond in cyano compounds[22, 28] and at the back of the carbonyl oxygen in benzene-chromium tricarbonyl[49], indicating sp hybridisation of the nitrogen and oxygen lone pairs. A very pronounced peak extending perpendicular to the molecular plane occurs at the oxygen atom in tetracyanoethylene oxide[28]. It is similar in shape but more intense than the lone-pair peaks on the water molecules in oxalic acid dihydrate and ammonium oxalate monohydrate, and the general appearance is consistent with approximate $sp^3$ hybridisation of the two lone-pair orbitals.

(b) *Bond density maxima:* Distinct maxima of varying height are found on the internuclear axes in C—C, C—O, C—N, B—H, C—H and N—H bonds. The maxima are clearly shifted away from the bond axis, however, in strained three-membered ring systems (ethylene oxide[28], and also in x-ray maps of ethyleneimine[52] and cyclopropane[53], see next section) and in the three-centre

B—H—B bond[48] (Figure 2.5), the bond peaks being displaced outwards in the former and inwards in the latter case. Cross-sections perpendicular to the bond at bond mid-point show the cylindrical symmetry of the single and triple bonds, and the extension of double or partially double bonds in the direction associated with the $\pi$-orbital axis (Figure 2.6). These observations are very well confirmed by the larger number of x-ray studies discussed in the next section. Sections perpendicular to the molecular plane containing the bond axis further indicate a polarisation of the density towards the more electronegative atom of a heteroatomic bond (glycylglycine[60], *sym*-triazine[42]).

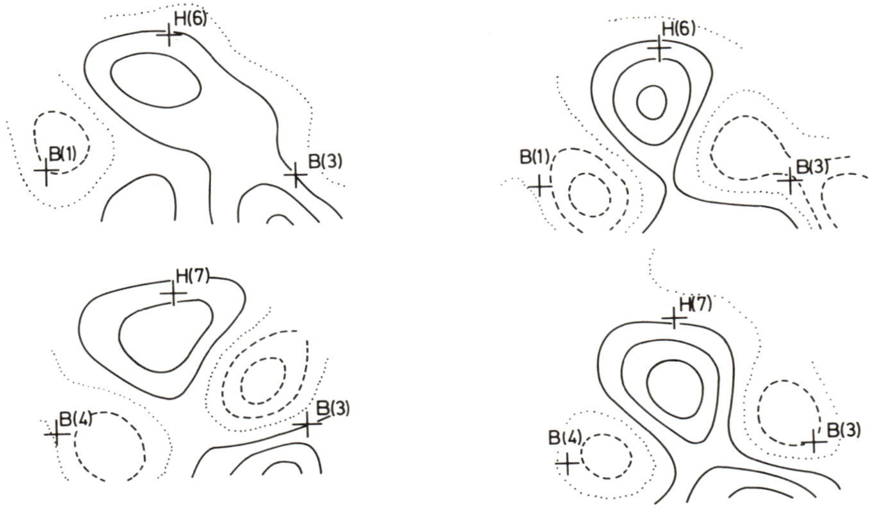


**Figure 2.5** X–N sections in the plane of the BHB atoms in decaborane. Contours at 0.1 e Å$^{-3}$. Zero counter dotted, negative contours dashed. (From Brill *et al.*[48], by courtesy of the International Union of Crystallography)

(c) *Other features*: A large residual peak is observed near the chromium atom in benzenechromium tricarbonyl. Assessment of its physical significance requires additional studies of metallo-organic complexes. A complication arises from the low neutron scattering power of Cr, compared with its dominance in the x-ray scattering. In such a situation, errors in the neutron measurement of the Cr parameters are magnified in the X–N map.

The densities on sucrose[19] and glycine[46] suggest that the polarising influence of hydrogen bonding on the molecular density is measurable, in agreement with theoretical studies of the formamide dimer[54] and the HF$\cdots$ $H_2O$ complex[55], which predict density changes of 0.1–0.2 e Å$^{-3}$.

A survey of peak heights reveals considerable fluctuations among chemically comparable features[43]. For example, C—C bond peak heights range from 0.25 e Å$^{-3}$ in oxalic acid to 0.6 e Å$^{-3}$ in tetracyanoethylene oxide, and lone-pair densities from 0.15 to 0.55 e Å$^{-3}$ in the same two compounds. The chemical significance of these differences is largely obscured at present by the effects of thermal averaging (Section 2.4.2), by an unnecessary lack of knowledge of the absolute scale of the x-ray data and an, as yet, incomplete

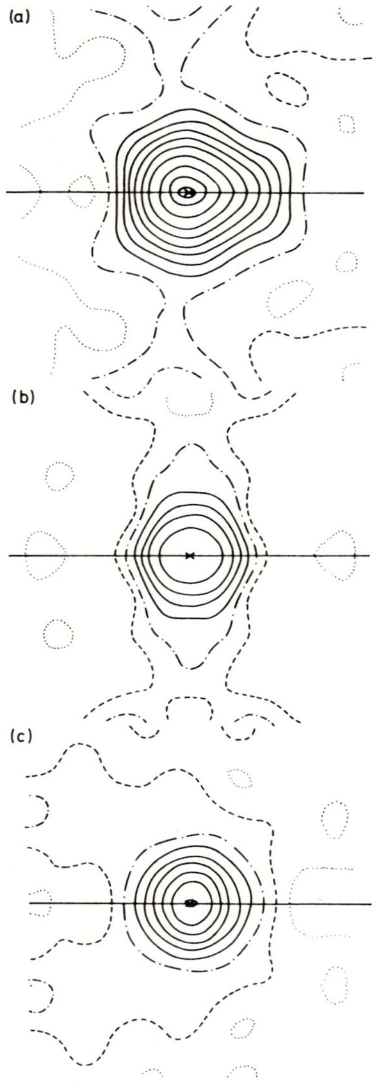


**Figure 2.6** X–N sections through mid-point, perpendicular to C≡N (a), C═C (b) and C—C (c) bonds in tetracyanoethylene. Contours at 0.1 e Å$^{-3}$ Negative contours dashed. (From Becker *et al.*[22], by courtesy of the American Chemical Society)

analysis of the effect of thermal diffuse scattering on the x-ray and neutron intensities.

But the prospects for more quantitative work seem good as further advances in experiment and interpretation can be expected. In the results obtained so far, density peaks and hollows appear where expected from advanced, but not from approximate, theoretical calculations (Section 2.4).

### 2.3.3 X–X densities

#### *2.3.3.1 Molecular crystals*

Combining x-ray and neutron results implies a doubling of experimental effort, and introduces errors inherent in the neutron data collection. To what extent can the neutron experiment be eliminated? A series of studies, listed in Table 2.2, provide a qualified optimistic answer. O'Connel, Ray and Maslen examined a number of previously published x-ray structures and found convincing evidence for bonding density in C—C bonds[23].

A systematic study of a number of related hydrocarbons[21] confirmed that even with low-order x-ray data alone certain aspects of the electron distribution can be studied. In general, the $\pi$-bond density, defined as the distortion from cylindrical bond symmetry, is evident in the direction perpendicular to the benzene rings at about 0.5 Å from the bond centres. But the apparent increase of the thermal vibrations of acetylenic and cyano[21] carbon atoms in the triple-bond direction and the atom shifts toward the bond centre[21, 22, 84] eliminate the density on the triple-bond axis so that the section perpendicular to the bond shows a trough surrounded by a ring of $\pi$ density (Figure 2.7).

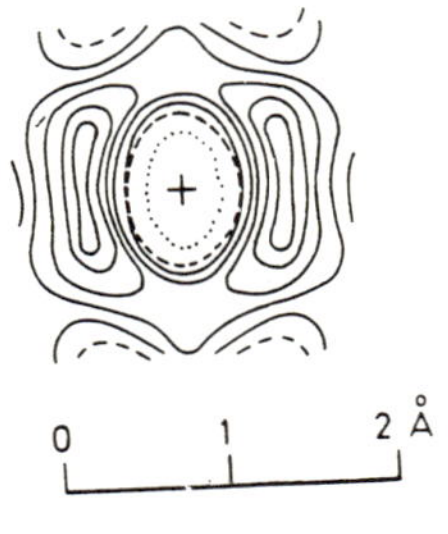


**Figure 2.7** X-ray difference density section perpendicular to C≡C bond at its centre in acetylenedicarboxylic acid showing an artificial deficiency on the bond axis. Contours at 0.02 e Å$^{-3}$. Negative contours dashed. The difference density at the centre of the bond is $-0.2$ e Å$^{-3}$. (From Irngartinger *et al.*[21], by courtesy of The Chemical Society)

It is clearly desirable to use unbiased parameters from a modified treatment of the x-ray data (Sections 2.2.2 and 2.5) in the calculation of difference densities. Several studies listed in Table 2.3 have appeared in which parameters from a high-order refinement are used in equation (2.5). In the

**Table 2.3 Recent x-ray charge-density studies of molecular crystals***

| *Compound* | $(\sin\theta/\lambda)_{max}$ | $(\sin\theta/\lambda)_{min}$ *of high-order refinement*/$Å^{-1}$ | *Temp.*/K | *Ref.* |
|---|---|---|---|---|
| Hexamethylenetetramine | 1.1 | 0.65 | 300 | 39 |
| Penta-*m*-phenylene | 0.64 | — | 300 | 21 |
| Hexa-*m*-phenylene | 0.64 | — | 300 | 21 |
| Cyclododeca-1,5,9-triene-3,7,11-triyne | 0.64 | — | 300 | 21 |
| 1,8-Bis(phenylethyl)naphthalene | 0.64 | 0.4 | 300 | 175 |
| *cis*-2,5-Dimethylhex-3-ene-2,5-diol | 0.96 | — | 110 | 57 |
| *trans*-2,5-Dimethylhex-3-ene-2,5-diol | 0.96 | 0.7, 0.8 | 110 | 56, 57 |
| *trans*-2,5-Dimethylhex-3-yne-2,5-diol | 0.96 | — | 110 | 56, 57 |
| Acetylenedicarboxylic acid | 0.64 | 0.4 | 300 | 21, 175 |
| Fumaramic acid | >0.64 | — | 300 | 58 |
| 2,5-Dimethylbenzoquinone | 0.64 | — | 300 | 59 |
| Cyanuric acid | 0.8 | 0.6 | 100 | 31 |
| Ethyleneimine quinone | 0.6 | — | 300 | 52 |
| | 0.7 | — | 253 | 52 |
| | 0.7 | — | 143 | 52 |
| Uracil | 0.902 | 0.648 | 300 | 33 |
| Sucrose | 0.81 | 0.64 | 300 | 19 |
| α-Glycylglycine | 0.85 | 0.65 | 300 | 60 |
| Ammonium tetraoxalate | 1.3 | 0.65 | 300 | 39 |
| Tetracyanoethylene | 0.94 | 0.6 | 300 | 22 |
| Tricyanocyclopropane | 0.64 | — | 300 | 53 |
| Tetracyanocyclobutane | 1.08 | — | | 66 |
| Cyanoguanidine | 1.08 | 0.64 | 300 | 38 |
| 2 Aminoethylsulphonic acid (taurine) | 0.91 | 0.65 | 300 | 34 |
| Thiocytosine | 1.0 | 0.64 | 300 | 61 |
| Hexaminecobalt(III) hexacyano-cobaltate(III) | 1.3 | 0.6† | 300 | 62 |
| Sodium azide | 1.4 | 0.65 | | 39 |
| 3-Peroxotitanium(IV) chelates | 0.705 | — | 300 | 63 |
| Kernite | 0.72 | 0.60 | 300 | 168 |
| Cyclotriborazane | 0.55 | — | 300 | 170 |

* Compounds of Table 2.2 are cross-listed here if an analysis of the x-ray data alone is available. Only studies using diffractometer data are included. Data correct as at January 1974.
† Data between 0.6 and 1.0 $Å^{-1}$ used in the high-order refinement.

difference function $\rho_{X-X(\text{high angle})}$, bond densities appear which are quite comparable with those in X–N maps, except that lone-pair peaks are underestimated when a cut-off of 0.65 $Å^{-1}$ is used[14, 19, 60]. This can be understood if lone-pair scattering persists beyond 0.65 $Å^{-1}$, an explanation which is supported by other evidence (Section 2.2.2).

The difference density in the azide ion[39] (Figure 2.8) is a quite interesting exception which indicates the usefulness of a careful collection of high-order data (<1.3 $Å^{-1}$) in a case where many of the positional and thermal parameters are fixed by the high crystallographic site-symmetry of the azide ion. In general, a sufficient number of reflections may be measured at low temperature so that the high-order refinement can be restricted to a more extreme range of sin $\theta/\lambda$. But the majority of x-ray difference maps are informative mainly in parts of the molecules not involving atoms with lone-pair electrons.

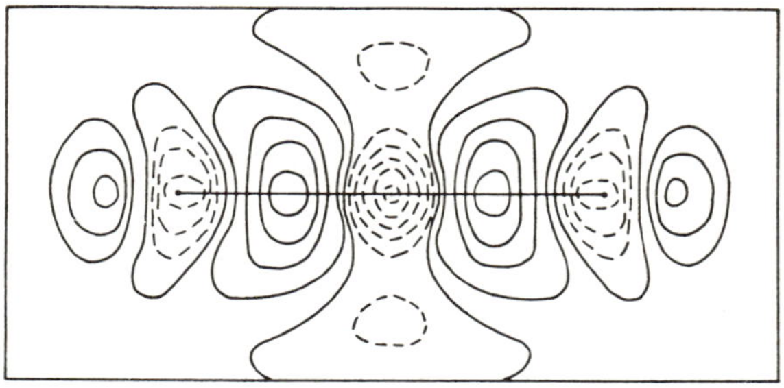

**Figure 2.8** X–X (high angle) density in the azide ion of sodium azide. Contours at 0.1 e Å$^{-3}$. Negative contours broken. (From Stevens[39], by courtesy of E. D. Stevens)

For example, even though the difference density is flat in the C=O region of the quinone part of ethyleneimine quinone[52], the bending of the bonds in the ethyleneimine ring is shown very convincingly at all three temperatures (Figure 2.9). This feature agrees well with observations made for the three-membered rings in tricyanocyclopropane[53], 2,5-dimethyl-7,7-diacyanonor-caradiene[64], in tetracyanoethylene oxide[28] and even for the four-membered

**Figure 2.9** X-Ray difference density in the plane of the three-membered ring of ethyleneimine quinone at 143 K. Contours at 0.05 e Å$^{-3}$. Negative contours omitted. Open circles represent carbon atoms. (From Ito and Sakurai[52], by courtesy of the International Union of Crystallography)

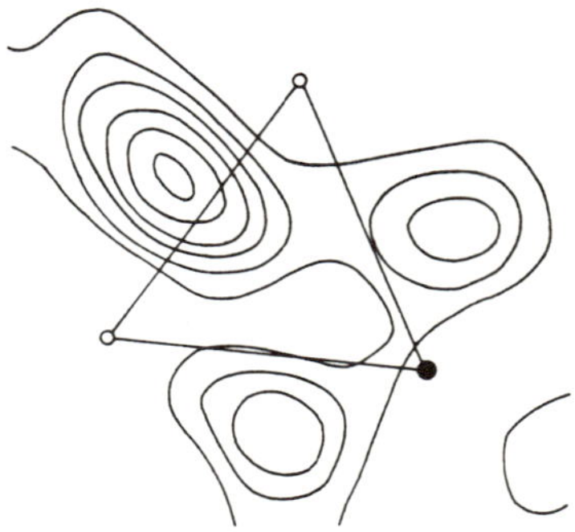

ring in tetracyanocyclobutane[66]. It represents a direct experimental proof for the bending of bonds in small-ring compounds as proposed first by Coulson and Moffit[65]. A similar bond-bending in the TiOO ring in titanium peroxo complexes is suggested by the work of Schwarzenbach[63].

**Figure 2.10** X-Ray difference density perpendicular to a double (a) and a triple (b) bond at 110 K. Contours at 0.1 e Å$^{-3}$. Zero contour dotted. (From Helmholdt *et al.*[56], by courtesy of the International Union of Crystallography)

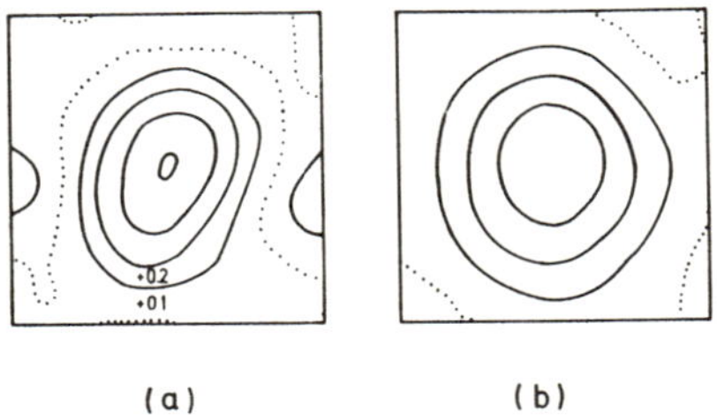

Sections perpendicular to the double and triple bonds in *trans*-2,5-dimethylhex-3-ene-2,5-diol and *trans*-2,5-dimethylhex-3-yne-2,5-diol[56,57], based on the low-order data only, with parameters from the full refinement are shown in Figure 2.10.

The improvement over the earlier low-order results (Figure 2.7) is striking. The shape of the triple-bond cross-section now agrees well with the X–N sections on tetracyanoethylene[22] (Figure 2.6) and tetracyanoethylene oxide[28].

### 2.3.3.2 *Difference densities for some metallic, ionic and covalently bonded crystals*

Ionic or 'network' crystals have high Debye temperatures so the thermal smearing of the density is smaller than in molecular crystals. Furthermore, as the atomic positions are often fixed by symmetry considerations, positional parameters are not biased by assumptions inherent in the selection of a refinement model. A major drawback is the small unit-cell size of materials such as diamond, silicon, LiH, NaCl, Al, etc. which makes the low-order region, in which the valence scattering is concentrated, poorly accessible. In aluminium, there are only five reflections below $\sin\theta/\lambda = 0.5$ Å$^{-1}$, and discrepancies between the structure factors calculated with isolated-atom form factors and those from a pseudo-potential calculation by Ascarelli and Racah exist only for the (111) and (200) reflections[67]. They both indicate that the valence electrons are more diffuse in the metal than in the isolated atom, as may be expected. The observed values (on an absolute scale) are within the experimental error identical with the results of the Ascarelli and Racah calculation, but in the case of the (220) reflection disagree with a second pseudo-potential calculation by Walter *et al.*[68]. The valence charge density in aluminium[67,68], and also in vanadium metal[69], shows its largest maxima between nearest-neighbour atoms. Experimental evidence indicates that this is not the case for metallic beryllium for which accurate data were collected by Brown[70]. The difference between the OPW (orthogonalised plane wave) scattering factors and the experimental measurements in the region below 0.45 Å$^{-1}$ indicate tight binding, i.e. covalent bond formation. Further analysis by Stewart[71], using methods described in Section 2.5, indicates that the bonding is of the multicentre type. This is confirmed by the valence-density map which shows a diffuse distribution with distinct maxima of about 0.08 e Å$^{-3}$ in the centre of a tetrahedron formed by four beryllium atoms (Figure 2.11)[72].

Very accurate structure factors from Pendellösung-fringe measurements are available for Si[73,74]. Their theoretical interpretation has been discussed in detail by Aldred and Hart[75], and excellent agreement with form factors from an SCOPW (self-consistent orthogonalised plane wave)[178] is obtained, especially if the Kohn–Sham–Caspar exchange term is used. A Fourier transform of the Si measurements (Figure 2.12) shows reasonably good agreement for the electron density in the bond region when comparison is made with pseudo-potential calculations[72,76]. The comparison in the nuclear regions is hampered by uncertainties in the dispersion correction for Si,

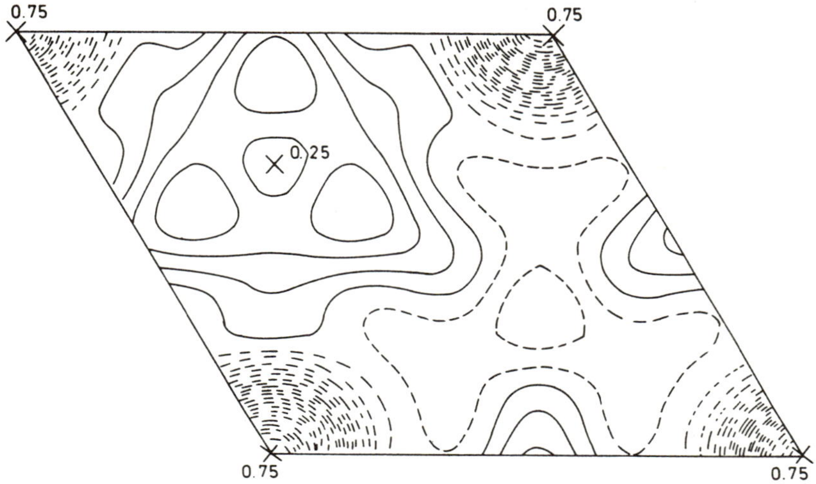


**Figure 2.11** Difference density in beryllium metal. Section parallel to the hexagonal *ab* plane at $z = 0.675$ containing the interatomic density maximum. The projection of the Be atoms and their $z$ coordinates are indicated. Contours at 0.025 e Å$^{-3}$. Negative contours dotted. (From Yang and Coppens[72])

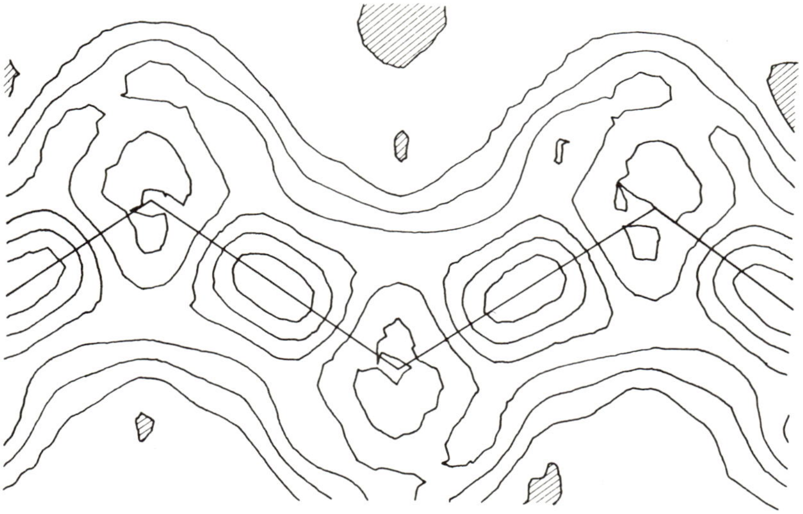

**Figure 2.12** Valence density in the plane of the Si—Si bonds in silicon. Contours at 0.1 e Å$^{-3}$. Negative areas cross-hatched. (From Yang and Coppens[72])

small variations of which have a pronounced effect on the density near the nucleus.

Qualitative agreement with pseudo-potential form factors has also been reported for InSb by Collela and co-workers[77]. This study, and related work on GaAs[78], GaP and GaSb[78, 173], confirm the covalent character of the bonding in Group III–Group V compounds. Accumulation of charge between nearest neighbours has also been observed in NiAs[79], NiSb[79], NiSe[79], NiO[80] and even in AgBr[81], but not in NaCl[181]. A survey of much of the earlier work on metals, alloys and covalent compounds has been given by Sirota[82], while Hosoya has summarised the results for ionic crystals and metallic oxides, carbides and nitrides up to 1969[83].

For the more recent studies, agreement between theory and experiment is quite good, and, more important, in several cases the measurements have been able to discriminate in favour of particular theoretical treatments (the Ascarelli and Racah calculation on Al, the Kohn–Sham–Caspar exchange term for Si, multicentre bonding in Be). The calculational methods applied to molecular crystals are quite different, but as discussed in the following sections, a similar series of sensitive tests is possible using the experimental charge densities. A series of ionic crystals (LiF, NaCl, KBr, NiO, MnO, $CaF_2$ and others) has been carefully analysed by methods developed by Kurki Suonio and his co-workers (Section 2.6.2).

## 2.4 THEORETICAL DIFFERENCE DENSITIES AND COMPARISON WITH EXPERIMENT

### 2.4.1 Molecular difference densities from theoretical calculations

Although the crystals discussed in the preceding section form infinite lattices of bonds, crystal forces in molecular solids may be considered as a perturbation on the electronic structure of the molecular unit. Thus, to a first approximation, experimental results may be compared with theoretical densities for the isolated molecular unit. Any such comparison on a quantitative level requires an appropriate averaging of the molecular electron density over the vibrational states of the molecule, a problem of considerable complexity which will be considered in the next section. The number of theoretical molecular *difference* densities that has been reported is small in comparison with the large existing body of molecular calculations. Nevertheless, several workers have come to realise that the difference function (or Roux function) is an exceedingly sensitive measure of comparison[85-92]. Indeed, many approximate calculations are known to give good values of the dipole moment, quadrupole moment and other quantities derived from the first-order density matrix, but fail to reproduce the details of the difference density as predicted by more sophisticated calculations or measured by diffraction methods. A systematic study of the improvement of the diatomics HF and $O_2$ has been made by Cade[93]. The very illuminating results obtained show that the difference densities are inadequate especially in the bond regions for minimal basis set, double zeta (two basis functions per electron) or even saturated sp basis sets (an adequate number of s- and p-type functions to obtain convergence of the molecular energy). In particular, the addition of polarised functions (i.e. functions with a higher principal quantum number, d orbitals in the case of first-row atoms) is required to build up the density in the bonding regions. Similar conclusions have been drawn for $N_2$ by Smith and Richardson[94] and for diborane and $N_2$ by Lipscomb *et al.*[95-97]. Extended Hückel difference densities on adenine show an analogous deficiency of charge in the bonding region[98]. The inadequacy of the semi-empirical INDO method[99] and of minimal basis set calculations is also apparent from comparison with the experimental difference densities on cyanuric acid (Figure 2.13)[100, 101]. In the theoretical maps, the overlap density

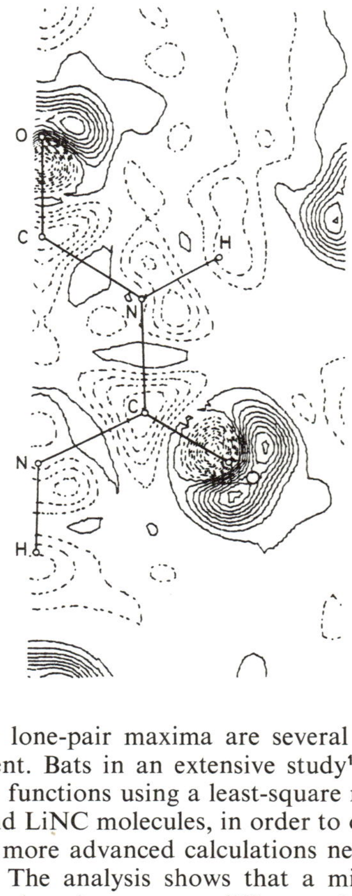


**Figure 2.13** Theoretical difference density in the plane of the cyanuric acid molecule according to a minimal basis set calculation. Contours at 0.1 e Å$^{-3}$. Negative contours dotted. The experimental map is shown in Figure 2.4. (From Jones *et al.*[100], by courtesy of the International Union of Crystallography)

is almost completely absent while the lone-pair maxima are several times higher than indicated by the experiment. Bats in an extensive study[102] has varied the orbital coefficients and basis functions using a least-square refinement on the $CN^-$ ion and the LiCN and LiNC molecules, in order to obtain a best fit to the electron density from more advanced calculations near the restricted Hartree–Fock (RHF) limit. The analysis shows that a minimal basis set (MBS) is inherently too restricted to give a proper difference density, even though the deficiencies may be less apparent when the gross features of the total charge distribution are compared. Still, the best possible fit to the $CN^-$ density with a MBS formalism gives an average deviation in the valence density of 10%, or about one electron. The fit will be worse in the usual minimisation of the total energy, in which the electron distribution near the nuclei is necessarily emphasised at the expense of the more remote parts of the electronic charge.

Consistency demands that to obtain a difference density both the atomic and the molecular density be calculated with comparable accuracy. But since the semi-empirical and approximate *ab initio* calculations are clearly insufficient, the nature of the basis set to be used for the separated atoms which are subtracted becomes less important, RHF atoms being needed in all useful cases. Bader *et al.*[89-93] in their pioneering analysis of the redistribution of charge upon the formation of diatomic molecules have subtracted out a non-spherical ground-state atom 'prepared' to undergo bond formation in the molecule. Subtraction of prepared atoms will accentuate the $\sigma$-bonding electrons in the difference map, as is evident from difference densities on $NF_3$, $NF_3O$, NSF and $NSF_3$[103]. An approach compatible with the experimental difference densities has been taken in an analysis of the electrostatic aspects of the bonding in diatomic molecules by Hirshfeld and Rzotkiewicz[104]. The 'pro-molecule' subtracted out by these authors is a superposition of spherically averaged ground-state atoms. Surprisingly, the net *electrostatic* force in the pro-molecule is always binding. A further analysis of the redistribution of the charge density in terms of the electrostatic forces acting upon the nuclei (which, according to the Hellmann–Feynmann principle must be zero for a stable molecule, but are non-zero in the hypothetical pro-molecule) show the effect of the $\sigma$-electron redistribution to be weakened by reversed polarisation inside the 2s nodal sphere, and further opposed by a small core polarisation. Such effects in the inner regions may not be experimentally detectable with diffraction methods because of thermal motion and the limited resolution of an x-ray experiment employing wavelengths in the 0.5–1 Å range[151]. Yet, they constitute a warning that the assumption of an unperturbed core distribution may not be warranted under all circumstances.

Theoretical densities have been analysed in terms of the transferability of segments of the charge density between different bonding environments, and the assignment of a fundamental unit in density space[105-108]. Ultimately, such a partitioning may achieve the synthesis of the electron density of large molecules from their constituent density fragments, and a subsequent derivation of other physical properties from the assembled charge density. A discussion of these theoretical analyses is outside the scope of this review, but the additivity, or lack of it, is clearly of importance for determining experimental strategy in selecting systems for future diffraction analysis. Difference maps on HC≡CH, LiC≡CH, HC≡N and N≡C—C≡N calculated by Hirshfeld[58] from molecular HF wave functions by McLean and Yoshimine[109] tend to support the transferability of features of the *difference* density. The regions of the C≡N groups in the last two molecules, for example, look remarkably similar (Figure 2.14) to each other and to difference densities for $C{\equiv}N^-$, LiCN and LiNC[102]. This conclusion seems to be confirmed by the comparison of the experimental X–N maps on tetracyanoethylene[22] and tetracyanoethylene oxide[28], and x-ray maps on tetracyanocyclobutane[66]. On the other hand, the transferability is only roughly valid in the series CO, SCO and $CO_2$ analysed by Hazelrigg and Politzer[88] from the McLean and Yoshimine wave functions. In particular, the lone pair on

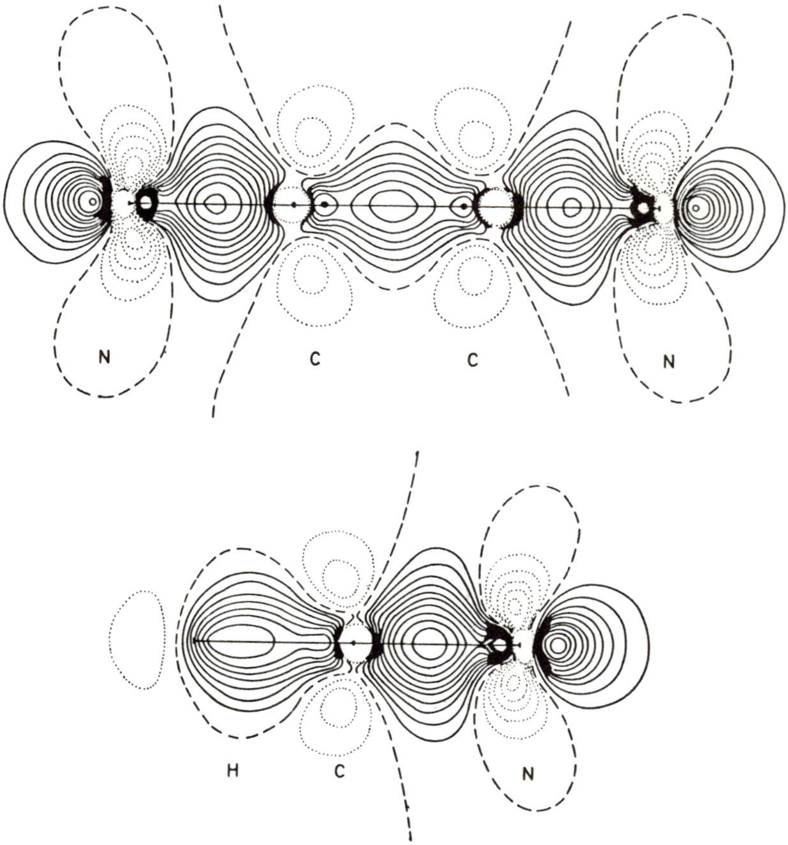


**Figure 2.14** Theoretical difference densities in HC≡N and N≡C—C≡N. Contours at 0.1 e Å$^{-3}$. Negative contours dotted, (negative contours near nuclei have been omitted). (From Hirshfeld[58], by courtesy of the International Union of Crystallography)

the oxygen is seen to vary with the nature of the substituents on the carbon atom. Furthermore, the concentration of electron density along the CO bond is much less in carbonyl sulphide than on carbon monoxide, in agreement with bond energies and slight differences in bond lengths. Such chemical differences are quite general, so electron-density additivity relationships can only be a first approximation, whether the basis functions are segmented density pieces or based on localised molecular orbitals[110].

The difference densities discussed up to this point are at best from calculations in the restricted Hartree–Fock limit[111], i.e. correlation effects have not been considered, and orbitals are assumed to be symmetry-restricted. What, if any, changes can be expected in the difference density if calculations are pursued beyond this limit? Only limited information is available at present. Calculations by Bader and Chandra[112], using wave functions by

Das and Wahl[113] on $H_2$ and $Li_2$, suggest the improvement is small in comparison with the differences between minimal basis set and RHF calculations. A similar result has been reported by Laws and Lipscomb for $H_2$[96]. But it is not obvious that such conclusions can be generalised to other molecular systems. Some evidence to the contrary is provided by the behaviour of the dipole moment upon approach to the RHF limit. Very large basis set calculations for the water molecule lead to a deterioration of the agreement between the calculated and observed dipole moments upon improvement of the basis set[114, 115]. The dipole moment of carbon monoxide has an incorrect sign in the RHF limit and the introduction of very many excited states in the calculation is necessary in order to obtain agreement with experiment[116]. The absolute error in the RHF dipole moment is 0.4 D for CO, CS and BH, and about 0.1–0.2 D for OH, FH, CH[111] and $H_2O$[115], the correct dipole moment being always lower than the RHF estimate. It is very desirable that difference densities corresponding to these wave functions become available so that the nature of the density change can be analysed and its experimental accessibility be assessed.

### 2.4.2 Thermal averaging of the charge density

A quantitative comparison of theoretical and experimental densities requires an accounting for the effect of thermal motion in the crystal. Low-frequency lattice modes give the main contribution to the atomic displacements, while internal modes of the molecules are barely excited at room temperature so that the associated atomic motions are practically temperature-independent when the sample is cooled. For the lattice modes, de-excitation is incomplete even at 10 K when the frequency is lower than about 50 $cm^{-1}$ [31]. For all modes zero-point motion remains, so that the static density is a hypothetical, unobservable quantity.

The effect of thermal smearing will be especially pronounced where the density function varies rapidly. The electron density and the difference function are respectively very positive and very negative at the nuclear positions, and an analysis of the temperature dependence of spherical-atom densities has led Stewart to conclude that 'with present-day precision, Fourier difference maps suffer an unmanageable bias near the time-average nuclear positions' [117]. This error is compounded by uncertainties in the experimental scale of the x-ray data[118] so that prospects for quantitative comparison in the regions around the nuclei are less promising than in the bonding and lone-pair regions. An extrapolation of the experimental densities to zero thermal motion requires a resolution not present in the experimental data; instead, the thermal averaging of the theoretical density may be considered. This is relatively straightforward for modes in which the bond lengths are not changing, such as lattice vibrations in a molecular crystal. But for the internal vibrational motion, the charge density in the Born–Oppenheimer approximation will be somewhat different for each molecular geometry along the normal mode displacement coordinate. Evaluation of a statistical average over the instantaneous difference densities requires a very large number of calculations, so approximate methods are needed. In the 'convolution approximation' proposed by Coulson and Thomas[119-121], the thermal average is calculated by assuming each of the atomic basis functions to vibrate with its nucleus, while the coefficients of the molecular orbital remain the same during the vibration. The statistical average of the two-centre terms in the charge density can then be evaluated from the vibrational wave function.

The convolution approximation agrees well with the exact 'Born–Oppenheimer' density for the $H_2$ molecule. Its consequences have been further explored in an important study of acetylene by Ruysink[57]. The application to a polyatomic molecule is not strictly justified by the fit obtained for $H_2$ because the $\pi$-type vibrations which occur in acetylene in addition to the $\sigma$ modes always lengthen the covalent bonds, but the approximation should still be quite reasonable. Figure 2.15 shows the difference density in acetylene

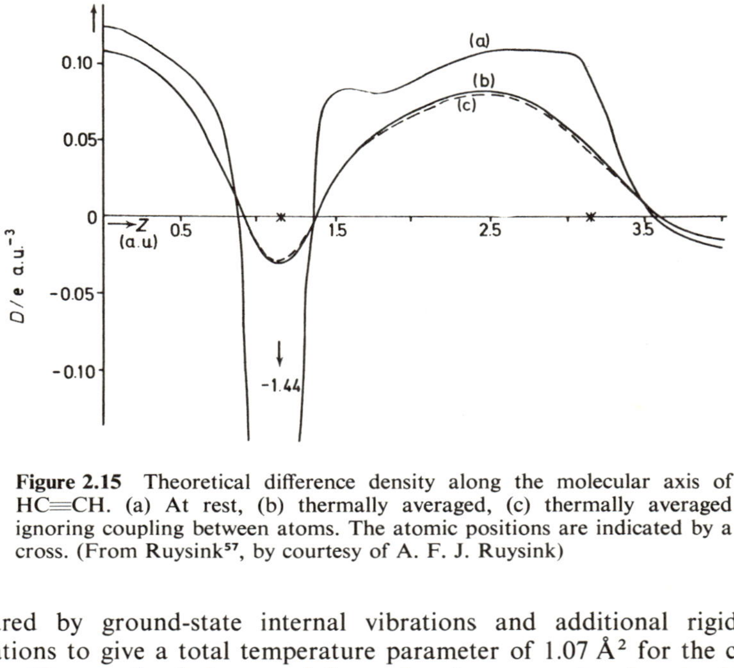


**Figure 2.15** Theoretical difference density along the molecular axis of HC≡CH. (a) At rest, (b) thermally averaged, (c) thermally averaged ignoring coupling between atoms. The atomic positions are indicated by a cross. (From Ruysink[57], by courtesy of A. F. J. Ruysink)

smeared by ground-state internal vibrations and additional rigid-body vibrations to give a total temperature parameter of 1.07 Å$^2$ for the carbon atoms. The negative hole around the nucleus is very much reduced, while the bond density is much less affected. When the coupling between the atomic motions is ignored in the smearing, an almost identical curve is obtained (curve (c) in Figure 2.15). The authors conclude that a knowledge of the normal vibrations is not required and that mean-square amplitudes from other sources (such as neutron diffraction or modified x-ray refinements) may be used for the calculation of the smeared-density function.

The temperature dependence of the bond scattering has been studied experimentally in the unique case of silicon, for which the class of reflections with $h + k + l = 4n + 2$ is only observable because of deviations from spherical atomic symmetry, either as a result of bonding or anharmonic thermal motion. Careful studies of the (222) and (442) reflections, both with x-ray and with neutron diffraction, suggest that the temperature dependence of the bond scattering is appreciably smaller than that of the core electrons, or perhaps even temperature-independent[122-124] *. The suggestion has been made that the bonding charge distribution itself changes with temperature[125]. Such a change could also be predicted in the convolution approximation, according to which internal molecular modes may even cause an increase in time-averaged bonding density if the strengthening during bond contraction outweighs the weakening at the times the bond is being stretched. But in molecular crystals external rigid-body modes are dominant, and they should lead to a thermal smearing of both core and valence electrons.

A surprisingly good fit to the peak heights in the thermally smeared density according to the convolution approximation can be obtained by a simple analytical model, in which a single Gaussian is fitted to the bond peak-shape in the direction perpendicular to the bond axis[43]. In acetylene, the convolution approximation predicts a reduction of the peak height by a factor of 0.87, with thermal (C≡C) parameters of 1.07 Å$^2$ on the carbon atoms[57]. The thermal behaviour of the Gaussian describing the theoretical bond density reproduces this reduction to within 1%[126]. Such an approximation, if generally valid, would greatly facilitate the routine comparison of theoretical and experimental peak heights in the difference density function.

### 2.4.3 Comparison of experimental and theoretical difference densities of molecules

In Section 2.3.3.2, we discussed the agreement between theory and experiment for some inorganic systems. Even though a fair number of molecular crystals has now been studied, quantitative comparisons with theory have been scarce, in part because theoretical calculations of sufficient sophistication can only be made for very small molecules, which are usually in the gaseous state at room temperature and therefore difficult to study accurately with x-ray diffraction. For this reason, comparison is often made with theoretical results on very small molecules with similar chemical groups, a procedure justified partly by the approximate transferability of molecular segments (Section 2.4.1).

Qualitative agreement exists between optimised MBS calculations on diborane[95-97] and cyclopropane[127] and experimental results on decaborane[48] and three-membered systems[28,52,53] (Figures 2.9 and 2.16) and between experiment and double-zeta-type MO calculations on the molecule of glycine[46]. This is to some extent due to cancellation of the effects of the inadequacy of the calculation (peak heights too low) and the neglect of thermal smearing (peak heights too pronounced). But, as in the other studies cited, features such as lone-pair electrons, polarisation of the H-atom charge density and the bonding charge appear in both theoretical and experimental maps.

* A later study by the same workers indicates a more equal temperature dependence for the bond and core scattering[184].

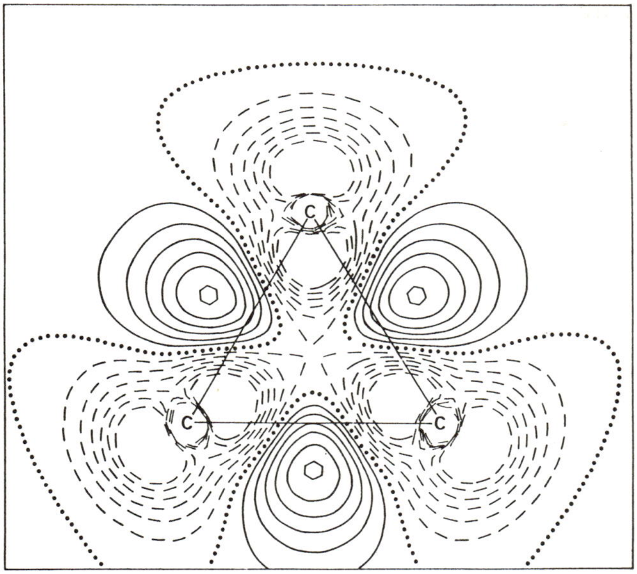


**Figure 2.16** Theoretical difference density in cyclopropane. Contours at 0.05 e Å$^{-3}$. Negative contours broken. (From Stevens *et al.*[127], by courtesy of the American Chemical Society)

Thermal smearing of the density in N≡C—C≡N in the single Gaussian approximation (Section 2.4.2) indicates the low experimental lone-pair peak height in tetracyanoethylene to be due to the lone-pair density being a much steeper feature, which is lowered more drastically by the molecular vibrations than the overlap peaks[43]. The relative heights of the bond and lone-pair peaks agree with the X–N results after the thermal averaging is accounted for (Table 2.4). The discrepancies in the absolute values may be attributed to uncertainties in the experimental scale and the extinction correction.

**Table 2.4 Peak heights in tetracyanoethylene (experimental) and dicyanogen (theoretical)[22]**

| | C—C | C═C | C≡C | *Lone pair* |
|---|---|---|---|---|
| Tetracyanoethylene | 0.6 | 0.4 | 0.9 | 0.4 |
| Dicyanogen (B═O) | 0.6 | — | 1.0 | 1.1 |
| Dicyanogen ($B_{exp}$) | 0.4 | — | 0.65 | 0.35 |

In studies of fumaramic acid[58] and tetracyanocyclobutane[66], the deformation from the spherical-atom density has been determined by a least-squares fit of a set of non-spherical density functions (see Section 2.5.2). The deformation density map for tetracyanocyclobutane shown in Figure 2.17 corresponds in

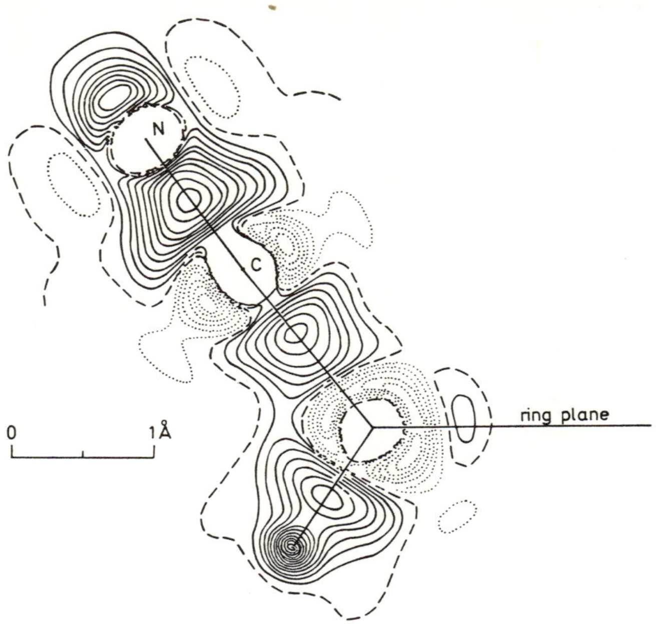


**Figure 2.17** Deformation density in a section through the C—C≡N group in tetracyanocyclobutane. Contours at 0.1 e $Å^{-3}$. Negative contours dotted. (From Harel and Hirshfeld[66], preliminary results)

principle to the difference density in the molecule at rest and is therefore to be compared with unsmeared theoretical calculations. The agreement with other theoretical (NCCN, $CN^-$, etc.) and experimental (tetracyanoethylene, tetracyanoethylene oxide) results is encouraging.

## 2.5 THE INTRODUCTION OF ATOMIC ASPHERICITY IN THE ELECTRON-DENSITY FORMALISM

### 2.5.1 Aspherical atoms

#### *2.5.1.1 Atoms with high site-symmetry: diamond, silicon, beryllium and ionic crystals*

The atomic deformations can be described by electron-density functions, centred at the atomic positions and adapted to its site-symmetry. Such density functions, applied by DeMarco and Weiss[128, 129], by Dawson *et al.*[130-133] and by Kurki Suonio *et al.*[134-140] apply to the thermally smeared density so that any deviation from isotropic thermal motion is incorporated in the atomic deformation terms. The angular dependence of the deformation terms is treated in terms of spherical harmonics which have the symmetry of the familiar s, p, d, f, etc. functions describing atomic orbitals and correspond to monopole, dipole, quadrupole, octupole, etc. deformation terms. The radial functions employed differ considerably. The earliest treatment contains the restriction that the radial function of the valence shell is not modified compared with the isolated atom[128, 129] while the Dawson formalism[130] allows a modified, Gaussian radial function for all but the leading spherically symmetric term, which is assumed to have the same radial dependence as the isolated atom. The analysis of powder diffraction data on diamond shows the antisymmetric third-order octupole (f) term to account quantitatively for the grosser non-spherical features in the electron density. Together with a fourth-order component of hexadecapole (g) symmetry, it corresponds to a mid-bond peak of 0.64 e $Å^{-3}$, a number in satisfactory agreement with peak heights discussed earlier in this review[131, 132]. A later treatment, in which the contributions of the deformation terms are least-squares-adjusted, confirms the earlier conclusions[141]. A similar, more recent, analysis of the very accurate Pendellösung data on silicon indicates the necessity for a radial variation of the spherical term of the charge density[75]. Aldred and Hart found the fit with experiment to be satisfactory only when the valence-shell form factor is radially compressed by 6.8%. A more flexible set of radial functions had been introduced earlier by Kurki Suonio and Meisalo[134], who employed an expansion of Fourier-invariant harmonic-oscillator wave functions. In this work, the adequacy of the experimental information is judged by the physical requirement of low-order harmonic behaviour, i.e., the amplitudes of the high-order radial functions should be relatively small. Applications have been described to experimental data on diamond, silicon[136, 137], NaCl[139], AgCl[139] and $CaF_2$[138]. The results again indicate the necessity of a spherical deformation component in the description of the atoms. For $CaF_2$, it is found that both fluorine and calcium are well separated from their surroundings by distinct low minima in the charge distributions, while the $Ca^{2+}$ ion is compressed and the $F^-$ ion extended in comparison with theoretical functions for the isolated ions.

In the generalised x-ray scattering model developed by Stewart[142], the density basis functions apply to the atom at rest, so that the results may be used to calculate electric-field gradients, gradient of gradients, local and molecular dipole and higher moments, coupling tensors and many other physical properties of the crystal[143]. The agreement of such quantities with other experimental measurements has been suggested as a criterion in judging the reliability of the x-ray diffraction data[143]. The thermal motion parameters have to be obtained in a separate analysis, while the limited resolution of the diffraction experiment imposes a fundamental limitation on the measurement of physical properties. The radial functions in Stewart's analysis are products of Slater-type functions, with orbital exponents being considered as variables in the least-squares treatment. For the powder data on diamond, the exponential value is 1.68(3) $bohr^{-1}$ when only (spherical) monopole terms are included[144]. This value does not differ significantly from the recommended theoretical standard molecular value (1.72) for the L-shell orbital exponent[145], but with spherical deformation terms a smaller value of the orbital exponent results (1.56–1.58). This is as expected, because the density in the C—C bond is represented by a two-centre orbital product ($\phi_A\phi_B$ for the bond between atoms A and B) in the theoretical one-electron-density function, and its description by a term centred at the nuclear position naturally leads to a more diffuse density function.

The efficiency of the deformation model and its superiority over the use of a single-term Gaussian for the description of the bond charge in diamond has been confirmed by an analysis of single-crystal data[146]. The very-high-order deformation terms[147] needed in the analysis of the diffraction data of beryllium have been interpreted in terms of multicentre bonding in this metal[71], a conclusion which is supported by an analysis of the valence charge density (see Section 2.3.3.2).

### *2.5.1.2 Application to organic molecules*

The density functions described in the previous section are especially suited to treat atoms at highly symmetric sites. In organic molecules, atomic site-symmetry is generally low and bond angles deviate from the idealised values such as 180, 90, 109.5°, which are the angles between the maxima of the spherical harmonic density functions. Nevertheless, considerable successs has been achieved in the application of a related charge density model to fumaramic acid[58] and tetracyanocyclobutane[66] by Hirshfeld and co-workers. Here the deformation of the density in the molecule at rest is represented by a series of orthogonalised cylindrically symmetric functions of the type $\cos^n\theta_K$ (with $0 \leqslant n \leqslant 3$) directed along the edges, face and body diagonals of a cubic reference frame. A simple Slater-type exponential is chosen for the radial function, with the exponent selected such as to place the maximum at about 1/3 to 1/2 of a typical bond length from the atom. The number of parameters is limited by imposing a rigid-body constraint on the molecular vibrations and by assumptions of chemical equivalence and local molecular symmetry. Though individual deformation terms may be of limited physical significance, the validity of the model is supported by the reduction in the crystallographic agreement index, and, above all, by the appearance of the deformation density (Figure 2.17) which show considerable similarity with theoretical and experimental maps on cyano compounds discussed earlier. The height of the lone-pair peak (0.8 e Å$^{-3}$) is intermediate between the theoretical (1.0)[58] and experimental values (0.35)[22], indicating the effect of the extrapolation to zero thermal motion. Finally, the positional parameters from this refinement reproduce the asphericity shift of the cyano carbon atom and the lack of bias in the nitrogen position[148], in agreement with the x-ray and neutron results on tetracyanoethylene[22].

### 2.5.2 Multicentre formalisms

The theoretical expression for the electron density in an LCAO (linear combination of atomic orbitals) framework contains explicit terms which are centred on two rather than on one nucleus:

$$\rho(x,y,z) = \sum_{\mu}\sum_{\nu} P_{\mu\nu}\phi_{\mu}\phi_{\nu} \tag{2.8}$$

The explicit introduction of the bond-scattering term in equation (2.8) by McWeeny[149, 150] was further generalised by Stewart[142]. Theoretical studies by Groenewegen *et al.*[151] support the assumption that for x-ray scattering the core density can be considered unperturbed by bonding. To limit the number of functions in the density formalism (2.8) further, a minimal valence basis set (MBS) of atomic s and p orbitals has been explored, leading to density functions with s, p and d symmetry[100, 152-154]. Applications have been reported in which all two-centre terms have been neglected[155, 156] to give a model more akin to those described in the previous section, but without the flexibility in the radial functions. In the treatment of theoretical diffraction data on diborane and experimental data on cyanuric acid and tetracyanoethylene oxide, only the two-centre terms in long bonds between non-adjacent atoms, which contain very little density, were neglected[100]. Electron-density deformation maps, based on the least-squares population parameters, indicate that a successful parametrisation of the electron density has been achieved, as the features of X–N and X–X maps are well reproduced. The adequacy of 'space-filling' properties of this MBS formalism compared with those of theoretical MBS calculations is due to the absence of constraints other than molecular neutrality. Massa and Clinton[157] have proposed a direct determination of the population coefficients $P_{\mu\nu}$ in equation (2.8) to fit the coherent diffraction data *and* the requirement of an antisymmetric *N*-electron wave function. No applications have yet been reported.

Individual values of the population parameters have little meaning because of the high correlation between parameters. It is therefore preferable to base the analysis on uncorrelated linear combinations of population parameters (eigenparameters)[100]. But comparisons with theory within a minimal basis set formalism are necessarily of limited value, and other type basis functions $\phi_{\mu}$ in equation (2.8) are needed for the comparison with theoretical treatments.

### 2.5.3 The use of molecular form factors

Because the molecule is the basic building block of an organic crystal, its scattering may be calculated with theoretical molecular form factors. Groenewegen and Feil[158] applied four alternative molecular form factors to the $NH_4^+$ molecular ion in the $NH_4F$ structure and found a superior fit with the diffraction data for the only multicentre calculation examined. An improved fit, compared with spherical atoms, has also been reported using an optimised MBS density and low-temperature diffraction data on diborane[159, 160]. The fit is best with a B—H bond length about 0.05 Å longer than those obtained from a spherical-atom refinement, which confirms the origin of the apparent shortening of x-ray bond lengths involving hydrogen atoms. A molecular refinement for sodium azide has been compared with spherical-atom models using different form factors[39]. Most calculations reported here have been restricted to the rigid-body approximation. The introduction of thermal smearing of internal modes in the convolution approximation (Section 2.4.2) is a logical future extension.

## 2.6 THE MEASUREMENT OF NET ATOMIC CHARGE

### 2.6.1 What constitutes an atom in a crystal?

Many of the concepts of chemistry are based on the existence of positively and negatively charged atoms. Thus, electrophilic and nucleophilic reactions are distinguished, ESCA spectra, and other physical chemical measurements are correlated with net atomic charges, and the three-dimensional structure and biological activity of protein molecules is partly explained by interactions of the positively charged $NH_3^+$ and the negatively charged $CO_2^-$ groups. Yet, atoms in molecules are not separate entities, and even in ionic crystals the charge density does not fall off to zero between the ions. The definition of an atomic charge is therefore by necessity controversial. The most widely used theoretical-chemical definition is implied in the Mulliken population analysis[161] in which the electron density in the two-centre terms in equation (2.8) is equally divided between the adjacent atoms. This is not always realistic as the overlap density will in general not be symmetrically distributed. Furthermore, atomic orbitals are often so diffuse as to place a substantial part of their charge in the vicinity of a neighbouring nucleus. The ambiguity is very well illustrated by the calculations on LiCN, LiNC and $CN^-$ [102]. The net atomic charges on the nitrogen atoms in these three molecules are, according to the population analysis, −0.180, −0.812 and −0.499 electrons, respectively, but the difference density maps are almost indistinguishable. Several authors have proposed alternative theoretical definitions[162-164] but none of these seems as yet to have found general acceptance.

Two definitions have been used in the diffraction work. The first, applied to ionic crystals, is based on the 'requirement of locality', which states that parts of the electron distribution cannot be assigned to a distant atom[136, 165]. In the second treatment, the valence density is projected into atomic density functions of presupposed or adjustable shape[47, 156, 166].

### 2.6.2 Ionic crystals with high symmetry

Because of the proximity of adjacent ions in a crystal, appreciable overlap exists when isolated atoms are placed at the ionic positions in the crystal. As a result, there is little difference between the charge density from a superposition of atoms or a superposition of ions in, for example, a KBr crystal[136, 165]. But these hypothetical models do not necessarily correspond to the charge density in the crystal. LiH is a particularly difficult case because the free-space hydrogen orbital is so diffuse as to place an appreciable part of its electrons in the neighbourhood of the adjacent lithium atom. Nevertheless, it was shown that the charge distribution is different from a superposition of isolated atoms and ions, and supports a net ionic charge of 0.8–1.0 electron units[5]. In the work of Kurki Suonio, the separation of the component densities from the diffraction data is defined through an analysis of the radial charge density $4\pi r^2 \rho_0(r)$ around the nuclear positions[136, 165]. A plot of the radial density, calculated with a theoretical correction for series termination effects, shows a clear minimum at a value of $r$ which is defined as the radius of best separation. Its value at this point is a measure of the conceptual indeterminacy, which is much higher for some metal oxides than for the alkali halides listed in Table 2.5, indicating that the ionic definition is more appro-

**Table 2.5 Net ionic charge in alkali halides according to Kurki Suonio[165] * †**

| | *Neutral model* | *Ionic model* | *Experimental* |
|---|---|---|---|
| KBr | +0.8, −0.2 | +1.0, −0.4 | +0.45, −1.15 |
| NaF | +0.9, −0.2 | +0.9, −0.6 | +0.85, −0.75 |
| LiF | +0.9, −0.3 | +1.0, −0.6 | +0.95, −0.70 |

* Neutral and ionic values from analysis of synthetic density distributions.

† Charges do not cancel because of spherical definition of atoms; density in internuclear region may be assigned to both atoms, while density in voids between spheres is not assigned.

priate in the latter case. According to this analysis, metal oxides and alkali halides are approximately described by the doubly and singly ionised states. This conclusion is further supported by Sanger's analysis of single-crystal data on MgO in which various form factors were compared[180].

### 2.6.3 Experimental charges from spherical-atom refinements: valence-shell projection

For less symmetrical crystals, the choice of a spherical atomic region becomes more difficult as large variations in bond lengths occur. Furthermore, the indeterminacy of the atomic regions will be considerable where density is accumulated in a covalent bond between atoms. Overlap of the atoms is allowed for in a projection of the valence density into spherical density functions centred at the atoms, assuming negligible perturbation of the core electrons. The population coefficients of the spherical functions may be adjusted by least squares, either after structure refinement[166], or simultaneously with the other structural parameters[156] (as in expression (2.3), but not necessarily with different core and valence positional parameters). The conceptual indeterminacy is present in the choice of atomic functions. Both Stewart[144, 167] and Coppens[36, 156] have reported that functions based on Slater-type atomic orbitals with standard molecular exponents are more appropriate and lead to thermal motion parameters in better agreement with neutron-diffraction values than those from isolated-atom Hartree–Fock orbitals. The choice of Slater orbitals is also supported by the experimental determination of the scale factor in kernite[118, 168]. With good quality data, not only the occupancy but also the shape of the density functions (spherical deformation) as described by the orbital exponent can be least-squares adjusted[47, 169]. In both glycylglycine[60] and oxalic acid[61] this leads to slight expansion of the oxygen atoms and a contraction of the carbon atoms relative to isolated-atom functions. The resulting charges generally agree in sign and approximate magnitude, independent of the basis functions selected, while good agreement is found between the charges on chemically equivalent atoms even when different structures are compared (such as water molecules in different crystalline hydrates)[156]. The x-ray diffraction dipole moment for uracil calculated with a point-charge approximation agrees well with theoretical values and with a measurement of 1,3-dimethyluracil[33]. The results may therefore be considered to give a qualitative indication of the net atomic charges, though aspherical density functions may sometimes be required. Generally, it is found that (a) net atomic charges are fractions of electron units, even in a mineral such as kernite[168], a Ni complex[156], or a zwitterion-like glycylglycine[60], (b) oxygen atoms are negatively charged by 0.2–0.5 electrons while carbon atoms are positive in groups like carboxyl and carbonyl but negative in an aliphatic environment[156], and boron atoms in kernite[168] and borazine[170] are positive by about 0.4 and 0.8 electrons respectively, and (c) atomic charges correlate with atomic electronegativity rather than with the number of bonds, no significant differences being observed between trigonal and tetrahedral borons, and between bridging and hydroxylic oxygen atoms in the borate polyanion in kernite.

Approximate electroneutrality is also supported by the work of Saito *et al.*[62, 171] on two cobalt(III) complexes in which the total density within 1.22 Å (the covalent radius) of three different Co atoms was counted as 25.8[171], and, in the more recent work, 26.3(3), 26.8(3)[62], compared with 27 electrons for the isolated atom.

In general this evidence supports Slater's conclusion that atoms tend to be much more neutral in a crystal than a straight ionic interpretation would indicate[172]. It also agrees remarkably well with Pauling's postulate of electroneutrality, according to which the atomic charges in a stable crystal vary at most between $+1/2$ and $-1/2$[173]. However, the ionic crystals discussed in the preceding section are to be considered as exceptions to the electroneutrality postulate.

## 2.7 CONCLUDING REMARKS

The work summarised above is only a first step on the way to more widespread measurements of charge densities, in the same sense that structure determination today is a generally accepted pre-requisite for other research. To quote Platt[174], 'a theory of chemistry and the chemical bond is primarily a theory of electron density'. Experimental studies of the electron density are therefore of central importance in our understanding of the nature of molecules and solids.

### References

1. Debye, P. and Scherrer, P. (1918). *Phys. Z.*, **19**, 474
2. Black, P. J. and Taylor, W. H. (1958). *Rev. Modern Phys.*, **30**, 55
3. Brill, R. (1967). *Solid State Phys.*, **20**, 1
4. Witte, H. and Wölfel, E. (1955). *J. Phys. Chem.*, **53**, 296; Weiss, A., Witte, H. and Wölfel, E. (1957). *Z. Phys. Chem.*, **10**, 98; Göttlicher, S. and Wölfel, W. (1959). *Z. Electrochem.*, **63**, 891

5. Calder, R. S., Cochran, W., Griffiths, D. and Lowde, R. D. (1962). *J. Phys. Chem. Solids*, **23,** 621
6. Bonham, R. A. (1969). *Rec. Chem. Progr.*, **30,** 185
7. Bartell, L. S. (1972). *Trans. Amer. Crystallogr. Assn.*, **8,** 37
8. Epstein, I. R. (1973). *Accounts Chem. Res.*, **6,** 145; Cooper, M. (1971). *Advan. Phys.*, **20,** 453
9. Moon, R. M. (1972). *Trans. Amer. Crystallogr. Assn.*, **8,** 59
10. (1972). *Proceedings NATO Summer Institute, University of Aarhus*
11. (1968). Proceedings of the Cambridge Conference on Accurate Determination of X-ray Intensities and Structure Factors, *Acta Crystallogr.*, **A25,** Part 1
12. Coppens, P. (1968). *Acta Crystallogr.*, **B24,** 1272
13. O'Connor, B. H. and Moore, F. H. (1973). *Acta Crystallogr.*, **B29,** 1903; O'Connor, B. H. (1973). *Acta Crystallogr.*, **B29,** 1893
14. Ross, F. K. and Coppens, P. (1972). *Abstracts Amer. Chem. Assn. Winter Meeting*, Paper D3; Wang, Y., Blessing, R. H., Ross, F. K., Williams, J. and Coppens, P., to be published.
15. Dawson, B. (1965). *Aust. J. Chem.*, **18,** 595
16. Coppens, P. and Coulson, C. A. (1967). *Acta Crystallogr.*, **23,** 718
17. Jensen, L. H. and Sundaralingam, M. (1964). *Science*, **145,** 1185
18. Halgren, T. A., Anderson, R. J., Jones, D. S. and Lipscomb, W. N. (1971). *Chem. Phys. Lett.*, **8,** 547
19. Hanson, J. C., Sieker, L. C. and Jensen, L. H. (1973). *Acta Crystallogr.*, **B29,** 797; Hanson, J. C., Sieker, L. C. and Jensen, L. H. (1972). *Trans. Amer. Crystallogr. Assn.*
20. Coppens, P. and Hirshfeld, F. L. (1964). *Israel J. Chem.*, **2,** 117
21. Irngartinger, H., Leiserowitz, L. and Schmidt, G. M. J. (1970). *J. Chem. Soc.* (*B*), 497
22. Becker, P., Coppens, P. and Ross, F. K. (1973). *J. Amer. Chem. Soc.*, **95,** 7604
23. O'Connel, A. M., Rae, A. I. M. and Maslen, E. N. (1966). *Acta Crystallogr.*, **21,** 208
24. Rietveld, H. M., Maslen, E. N. and Clews, C. J. B. (1970). *Acta Crystallogr.*, **B26,** 693
25. Hazel, A. C., Larsen, F. K. and Lehman, M. S. (1972). *Acta Crystallogr.*, **B28,** 2977
26. Kay, M. I., Okaya, Y. and Cox, D. E. (1971). *Acta Crystallogr.*, **B27,** 26
27. Stewart, R. F., Davidson, E. R. and Simpson, W. T. (1965). *J. Chem. Phys.*, **42,** 3175
28. Matthews, D. A. and Stucky, G. D. (1971). *J. Amer. Chem. Soc.*, **93,** 5954; Matthews, D. A. and Stucky, G. D. (1972). *Trans. Amer. Crystallogr. Assn.*, **8,** 113; Matthews, D. A., Swanson, J. and Stucky, G. D. (1971). *J. Amer. Chem. Soc.*, **93,** 5945
29. Coppens, P., Sabine, T. M., Delaplane, R. G. and Ibers, J. A. (1969). *Acta Crystallogr.*, **B25,** 2451
30. Taylor, J. C. and Sabine, T. M. (1972). *Acta Crystallogr.*, **B28,** 3340
31. Coppens, P. and Vos, A. (1971). *Acta Crystallogr.*, **B27,** 146; Verschoor, G. C. and Keulen, E. (1971). *Acta Crystallogr.*, **B27,** 134
32. Stewart, R. F. (1968). *J. Chem. Phys.*, **48,** 4882
33. Stewart, R. F. and Jensen, L. H. (1969). *Z. Kristall.*, **128,** 133
34. O'Connel, A. M. (1969). *Acta Crystallogr.*, **B25,** 1273
35. Little, R. G., Pautler, D. and Coppens, P. (1971). *Acta Crystallogr.*, **B27,** 1493
36. Coppens, P. (1971). *Acta Crystallogr.*, **B27,** 1931
37. Dunitz, J. D. and Seiler, P. (1973). *Acta Crystallogr.*, **B29,** 589
38. Hope, H. and Kim, N. E. (1971). *Abstracts Amer. Chem. Assn. Winter Meeting*, Paper D2
39. Stevens, E. D. (1973). Experimental Determination of Electron Density Distributions by X-ray Diffraction, *Thesis*, University of California at Davis
40. Dawson, B. (1964). *Acta Crystallogr.*, **17,** 997
41. Coppens, P. (1969). *Acta Crystallogr.*, **A25,** 180
42. Coppens, P. (1967). *Science*, **158,** 1577
43. Coppens, P. (1974). *Acta Crystallogr.*, **B30,** 255
44. Cruickshank, D. W. J. and Rollett, J. S. (1953). *Acta Crystallogr.*, **6,** 705
45. Harkema, S. (1971). Electron Density and X-ray Diffraction, *Thesis*, Technische Hogeschool Twente, The Netherlands
46. Almlöf, J., Kvick, Å. and Thomas, J. O. (1973). *J. Chem. Phys.*, **59,** 3901; *Abstracts, 1st Eur. Crystallogr. Conf. Bordeaux*, Group B3
47. Blessing, R. H., Cooper, W. F., Yang, Y. W. and Coppens, P. (1973). *Amer. Crystallogr. Assn. Conference Program and Abstracts, Series 2*, **1,** 200

48. Brill, R., Dietrich, H. and Dierks, H. (1971). *Acta Crystallogr.*, **B27**, 2003
49. Rees, B. and Coppens, P. (1973). *Acta Crystallogr.*, **B29**, 2515
50. Brown, G. M. and Levy, H. A. (1973). *Acta Crystallogr.*, **B29**, 790
51. Duckworth, J. A. K., Willis, B. T. M. and Pawley, G. S. (1970). *Acta Crystallogr.*, **A26**, 263
52. Ito, T. and Sakurai, T. (1974). *Acta Crystallogr.*, **B29**, 1594
53. Hartman, A. and Hirshfeld, F. L. (1966). *Acta Crystallogr.*, **20**, 80
54. Dreyfus, M., Maigret, B. and Pullman, A. (1970). *Theor. Chim. Acta*, **17**, 109; Dreyfus, M. and Pullman, A. (1970). *ibid*, **19**, 30
55. Kollman, P. A. and Allen, L. C. (1970). *J. Chem. Phys.*, **52**, 5085
56. Helmholdt, R. B., Ruysink, A. F. J., Reynaars, H. and Kemper, G. (1972). *Acta Crystallogr.*, **B28**, 318
57. Ruysink, A. F. J. (1973). Electron Distribution in Crystals, *Thesis*, University of Groningen, The Netherlands; Ruysink, A. F. J. and Vos, A. (1974). *Acta Crystallogr.*, **A30**, 463
58. Hirshfeld, F. L. (1971). *Acta Crystallogr.*, **B27**, 769
59. Hirshfeld, F. L. and Rabinovich, D. (1969). *Acta Crystallogr.*, **23**, 989
60. Griffin, J. F. and Coppens, P. (1974). To be published
61. Furberg, S. and Jensen, L. H. (1970). *Acta Crystallogr.*, **B26**, 1260
62. Iwata, M. and Saito, Y. (1973). *Acta Crystallogr.*, **B29**, 822
63. Schwarzenbach, D. and Manohar, H. (1973). *Abstr. 1st Eur. Crystallogr. Conf., Bordeaux*, Group A6
64. Fritchie, C. J., Jr. (1966). *Acta Crystallogr.*, **20**, 27
65. Coulson, C. A. and Moffit, W. A. (1959). *Phil. Mag.*, **40**, 1
66. Harel, M. and Hirshfeld, F. L. (1973). *Abstr. 1st Eur. Crystallogr. Conf., Bordeaux*, Group B3
67. Temkin, R. J., Henrich, V. E. and Racah, P. M. (1973). *Solid State Commun.*, **13**, 811
68. Walter, J. P., Fong, C. Y. and Cohen, M. L. (1973). *Solid State Commun.*, **12**, 303
69. Linkoaho, M. V. (1972). *Physica Scripta*, **5**, 271
70. Brown, P. J. (1972). *Phil. Mag.*, **26**, 1377
71. Stewart, R. F. (1973). *Amer. Chem. Assn. Conference Program and Abstracts, Series 2*, **1**, 200
72. Yang, Y. W. and Coppens, P., unpublished results
73. Tanemura, S. and Kato, N. (1972). *Acta Crystallogr.*, **A28**, 69
74. Aldred, P. J. E. and Hart, M. (1973). *Proc. Roy. Soc.* (*London*), **A332**, 223
75. Aldred, P. J. E. and Hart, M. (1973). *Proc. Roy. Soc.* (*London*), **A332**, 239
76. Walter, J. P. and Cohen, M. L. (1971). *Phys. Rev.*, **B4**, 1877
77. Collela, R. (1971). *Phys. Rev.*, **B3**, 4308; Bilderback, D. H. and Collela, R. (1973). *Amer. Crystallogr. Assn. Conference Program and Abstracts, Series 2*, **1**, 137
78. Uno, R., Okano, T. and Yukino, K. (1970). *J. Phys. Soc. Jap.*, **28**, 437
79. Dyck, M. van and DeRanter, C. J. (1973). *Abstr. 1st Eur. Cyrstallogr. Conf. Bordeaux*, Group B3
80. Uno, R. (1965). *J. Phys. Soc. Jap.*, **20**, 308
81. Vogl, E. and Waidelich, W. (1968). *Z. Angew. Phys.*, **2**, 98
82. Sirota, N. N. (1969). *Acta Crystallogr.*, **A25**, 223
83. Hosoya, S. (1969). *Acta Crystallogr.*, **A25**, 243
84. Tsuchiya, H., Marumo, F. and Saito, Y. (1972). *Acta Crystallogr.*, **B28**, 1935
85. Roux, M., Besnainou, S. and Daudel, R. (1956). *J. Chim. Phys.*, **54**, 218; Roux, M. (1958). *J. Chim. Phys.*, **55**, 754
86. Roux, M., Cornille, M. and Burnelle, L. (1962). *J. Chem. Phys.*, **37**, 933
87. Ransil, B. J. and Sinai, J. J. (1967). *J. Chem. Phys.*, **46**, 4050
88. Hazelrigg, M. J. and Politzer, P. (1969). *J. Amer. Chem. Soc.*, **73**, 1009
89. Bader, R. F. W., Henneker, W. H. and Cade, P. E. (1967). *J. Chem. Phys.*, **46**, 3341
90. Bader, R. F. W., Keaveny, I. and Cade, P. E. (1967). *J. Chem. Phys.*, **47**, 3381
91. Bader, R. F. W. and Bandrauk, A. D. (1968). *J. Chem. Phys.*, **49**, 1653
92. Cade, P. E., Bader, R. F. W., Henneker, W. H. and Keaveny, I. (1969). *J. Chem. Phys.*, **50**, 5315
93. Cade, P. E. (1972). *Trans. Amer. Crystallogr. Assn.*, **8**, 1
94. Smith, P. R. and Richardson, J. W. (1965). *J. Phys. Chem.*, **69**, 3346; Smith, P. R. and Richardson, J. W. (1967). *J. Phys. Chem.*, **71**, 924

95. Laws, E. A., Stevens, R. M. and Lipscomb, W. N. (1972). *J. Chem. Phys.*, **56,** 2029
96. Laws, E. A. and Lipscomb, W. N. (1972). *Israel J. Chem.*, **10,** 77
97. Laws, E. A., Stevens, R. M. and Lipscomb, W. N. (1972). *J. Amer. Chem. Soc.*, **94,** 4467
98. Boyd, D. B. (1972). *J. Amer. Chem. Soc.*, **94,** 64
99. Pople, J. A. and Beveridge, D. L. (1970). *Approximate Molecular Orbital Theory* (New York: McGraw-Hill)
100. Jones, D. S., Pautler, D. and Coppens, P. (1972). *Acta Crystallogr.*, **A28,** 635
101. McIver, J. W., Coppens, P. and Nowak, D. (1971). *Chem. Phys. Lett.*, **11,** 82
102. Bats, J. W. (1973). A Theoretical Study of the Electron-density in $CN^-$, LiCN and LiNC, *Thesis*, Technische Hogeschool Twente, The Netherlands
103. Cruickshank, D. W. J. and Weaver, R. F. (1973). *Abstr. 1st Eur. Crystallogr. Conf., Bordeaux*, Group B3
104. Hirshfeld, F. L. and Rzotkiewicz, S. (1974). *Mol. Phys.*, **27,** 319
105. Bader, R. F. W., Beddall, P. M. and Cade, P. E. (1971). *J. Amer. Chem. Soc.*, **93,** 3095
106. Bader, R. F. W. and Beddall, P. M. (1973). *J. Amer. Chem. Soc.*, **95,** 305
107. Bader, R. F. W., Beddall, P. M. and Peslak, J., Jr. (1973). *J. Chem. Phys.*, **58,** 557
108. Aslangul, C., Constanciel, R., Daudel, R. and Kottis, P. (1972). *Advan. Quant. Chem.*, **6,** 93
109. McLean, A. D. and Yoshimine, M. (1967). *Tables of Linear Molecule Wave Functions* (San Jose, Cal.: International Business Machine Corporation)
110. England, W., Salmon, L. S. and Ruedenberg, K. (1971). *Topics Current Chem.*, **23,** 31
111. Schaefer, H. F., III (1972). *The Electronic Structure of Atoms and Molecules* (New York: Addison-Wesley)
112. Bader, R. F. W. and Chandra, A. K. (1968). *Can. J. Chem.*, **46,** 953
113. Das, G. and Wahl, A. C. (1966). *J. Chem. Phys.*, **44,** 87
114. Clementi, E. and Popkie, H. (1972). *J. Chem. Phys.*, **57,** 1077
115. Kern, C. W. and Karplus, M. (1972). *Water, A Comprehensive Treatise*, Vol. 1, p. 21 (New York: Plenum Press)
116. Grimaldi, F., Lecourt, A. and Moser, C. (1967). *Int. J. Quant. Chem.*, **1S,** 153
117. Stewart, R. F. (1968). *Acta Crystallogr.*, **A24,** 497
118. Coppens, P. (1972). *Israel J. Chem.*, **10,** 85
119. Coulson, C. A. and Thomas, M. W. (1971). *Acta Crystallogr.*, **B27,** 1354
120. Thomas, M. W. (1971). *Acta Crystallogr.*, **B27,** 1760
121. Thomas, M. W. (1973). *Chem. Phys. Lett.,*, **20,** 303
122. Roberto, J. B. and Batterman, B. W. (1970). *Phys. Rev.*, **B2,** 3220
123. Trucano, P. and Batterman, B. W. (1972). *Phys. Rev.*, **B6,** 3659
124. Keating, D., Nunes, A., Batterman, B. W. and Hastings, J. (1971). *Phys. Rev.*, **B4,** 2472
125. Philips, J. C. (1971). *Phys. Lett.*, **37A,** 434
126. Coppens, P., unpublished results
127. Stevens, R. M., Switkes, E., Laws, E. A. and Lipscomb, W. N. (1971). *J. Amer. Chem. Soc.*, **93,** 2603
128. DeMarco, J. J. and Weiss, R. J. (1965). *Phys. Rev.*, **A137,** 1869
129. Weiss, R. J. (1966). *X-ray Determination of Electron Distributions* (Amsterdam: North Holland Publishing Co.)
130. Dawson, B. (1967). *Proc. Roy. Soc.* (*London*), **A298,** 255
131. Dawson, B. (1967). *Proc. Roy. Soc.* (*London*), **A298,** 264
132. Dawson, B. and Sanger, P. L. (1967). *Proc. Roy. Soc.* (*London*), **A301,** 195
133. Dawson, B. (1969). *Acta Crystallogr.*, **A25,** 12
134. Kurki Suonio, K. and Meisalo, V. (1967). *Ann. Acad. Sci. Fenn., Ser. A VI*, **241,** 3
135. Kurki Suonio, K. (1968). *Acta Crystallogr.*, **A24,** 379
136. Kurki Suonio, K. (1971). Analysis of Crystal Atoms on the Basis of X-ray Diffraction, *Italian Crystallographic Association Meeting, Bari, Italy*
137. Kurki Suonio, K. and Ruuskanen, A. (1971). *Ann. Acad. Sci. Fenn., Ser. A VI*, **358**
138. Ruuskanen, A. and Kurki Suonio, K. (1973). *J. Phys. Soc. Jap.*, **34,** 715
139. Linkoaho, M. V. (1969). *Acta Crystallogr.*, **A25,** 450
140. Linkoaho, M., Rantavuori, E., Korhonen, U., Kurki Suonio, K. and Ruuskanen, A. (1972). *Acta Crystallogr.*, **A28,** 260
141. McConnell, J. F. and Sanger, P. (1970). *Acta Crystallogr.*, **A26,** 83
142. Stewart, R. F. (1969). *J. Chem. Phys.*, **51,** 4569

143. Stewart, R. F. (1972). *J. Chem. Phys.*, **57,** 1664
144. Stewart, R. F. (1973). *J. Chem. Phys.*, **58,** 4430
145. Hehre, W. J., Stewart, R. F. and Pople, J. A. (1970). *J. Chem. Phys.*, **52,** 2769
146. Larson, A. C., Cromer, D. T. and Ryan, R. R. (1973). *Amer. Chem. Assn. Conference Program and Abstracts, Series 2*, **1,** 201
147. Stewart, R. F. (1973). *J. Chem. Phys.*, **58,** 1668
148. Hirshfeld, F. L. (1973). Private communication
149. McWeeny, R. (1952). *Acta Crystallogr.*, **5,** 463
150. McWeeny, R. (1953). *Acta Crystallogr.*, **6,** 631
151. Groenewegen, P. P. M., Zeevalkink, J. and Feil, D. (1971). *Acta Crystallogr.*, **A27,** 487
152. Coppens, P., Willoughby, T. V. and Csonka, L. N. (1971). *Acta Crystallogr.*, **A27,** 248
153. Matthews, D. A., Stucky, G. D. and Coppens, P. (1972). *J. Amer. Chem. Soc.*, **94,** 8001
154. Coppens, P. (1972). *Trans. Amer. Crystallogr. Assn.*, **8,** 93
155. Coppens, P., Csonka, L. and Willoughby, T. V. (1970). *Science*, **167,** 1126
156. Coppens, P., Pautler, D. and Griffin, J. F. (1971). *J. Amer. Chem. Soc.*, **93,** 1051
157. Massa, L. J. and Clinton, W. L. (1972). *Trans. Amer. Crystallogr. Assn.*, **8,** 149; Clinton, W. L. and Massa, L. J. (1972). *Phys. Rev. Lett.*, **20,** 1363
158. Groenewegen, P. P. M. and Feil, D. (1969). *Acta Crystallogr.*, **A25,** 444
159. Jones, D. S. and Lipscomb, W. N. (1969). *J. Chem. Phys.*, **51,** 3133
160. Jones, D. S. and Lipscomb, W. N. (1970). *Acta Crystallogr.*, **A26,** 196
161. Mulliken, R. S. (1955). *J. Chem. Phys.*, **23,** 1833
162. Politzer, P. and Harris, R. R. (1970). *J. Amer. Chem. Soc.*, **92,** 6451
163. Maclagan, R. G. A. R. (1971). *Chem. Phys. Lett.*, **8,** 114
164. Jug, K. (1973). *Theor. Chim. Acta*, **29,** 9
165. Kurki Suonio, K. and Salmo, P. (1971). *Ann. Acad. Sci. Fenn., Ser. A VI*, **369**
166. Stewart, R. F. (1970). *J. Chem. Phys.*, **53,** 205
167. Stewart, R. F. (1969). *J. Chem. Phys.*, **51,** 4569
168. Cooper, W. F., Larsen, F. K., Coppens, P. and Giese, R. F. (1973). *Amer. Mineral.*, **58,** 21
169. Coppens, P., unpublished results
170. Corfield, P. W. R. and Shore, S. G. (1973). *J. Amer. Chem. Soc.*, **95,** 1480
171. Kobayashi, A., Marumo, F. and Saito, Y. (1972). *Acta Crystallogr.*, **B28,** 2709
172. Slater, J. C. (1967). *Quantum Theory of Molecules and Solids*, Vol. 2, 108 (New York: McGraw-Hill)
173. Pauling, L. (1948). *J. Chem. Soc.*, 1461
174. Platt, J. R. (1961). *Handbuch Der Physik*, **XXXVII/2,** 188 (Berlin: Springer Verlag); Philips, J. C. (1969). *Covalent Bonding in Crystals, Molecules and Polymers* (Chicago: The University of Chicago Press)
175. Jungk, A. E. and Schmidt, G. M. J. (1972). *Chem. Ber.*, **105,** 2607; (1972). *ibid.*, **105,** 2623
176. Hall, S. R. and Maslen, E. N. (1967). *Acta Crystallogr.*, **22,** 216
177. Harkema, S. and Feil, D. (1969). *Acta Crystallogr.*, **A25,** S78
178. Stukel, D. J. and Euwema, R. N. (1970). *Phys. Rev.*, **B1,** 1635
179. Uno, R. and Ishigaki, A. (1972). *Acta Crystallogr.*, **A28,** S100
180. Sanger, P. L. (1969). *Acta Crystallogr.*, **A25,** 694
181. Göttlicher, S. (1968). *Acta Crystallogr.*, **B24,** 122
182. Craven, B. M. and Sabine, T. M. (1969). *Acta Crystallogr.*, **20,** 214
183. Freeman, H. C., Paul, G. L. and Sabine, T. M. (1970). *Acta Crystallogr.*, **B26,** 925
184. Roberts, J. B., Batterman, B. W. and Keating, D. T. (1973). *Report 2120*, Materials Science Center, Cornell University

# 3
# Chemistry and Crystallography of Caryophyllene

**J. MONTEATH ROBERTSON, F.R.S.**
University of Glasgow

---



---

## 3.1 INTRODUCTION

With the possible exception of cadinene, caryophyllene is perhaps the most widespread representative of the naturally occurring sesquiterpenes. The main source of caryophyllene is oil of cloves, but its isolation in a pure state is difficult. The hydrocarbon has the molecular formula $C_{15}H_{24}$, and as it contains two ethylenic linkages it must be bicyclic. However, one of the most remarkable features of caryophyllene is the ease with which it and its derivatives undergo further cyclisation and molecular rearrangement. This fact makes the study of these compounds by the ordinary methods of organic chemistry extremely difficult and sometimes uncertain, and makes structural investigation by the methods of x-ray crystallography vitally important. Nevertheless, the structural relationships of caryophyllene and many of its derivatives have been successfully elucidated by the classical degradative and synthetic methods of organic chemistry, with x-ray analysis occasionally supplementing the work with important stereochemical information. In the case of other derivatives, the structures have been solved, and in a few cases could perhaps only have been solved, by the methods of x-ray crystallography. The whole field is a good illustration of collaborative work between chemists and crystallographers.

In addition to the use of x-ray crystallography, it should be stated that recent advances in this difficult field have been due to a very large extent to the development of other physical methods with special applications to organic chemistry, such as mass spectrometry, nuclear magnetic resonance, refined gas–liquid chromatographic methods, and so on. However, for the complete and unambiguous determination of structure, the x-ray crystallographic method, when it can be applied, is unique because it provides the exact position of every atom in space.

In this article the extensive chemical work carried out in this field is referred to rather briefly, but an attempt is made to outline the parallel x-ray crystallographic investigations that have proved important in bringing the subject to its present state.

One of the early difficulties in caryophyllene chemistry arose from the presence of several closely related hydrocarbons in the main natural source, oil of cloves. These were formerly designated $\alpha$-, $\beta$- and $\gamma$-caryophyllene. It is now known that $\alpha$-caryophyllene is identical with humulene (from oil of hops) and $\gamma$-caryophyllene is an isomer of $\beta$-caryophyllene and is now referred to as isocaryophyllene. The main constituent, $\beta$-caryophyllene, may therefore be referred to simply as caryophyllene, and the important derivative $\beta$-caryophyllene alcohol is caryolan-1-ol. This derivative obtained from caryophyllene by hydration with a mixture of sulphuric and acetic acids, or in other ways, followed by steam distillation, is crystalline and has been used as a ready means of characterising caryophyllene and detecting it in various ethereal oils ever since Wallach's early work[1,2]. Another important derivative used in the early characterisation of caryophyllene is the blue crystalline nitrosite, whose structure has only recently been finally determined by x-ray analysis.

## 3.2 CARYOPHYLLENE

The structure of caryophyllene had baffled generations of workers and its final elucidation by Barton[3-7] and Šorm[8,9] was a notable triumph for organic chemistry. Oxidative degradation is difficult, but has yielded several important derivatives. Among them are norcaryophyllenic acid (1), caryophyllenic acid (2) and homocaryophyllenic acid (3). The structures of these compounds were conclusively established by the elegant syntheses of Rydon and others[10-13,51]. It follows that one of the two rings in caryophyllene is a dimethylcyclobutane derivative.

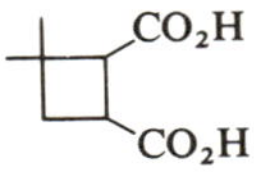


(1)

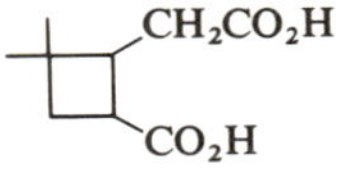


(2)

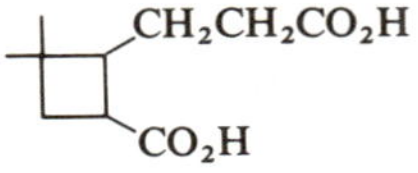


(3)

The size of the second ring in caryophyllene remained in doubt for many years and an extensive literature exists on attempts to solve this difficult problem. The question was finally settled by the intensive chemical studies of Barton *et al.*[3-7], and by the chemical and infrared studies of Šorm *et al.*[8,9]. These were based mainly on the reactions of a crystalline epoxide and epoxyketone (9) obtained from caryophyllene by oxidation (Treibs[14]). These chemical studies finally established formula (4) for caryophyllene, and so it became the first natural product known to have a nine-membered ring.

Isocaryophyllene can be obtained from the blue nitrosite of caryophyllene when it is heated with ethanol, and in other ways. The reactions of caryophyllene and isocaryophyllene show that they are simply geometrical isomerides about the endocyclic ethylenic linkage. Caryophyllene is less stable and more reactive, and the evidence tends to show that it is the *trans* isomer (4) while isocaryophyllene is the *cis* isomer (5). The various reactions and chemical arguments that led to these important conclusions have been described in some detail in a review by Barton and de Mayo[15]. At a later date, the total syntheses of *d,l*-caryophyllene and also *d,l*-isocaryophyllene were achieved by Corey *et al.*[50]. This cleared up any remaining doubts about the structure and the relations of the two isomers.

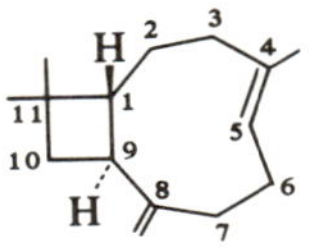


(4)

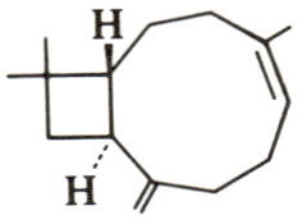


(5)

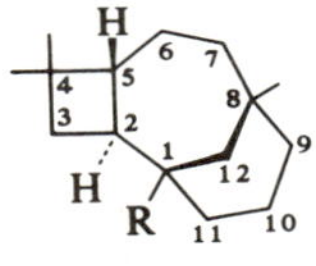


(6)

## 3.3 CARYOLAN-1-OL

β-Caryophyllene alcohol or caryolan-1-ol was recognised at an early date[16] to be a cyclisation product and not a simple hydroxyl derivative of the parent hydrocarbon. It is the first of a very long series of tricyclic derivatives of caryophyllene, whose nine-membered ring can be bridged and rearranged in a bewildering number of ways. A number of these structures have been elucidated and synthesised by the classical methods of organic chemistry, but many more have had to await the methods of x-ray crystallography for the final and precise determination of their structures.

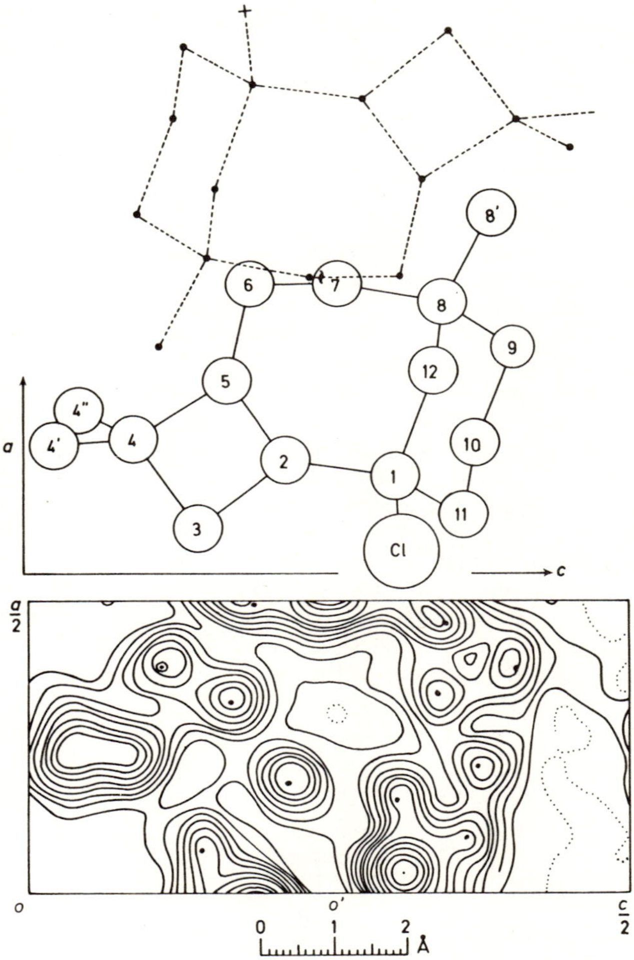


**Figure 3.1** Electron-density projection of caryolanyl chloride. (From Robertson[81], by courtesy of the Royal Institute of Chemistry)

Caryolan-1-ol forms a beautifully crystalline chloride and bromide which are isomorphous, and these were the subject of the first important x-ray crystallographic investigation in the caryophyllene series[17, 18]. It was clear that these derivatives might be employed to solve the difficult x-ray phase problems involved by the method of isomorphous substitution[19]. At the time of this work there were no computers, and the only feasible method of approach was by means of two-dimensional projections. By a gradual process of successive approximations, it was finally possible to locate the atomic centres in these poorly resolved projections, and the result for one of the projections is shown in Figure 3.1. From this and the other projections, the model shown in Figure 3.2 was constructed, and this clearly indicates the

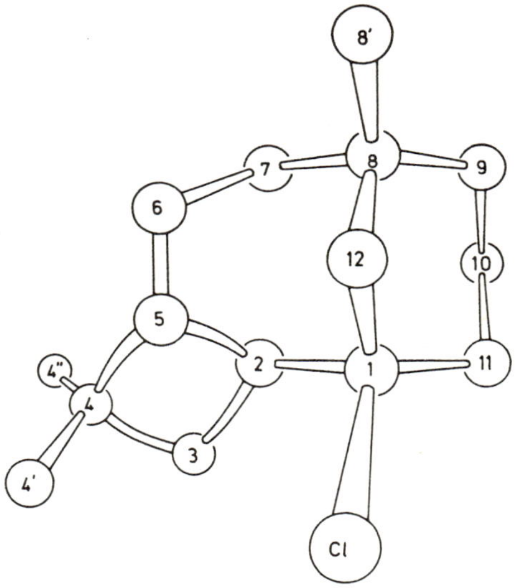


**Figure 3.2** Perspective drawing of the caryolanyl chloride molecule. (From Robertson[81], by courtesy of the Royal Institute of Chemistry)

relative positions of the atoms and the complete stereochemistry of the molecule. (The projection shown in Figure 3.1 does in fact correspond rather closely to this particular view of the molecule, although it is obscured by overlapping molecules at different levels in the unit cell.) The structural formula for the chloride is therefore (6; R = Cl) and for caryolan-1-ol (6; R = OH). These constitutions are in complete agreement with the chemical evidence of Barton *et al.*[3-7]. The x-ray work, however, enabled certain important points regarding the stereochemistry of these molecules to be elucidated for the first time. From the general shape of the molecule shown in Figure 3.2 it is clear that the cyclobutane ring is fused *trans* to the larger ring. Further, the hydrogen atom attached to C(5), which corresponds to C(1) in the formula for caryophyllene (4), evidently lies on the same side of the molecule as the methylene bridge, C(12). These facts completed the stereochemical description of caryophyllene and caryolan-1-ol.

## 3.4 THE STRUCTURE OF THE BLUE NITROSITE

The blue crystalline nitrosite of caryophyllene is an important derivative which has been used since early times to characterise and identify the sesquiterpene. It is also a convenient source of isocaryophyllene, which is formed when the nitrosite is heated in ethanol. A stable crystalline iodonitrosite can also be prepared by treatment of the nitrosite with iodine in chloroform solutions at room temperature[20]. Although known for many years, the structure of these compounds was never satisfactorily established[21,22]. A recent accurate crystallographic investigation[23] has now cleared this matter up, with rather surprising and unexpected results.

A weak chloroform–toluene solution of the iodonitrosite gave a very intense paramagnetic resonance spectrum[24]. The form of this spectrum and its magnetic parameters are almost identical with those given under the same conditions by the di-t-butyl nitroxide radical, $(Me_3C)_2NO$, showing that the caryophyllene compound belongs to this relatively new class of stable aliphatic nitroxide radicals[25]. Caryophyllene bromonitrosite behaves similarly. Further electron paramagnetic resonance studies[26] have shown that caryophyllene nitrosite is a very versatile source of nitroxide radicals, at least six further species being obtained when the solid or solutions of it are irradiated in different ways. Some of these radicals can be isolated as pure stable solids.

The presence of the iodine atom in the iodonitrosite provided an opportunity for a complete x-ray structure determination[23] by the well-known heavy-atom method[27]. This was an accurate analysis (discrepancy factor $R = 0.06$), and as the iodine atom scatters x-rays anomalously, it was possible at the same time to determine the absolute configuration of the molecule by Bijvoet's anomalous dispersion method[28]. The results of this analysis establish the structure and absolute stereochemistry of caryophyllene iodonitrosite as (7). It is a tricyclic molecule with the cyclobutane ring *trans*-fused to a seven-membered ring which is in turn *cis*-bridged to a six-membered ring as in caryolan-1-ol (6). This confirms that the absolute stereochemistry of caryophyllene itself must be as indicated in (4), a result that was deduced earlier by Barton and Nickon[7]. As a result of this analysis and further spectroscopic and chemical evidence[29], the structure of the parent caryophyllene nitrosite must be formulated as (8).

As mentioned above, caryophyllene iodonitrosite is a stable free radical. Its formation probably involves the addition of I· to the *exo*-methylene group in (8) and this tertiary alkyl radical then undergoes transannular addition to the nitroso group, thus generating the nitroxide radical[24]. Steric inhibition probably prevents the formation of a dimer, and the stability of the radical may be ascribed to the formation of a three-electron bond, as suggested by Hoffmann and Henderson[25] in the case of other aliphatic nitroxide radicals.

The dimensions of the molecule are shown in Figure 3.3, and these are on the whole in agreement with accepted values. The N—O bond length of 1.308 Å is what one might expect for a three-electron bond, Pauling's values for the N—O single and double bond being 1.44 and 1.20 Å, respectively[30].

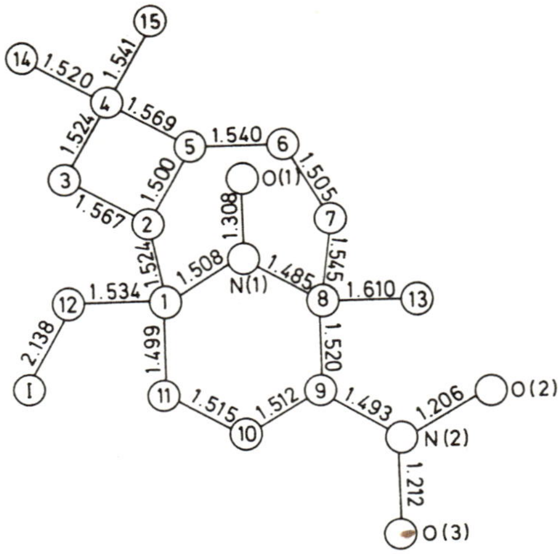


**Figure 3.3** Bond lengths in caryophyllene iodo-nitrosite. Standard deviations are about 0.02 Å. (From Hawley *et al.*[23], by courtesy of the Chemical Society)

The N—O bond makes an angle of 24° with the C(1)N(1)C(8) plane. The cyclobutane ring is severely puckered. The seven-membered ring and the six-membered ring both adopt chair conformations.

### 3.4.1 Caryophyllene chlorohydrin

During the discussion of the work by Barton and his collaborators on the structure of caryophyllene, it has already been mentioned that Treibs'[14] epoxy-ketone (9) played an important part in this elaborate chain of experiment and deduction. The action of hydrogen chloride converts this ketone into a chlorohydrin, $C_{14}H_{23}ClO_2$, which is not a simple addition product.

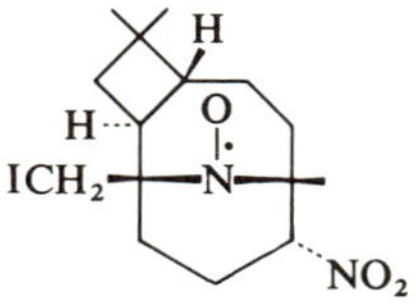


(7)

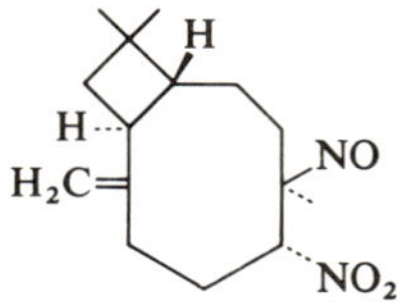


(8)

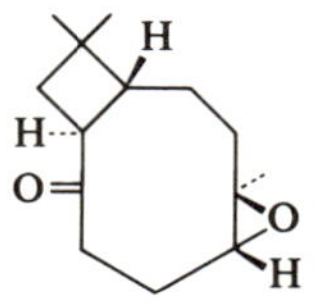


(9)

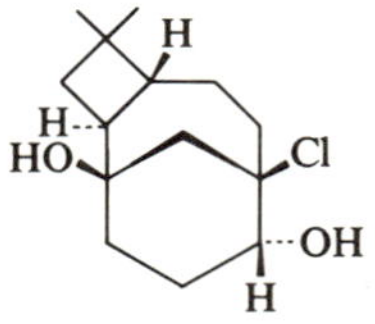


(10)

The structure of this chlorohydrin remained obscure for some time until an x-ray investigation by Rogers *et al.*[37, 82] and further chemical work by Sutherland *et al.*[38] established the structure as (10). The crystallographic work shows that the stereochemistry and conformational details of this molecule agree very closely with those of caryolan-1-ol (6; R = OH). The trigonal crystal structure is peculiar, the molecules being linked by hydrogen bonds and forming spirals around two of the $3_1$ screw axes, while leaving a large void around the third screw axis.

## 3.5 CLOVENE AND RELATED TRICYCLIC HYDROCARBONS

The first tricyclic derivative of caryophyllene is the alcohol caryolan-1-ol which has already been described. In addition, a very large number of tricyclic hydrocarbons are now known which can be obtained from caryophyllene or caryolan-1-ol in a variety of ways. The structures of these compounds are are shown in the formulae (11)–(18). They are all isomers of clovene, but trivial names and prefixes soon became exhausted, and the last two members are given the rather clumsy systematic names. The constitutions of clovene, epiclovene and neoclovene have all been determined and confirmed by the degradative and synthetic methods of organic chemistry without any help from x-ray analysis. The remaining structures, isoclovene, the pseudoclovenes and (17) and (18), were all elucidated by the methods of x-ray crystallography.

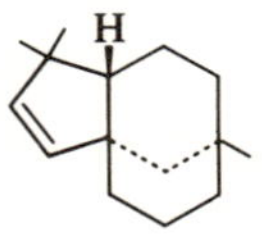

Clovene

(11)

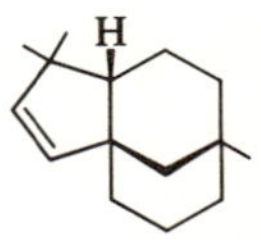

Epiclovene

(12)

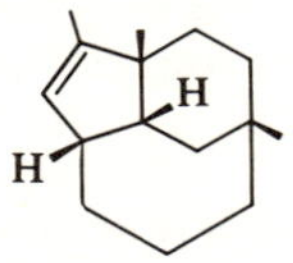

Isoclovene

(13)

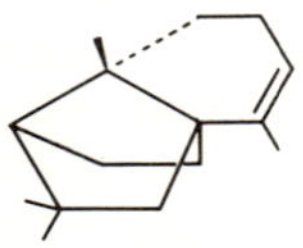

Neoclovene

(14)

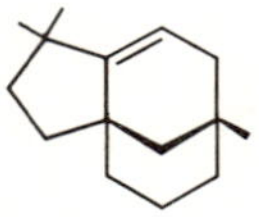

Pseudoclovene A

(15)

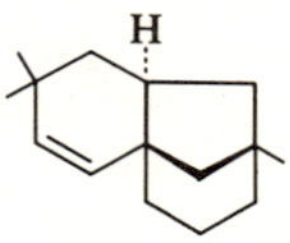

Pseudoclovene B

(16)

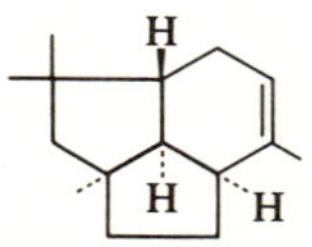

1,5,9,9-Tetramethyltricyclo[6,2,1,$0^{4,11}$]undec-5-ene

(17)

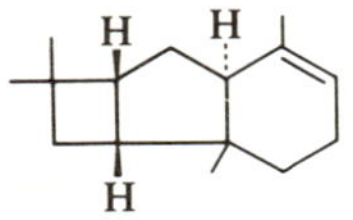

2,6,10,10-Tetramethyltricyclo[7,2,0,$0^{2,7}$]undec-5-ene

(18)

### 3.5.1 Clovene and epiclovene

Much of the early literature on clovene is confusing and sometimes erroneous. Using glacial acetic acid and dilute sulphuric acid, Wallach and Walker[1] were the first to obtain the oily hydrocarbon clovene together with caryolan-1-ol from caryophyllene. They also reported a hydrocarbon thought to be clovene obtained by dehydration of caryolan-1-ol with phosphoric oxide. Various other early preparations of clovene have been described[31, 32] but the hydrocarbon was never adequately characterised owing to the lack of crystalline derivatives. Its degradation to the dicyclic dibasic clovenic acid (19) by Ruzicka and Gibson[33] provides a means of identification, but not for the regeneration of pure clovene. In an important paper, Lutz and Reid[34] showed that the bromination of clovene yielded a mixture of solid isomeric dibromides that could be purified by distillation and fractional crystallisation, and from which very pure clovene could be regenerated by extended treatment with zinc dust. This very pure clovene gave a high yield of clovenic acid.

By careful repetition of the dehydration experiments[1] with caryolan-1-ol, Lutz and Reid[34] showed that the product was not clovene but a different dehydration product which they termed pseudoclovene (see below). Although clovene can readily be obtained from caryophyllene by an acid-catalysed reaction, it is not possible to convert the alcohol caryolan-1-ol into clovene in this way. They belong to different stereochemical families.

After characterisation and purification of clovene, the structure (11) was finally determined by Barton *et al.*[6] and by Lutz and Reid[34]. The chemical proof cannot be given here. Barton's work proceeded by an elaborate stepwise degradation of the glycol (20) obtained from caryophyllene and caryophyllene oxide to yield finally clovenic acid (19). Final confirmation of the structure was obtained by the total synthesis of clovene by Parker *et al.*[35]. This long and elegant synthesis commenced with the bicyclic acid (21) and after *ca.* 12 intermediate steps pure ($\pm$)-clovene was obtained, indistinguishable from naturally derived ($-$)-clovene.

While clovene is one of the major rearrangement products of caryophyllene, it is not obtained from caryolan-1-ol as Lutz and Reid have shown. The major acid-catalysed rearrangement products of caryolan-1-ol are isoclovene and the pseudoclovenes (see below). Recently, however, Parker *et al.*[36] have isolated and characterised another of these rearrangment products obtained by the action of $P_2O_5$ in DMF on caryolan-1-ol. This product has now been shown to be stereoisomeric with clovene, and hence is named epiclovene (12). Comparison of the formula for epiclovene (12) with that for caryolan-1-ol

(6) from which it is derived shows that they belong to the same stereochemical family. Taking the hydrogen atom at position 1 in caryophyllene, or position 5 in caryolan-1-ol (6) as the reference centre, then following Barton's terminology[6], substituents or bonds on the same side of the plane of the main ring as this hydrogen may be called $\beta$ (thickened bonds) and those on the opposite side $\alpha$ (broken bonds). Thus the methylene bridge in clovene is $\alpha$ while in epiclovene and caryolan-1-ol it is $\beta$.

### 3.5.2 Neoclovene

During a careful preparation of clovene by the action of concentrated sulphuric acid in ether on caryophyllene, Parker *et al.*[39] isolated a small quantity of a new and previously undetected higher boiling isomer which they named neoclovene (14). This new hydrocarbon was not obtained by this treatment from either clovene or caryolan-1-ol but only from caryophyllene.

The structure of neoclovene (14) was elucidated by these authors after a long and searching investigation involving many degradations and making the utmost use of mass spectrometry, i.r., $^1$H n.m.r. and other physical methods, but not x-ray crystallography. The presence of four methyl groups was soon established, indicating a new kind of carbon skeleton not encountered in most other caryophyllene rearrangement products. The formation of this unusual structure from caryophyllene has been rationalised by the authors[39] in terms of the mechanistic scheme shown in Figure 3.4.

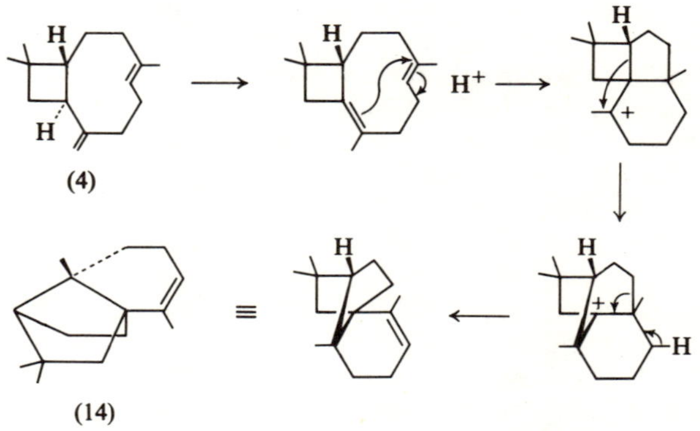


**Figure 3.4** Rearrangement of caryophyllene into neoclovene. (From Parker *et al.*[39], by courtesy of the Chemical Society)

Confirmation of this scheme and of the structure of neoclovene has now been obtained by a synthesis of neoclovene[40] based essentially on the steps shown in Figure 3.4. The starting point of this synthesis was the ketol (22), which is a known degradation product[3] of caryophyllene.

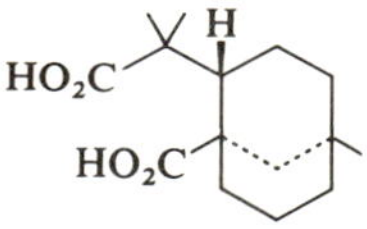


(19)

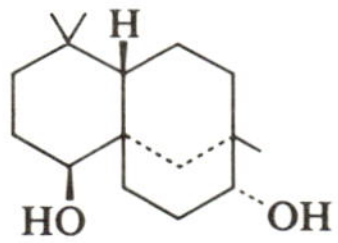


(20)

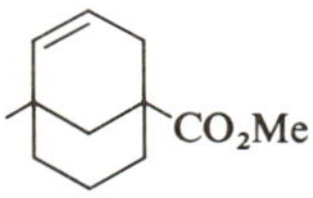


(21)

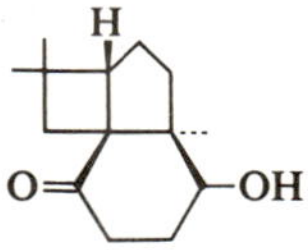


(22)

### 3.5.3 Isoclovene

During an early study[31] of the dehydration of caryolan-1-ol by a slight modification of Wallach's method[1], it was found that the product could be separated into two fractions by distillation under reduced pressure. The lower boiling mobile oil was at that time thought to be clovene, but the higher boiling, slightly more viscous, oil had different properties and was called isoclovene. It is characterised by the formation of a crystalline monohydrochloride and monohydrobromide from which the pure tricyclic sesquiterpene can be regenerated.

Although this member of the clovene family can thus be easily characterised and obtained in a pure state, all attempts to determine its structure were unavailing until the advent of x-ray analysis. Taking advantage of the heavy halogen atom in the hydrochloride, Clunie and Robertson[41, 42] then carried out a three dimensional crystal structure analysis which led unambiguously to the unexpected formula (13) for isoclovene. The bond lengths and angles found for the hydrochloride are shown in Figure 3.5, where the average

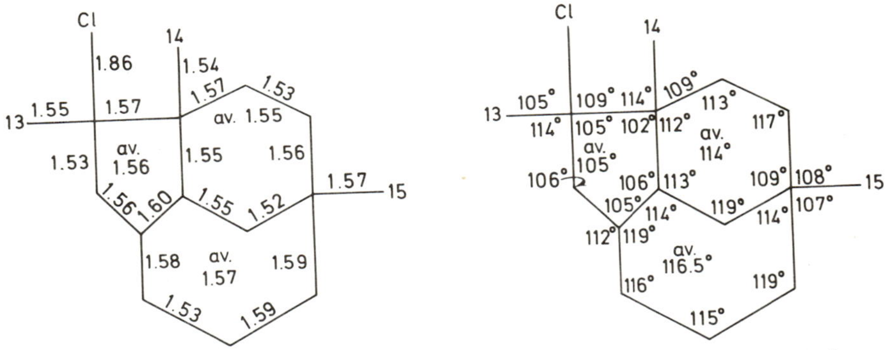


**Figure 3.5** Bond lengths and angles in isoclovene hydrochloride. (From Clunie and Robertson[42], by courtesy of the Chemical Society)

standard deviations are about 0.04 Å and about 2°. The fusion of the five-, six- and seven-membered rings causes considerable strain, the bond angles in the seven-membered ring, for example, being consistently greater than the tetrahedral value. Otherwise, the dimensions have about the expected values. The five-membered ring is hinge-shaped and the six-membered ring has the chair conformation. The seven-membered ring is also a distorted chair.

A mechanistic scheme which accounts for the formation of isoclovene, as well as the formation of the two pseudoclovenes from caryolan-1-ol, has been proposed by Parker *et al.*[36] and is given below in Figure 3.8.

### 3.5.4 Pseudoclovene A

It was shown by Lutz and Reid[34] that dehydration of caryolan-1-ol with phosphorus pentoxide does not yield clovene, but an isomeric tricyclic hydrocarbon which they found termed pseudoclovene. They suggested a tentative formula for this compound based on the formula for caryolan-1-ol, but there was very little evidence to support this formulation. In a more recent paper, Parker *et al.*[43] made a very careful gas–liquid chromatographic study of the products obtained by the treatment of caryolan-1-ol with $P_2O_5$ or polyphosphoric acid, and this revealed a complex mixture of hydrocarbons. The two main components of this mixture proved to be pseudoclovene and isoclovene.

Further investigation of the pseudoclovene fraction showed that it in turn consisted of a mixture of two components present in 1:1 ratio which were separable by silver nitrate–silica gel adsorption column chromatography. These two components were called pseudoclovene A and pseudoclovene B. A chemical and spectrometric study of pseudoclovene A showed that the structure proposed earlier by Lutz and Reid for pseudoclovene was not applicable to this compound. A lengthy treatment of pseudoclovene A with $OsO_4$–pyridine in ether then yielded a crystalline secondary–tertiary diol. From this it became possible to prepare a heavy atom compound suitable for x-ray analysis by converting the diol into the corresponding mono-*p*-bromobenzenesulphonate ester.

The crystal structure of this compound[44] proved to be rather complex. The space group is $P2_1$ and there are two molecules of $C_{21}H_{29}O_4SBr$ in the asymmetric unit, which are closely related by pseudo-fourfold screw axis. However, the phase-determining heavy-atom method[27] proved adequate,

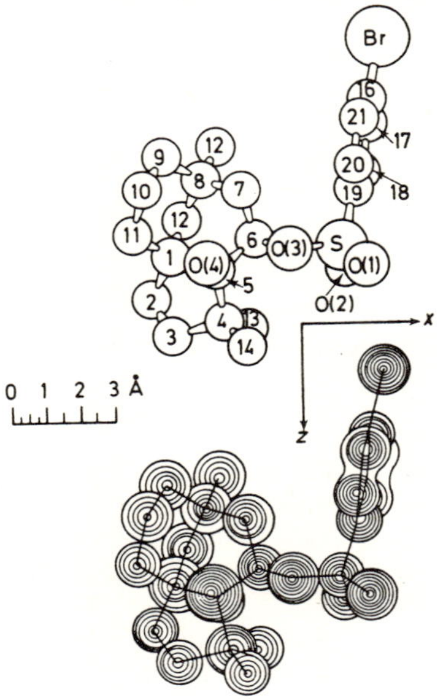


**Figure 3.6** Electron distribution in pseudoclovene A derivative. (From Hawley *et al.*[44], by courtesy of the Chemical Society)

and after least-squares refinement the final residual $R$ was reduced to 0.08 over 3100 independent reflections. The two independent molecules in the asymmetric unit proved to be identical within the limits of error. Superimposed electron-density sections for one molecule of the ester are shown in Figure 3.6, and the bond lengths averaged for the two molecules are given in Figure 3.7, where the estimated standard deviations are about 0.02 Å.

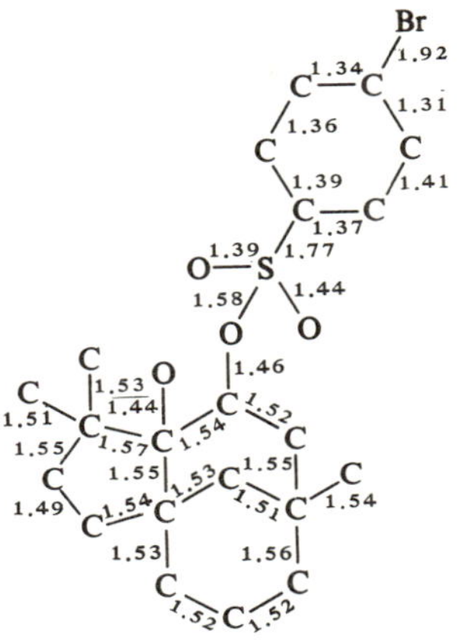


**Figure 3.7** Bond lengths in pseudoclovene A derivative (Calculated from Hawley *et al.*[44])

It was also possible to apply Bijvoet's anomalous dispersion method[28] to determine the absolute configuration of this derivative. The result shows that the mono-*p*-bromobenzenesulphonyl ester has the structure and absolute configuration (23), and hence the structure and absolute stereochemistry of pseudoclovene A is established as (15).

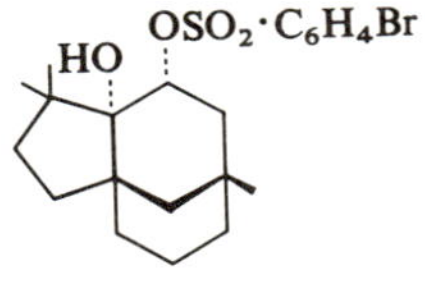


(23)

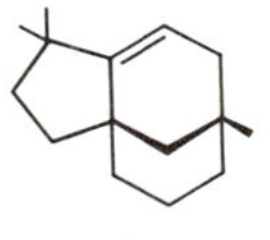

(15)

A successful synthesis of pseudoclovene A has also been described briefly by Parker *et al.*[43], starting from a precursor developed during their total synthesis of clovene[35].

### 3.5.5 Pseudoclovene B

A chemical and spectrometric study[45] of pseudoclovene B showed the presence of a dimethylcyclohexene fragment in the molecule, indicating a structure quite different from that of pseudoclovene A. For a complete structure determination, it proved possible to prepare a suitable heavy-atom derivative in the form of a crystalline dibromide, from which pseudoclovene B could be regenerated.

The monoclinic crystals of this dibromide, space group $P2_1$, proved very suitable for a complete structure determination, except that they decomposed on prolonged exposure to x-rays. Several specimens had to be employed for the data collection, but a reasonably accurate three-dimensional analysis was finally achieved ($R = 0.12$) which has been described briefly[45]. It was also possible to determine the absolute configuration from intensity differences of several Bijvoet pairs of reflections.

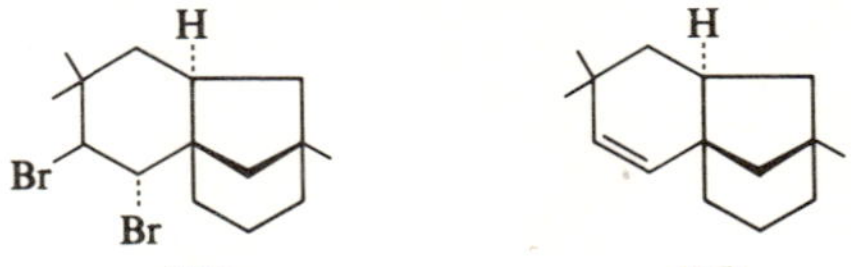


(24) (16)

The structure of the dibromide was thus determined as (24) and hence the structure and absolute stereochemistry of pseudoclovene B as (16).

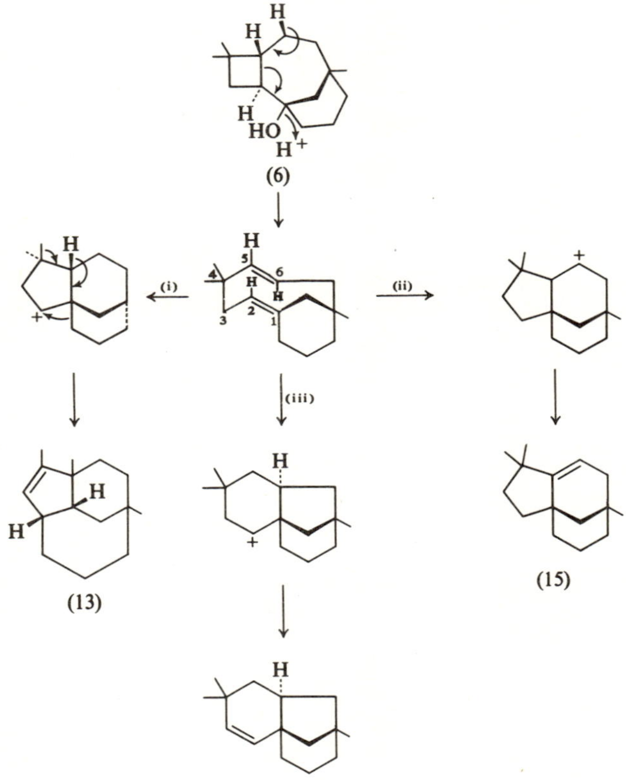


**Figure 3.8** Mechanism suggested for the rearrangement of caryolan-1-ol (6) to form isoclovene (13), pseudoclovene A (15) and pseudoclovene B (16). (i) Protonation at C(6), C(1)–C(5) closure. (ii) Protonation at C(2), C(1)–C(5) closure. (iii) Protonation at C(5), C(1)–C(6) closure. (From Baines *et al.*[36], and Parker *et al.*[45], by courtesy of Pergamon Press and the Chemical Society)

Several formal mechanisms have been proposed from time to time to account for the rearrangement of caryolan-1-ol into isoclovene[41, 42] and the pseudoclovenes[43,45]. However, Baines *et al.*[36] have now suggested a comprehensive scheme based on a bridged cyclononadiene intermediate which explains the formations of both the pseudoclovenes as well as isoclovene from caryolan-1-ol. This is shown in Figure 3.8.

### 3.5.6 1,5,9,9-Tetramethyltricyclo[6,2,1,$0^{4,11}$]undec-5-ene

The major products obtained from the acid-catalysed rearrangement of caryophyllene are caryolan-1-ol, clovene and neoclovene. Gollnick *et al.*[46] have now studied the rearrangement of isocaryophyllene (5) under the same conditions. When treated with a mixture of sulphuric acid and ether, isocaryophyllene yielded a complex mixture of hydrocarbons, but neither caryolan-1-ol nor clovene was detected. The two main components produced were neoclovene (*ca.* 35%) and another tricyclic hydrocarbon (also *ca.* 35%) which yielded a crystalline dibromide, $C_{15}H_{24}Br_2$.

This dibromide, although sensitive to light and easily decomposed, was suitable for x-ray examination by the heavy-atom method. The orthorhombic crystals, space group $P2_12_12_1$, have four molecules per unit cell. The final result of this analysis, which was of reasonable accuracy ($R = 0.14$), gave the structure of the dibromide as (25) and hence the structure of the new hydrocarbon as (17)

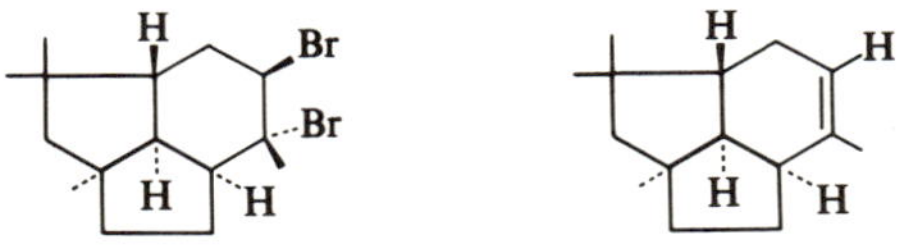


(25) (17)

The mechanism suggested[46] for the formation of this new hydrocarbon from isocaryophyllene is shown in Figure 3.9.

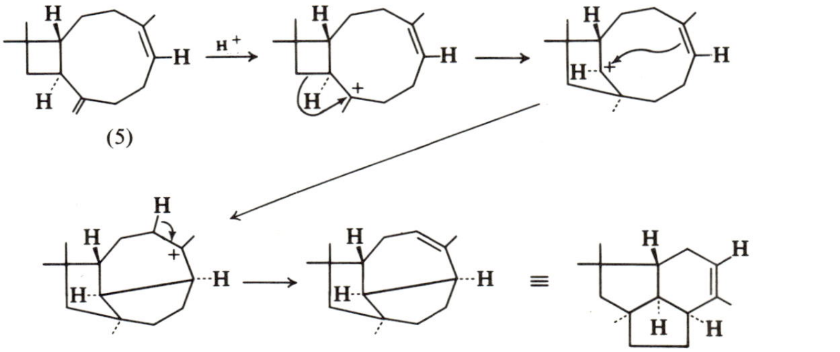


(17)

**Figure 3.9** The rearrangement of isocaryophyllene. (After Gollnick *et al.*[46], by courtesy of the Chemical Society)

### 3.5.7 2,6,10,10-Tetramethyltricyclo[7,2,0,$0^{2,7}$]undec-5-ene

The final hydrocarbon in this series has been obtained by Gollnick *et al.*[47] by the dechlorination of caryophyllene dihydrochloride. Various dechlorination experiments of this kind have been described in the earlier literature[31,48,49], but the products were never satisfactorily identified. Gollnick *et al.* employed one of the early procedures by refluxing in AcOH and isolated two main hydrocarbons from the product. One of them (34% yield) was identified as neoclovene (14). The other tricyclic hydrocarbon (45% yield) was difficult to identify from spectrometric evidence, but it formed a nicely crystalline dibromide, $C_{15}H_{24}Br_2$, which only decomposed slowly on exposure to light.

This derivative proved suitable for an accurate x-ray crystallographic examination by the heavy-atom method. There are four molecules in the unit cell, space group $P2_12_12$. The result of this analysis ($R = 0.10$) gave the structure and stereochemistry of the dibromide as (26) and hence the structure of the parent hydrocarbon as (18). This unexpected structure with four methyl groups consists of a four-, five- and six-membered ring fused in linear array.

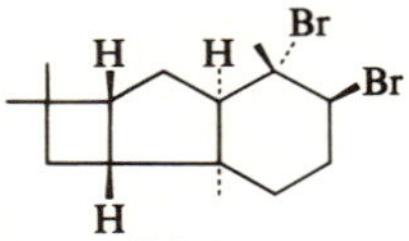

(26)

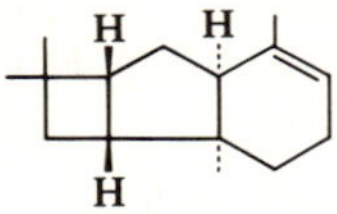

(18)

The steps involved in the formation of this hydrocarbon are suggested to be those shown in Figure 3.10, but it should be noted that the stereochemistry of caryophyllene dihydrochloride has not yet been verified by x-ray analysis.

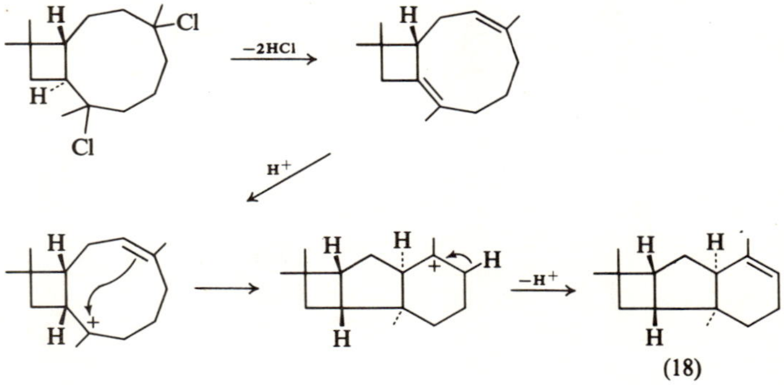


**Figure 3.10** The dechlorination of caryophyllene dihydrochloride. (After Gollnick *et al.*[47], by courtesy of the Chemical Society)

## 3.6 HUMULENE

Humulene was first recorded by Chapman[53] as the chief sesquiterpene constituent of oil of hops. It is also a minor constituent of the high-boiling terpene fraction of oil of cloves, and is found in other natural sources as well. It is included in this review because of its co-occurrence with caryophyllene in nature. It was originally described as α-caryophyllene, and only fairly recently has the true nature of α-caryophyllene alcohol as a humulene derivative been firmly established. The biogenesis of humulene and caryophyllene have been discussed by Halsall and Theobold[54], and by Parker *et al.*[55].

Like caryophyllene, humulene is characterised by the formation of a blue crystalline nitrosite[53]. It is a triply unsaturated sesquiterpene, $C_{15}H_{24}$, and is therefore monocyclic. Early work showed that it was a tetramethylcycloundecatriene, and formula (27) was deduced by oxidative degradation and synthetic studies[56-63], and confirmed by infrared and n.m.r. studies[64-66]. However, the *cis* or *trans* orientation of the double bonds remained undecided. The *trans,trans,cis* stereochemistry shown in (27) was predicted by Hendrickson[67], whereas the all-*trans* stereochemistry shown in (28) was favoured by Sutherland and Waters[68].

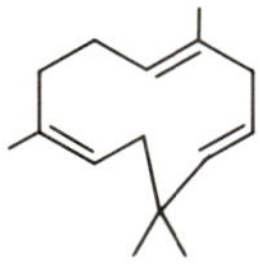

(27)

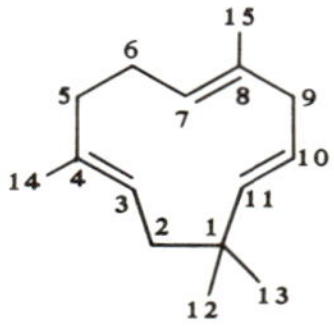


(28)

### 3.6.1 Humulene–silver nitrate adduct

The stereochemistry of humulene was finally settled by the x-ray analysis of the silver nitrate adduct[69-71], which crystallises in the monoclinic system, space group $P2_1/c$, with four $C_{15}H_{24}\cdot 2AgNO_3$ units in the unit cell. In this structure, each silver ion is bonded to one of the double bonds in humulene. Phasing on the heavy silver ions led to a reasonably accurate analysis ($R = 0.16$). The electron-density distribution is shown in Figure 3.11 and the corresponding atomic arrangement in Figure 3.12.

These results establish unambiguously the all-*trans* stereochemistry shown in (28), The average carbon–carbon double bond length is 1.34 Å, and the other bond lengths have their expected values. The silver ions clearly interact with the double bonds C(7)—C(8) and C(10)—C(11), the average $Ag^+ \cdots C$ separation being 2.38 Å. Owing to the ethylenic bonds, the 11-membered ring in humulene appears to be almost strain-free, the average C—C($sp^3$)—C valency angle in the ring being 111°.

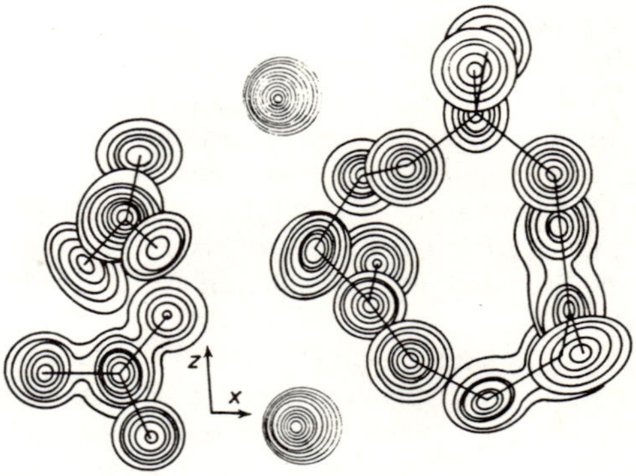


**Figure 3.11** Electron-density distribution in humulene bis(silver nitrate). (From McPhail and Sim[71], by courtesy of the Chemical Society) .

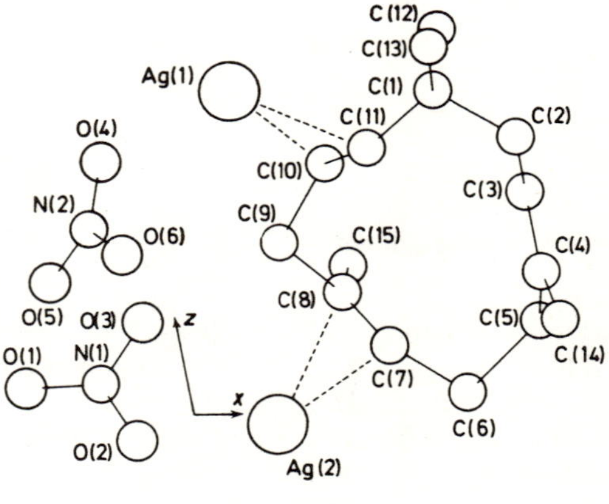


**Figure 3.12** Atomic arrangement in the humulene adduct corresponding to Figure 3.11. (From McPhail and Sim[71], by courtesy of the Chemical Society)

### 3.6.2 Humulene diepoxide

Although the conformation of the humulene molecule has been accurately defined in the analysis of the silver nitrate adduct, it seemed possible that the formation of this rather complex adduct might favour a conformation that was relatively unimportant in the free humulene molecule. It was therefore desirable to obtain the structure of another derivative not involving a heavy atom. Humulene itself is a liquid, but the diepoxide (29) is a solid, crystallising in the monoclinic system, space group $P2_1/c$.

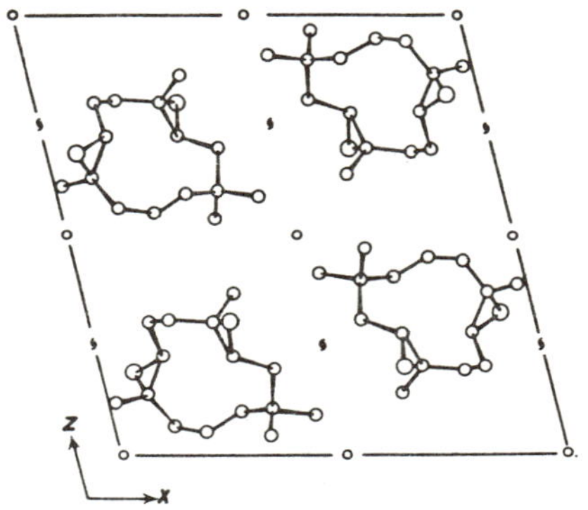


**Figure 3.13** Structure and arrangement of humulene diepoxide molecules in the unit cell. (From Cradwick *et al.*[72], by courtesy of the Chemical Society)

A very accurate structure analysis ($R = 0.07$) of this derivative, which does not involve any heavy atom, has been achieved by Cradwick *et al.*[72], using the symbolic addition procedure for phase determination[73]. The molecular structure and arrangement in the unit cell is shown in Figure 3.13. The conformation of the molecule approximates very closely to that shown by the silver nitrate adduct which is illustrated in formula (30). The torsion angles about the carbon–carbon bonds in the silver nitrate adduct and in the diepoxide have been calculated and the correspondence is extremely good, the mean deviation between equivalent angles for the 19 torsion angles calculated being only 8°. This shows that the conformations adopted by the two molecules are almost identical. All the evidence points to the fact that this is the conformation preferred by the free humulene molecule. The accurate bond lengths and angles obtained in this investigation are all close to the expected values.

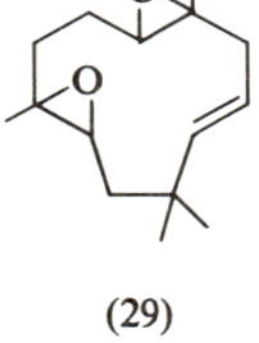


(29)

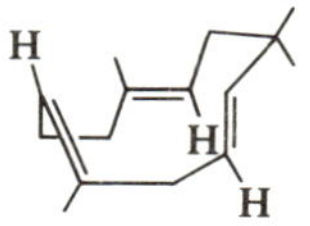


(30)

### 3.6.3 Humulene rearrangement products

The 11-membered ring in humulene, like the 9-membered ring in caryophyllene, can undergo a number of transannular rearrangements with the formation of tricyclic derivatives. Two of these products, α-caryophyllene alcohol and humulene bromohydrin, have now been the subjects of quantitative x-ray structure investigations. The first, α-caryophyllene alcohol, was known for many years as a hydration product of caryophyllene, being obtained along with β-caryophyllene alcohol, or caryolan-1-ol, and clovene when caryophyllene was treated with sulphuric acid. The yield of the α-alcohol obtained when commercial caryophyllene was treated with sulphuric acid in ether[74] was always small, usually about 4%, and its structure was never determined in these early studies. A careful re-investigation by Sim *et al.*[75,76] showed that when pure caryophyllene was treated with sulphuric acid and ether, no α-caryophyllene alcohol was obtained; but humulene on this treatment gave a 40% yield of the α-alcohol. The α-alcohol is therefore a humulene and not a caryophyllene derivative. The early confusion arose because commercial caryophyllene from oil of cloves always contains a small quantity (up to 10%) of humulene, which at that time could not be separated.

#### *3.6.3.1 α-Caryophyllene alcohol*

The structure of α-caryophyllene alcohol was established by Sim *et al.*[75,76] employing the *p*-bromobenzenesulphonate as a heavy atom derivative for their x-ray analysis, and simultaneously by Nickon *et al.*[77] who employed a chemical degradation sequence. An elegant synthesis of the alcohol was accomplished by Corey and Nozoe[78] shortly afterwards.

The monoclinic crystals of the bromobenzenesulphonate, space group $P2_1/c$, provided a reasonably accurate analysis ($R = 0.12$ over about 2000 reflections) from which structure (31) was derived without ambiguity. The molecular conformation is shown in (32). The cyclohexane ring adopts a pronounced chair conformation with the oxygen substituent axial. Both five-membered rings deviate markedly from planarity, adopting envelope conformations. All the bond lengths are found to be near their expected values.

It is clear that the molecular conformation (32) can easily be derived by transannular cyclisation of the humulene conformation (30).

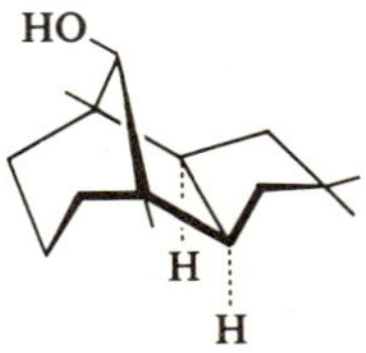


(32)

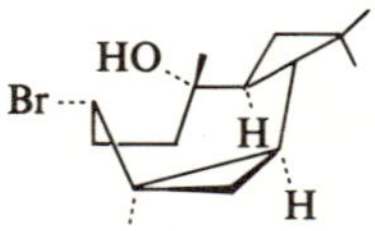


(34)

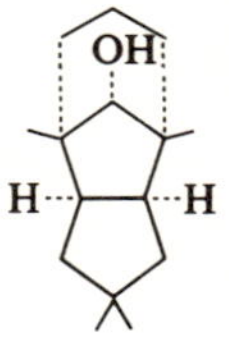


(31)

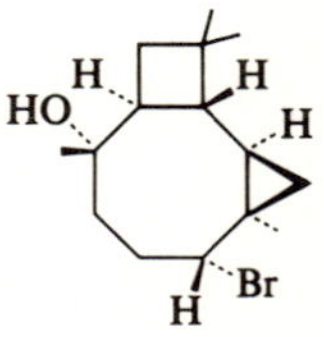


(33)

#### *3.6.3.2 Humulene bromohydrin*

Sutherland *et al.*[79] showed that humulene reacts with *N*-bromosuccinimide to give a monobromide and a bromohydrin, $C_{15}H_{25}OBr$. From the chemical investigation, these authors showed the structure of the bromohydrin to be (33) but without the full stereochemical details. These details were provided by the accurate ($R = 0.10$) x-ray crystallographic study of Allen and Rogers[80]. They showed that the large cyclo-octane ring has a distorted boat–chair conformation. The cyclobutane ring is buckled and is fused *trans* to the larger ring.

The conformation adopted by this bromohydrin can be represented by the formula (34) and from this formulation it is seen that it is readily derived by another transannular cyclisation of the humulene structure shown in formula (30). In this way, it is possible to explain the formation of the two tricyclic humulene derivatives, α-caryophyllene alcohol and humulene bromohydrin.

## 3.7 CONCLUSION

The structure of about 30 complex molecules derived from the sesquiterpene caryophyllene and its near relative humulene have been described in this article. The elucidation of these structures provided an outstanding challenge to organic chemistry over many years, and in the course of time many of them were solved by the degradative and synthetic methods of that science. All of them were isolated, purified and characterised by chemical and physico-chemical methods, but the detailed structure and stereochemistry of a number of these molecules had to depend on the methods of x-ray crystallography for their final elucidation. Any attempt to apportion credit between organic chemistry and x-ray crystallography in this and similar fields is an unrewarding task. Molecular science is best served by the active collaboration of chemists and crystallographers as exemplified throughout this chapter. Without chemistry, crystallographers would have no interesting structures to solve. Without crystallography, chemists would often be spending valuable time in trying to solve structures in a difficult way.

### References


1. Wallach, O. and Walker, W. (1892). *Justus Liebig's Ann. Chem.*, **271,** 285
2. Wallach, O. and Tuttle, F. E. (1894). *Justus Liebig's Ann. Chem.*, **279,** 391
3. Barton, D. H. R. and Lindsey, A. S. (1951). *J. Chem. Soc.*, 2988
4. Barton, D. H. R., Bruun, T. and Lindsey, A. S. (1952). *J. Chem. Soc.*, 2210
5. Aebi, A., Barton, D. H. R. and Lindsey, A. S. (1953). *J. Chem. Soc.*, 3124
6. Aebi, A., Barton, D. H. R., Burgstahler, A. W. and Lindsey, A. S. (1954). *J. Chem. Soc.*, 4659
7. Barton, D. H. R. and Nickon, A. (1954). *J. Chem. Soc.*, 4665
8. Šorm, F., Dolejš, L. and Plíva, J. (1950). *Collect. Czech. Chem. Commun.*, **15,** 186
9. Šorm, F., Jarolim, V., Streibl, M. and Dolejš, L. (1956). *Chem. Ind. (London)*, 154
10. Rydon, H. N. (1936). *J. Chem. Soc.*, 593; (1937). *ibid.*, 1340
11. Dawson, T. L. and Ramage, G. R. (1950). *J. Chem. Soc.*, 3523
12. Dawson, T. L. and Ramage, G. R. (1951). *J. Chem. Soc.*, 3382

13. Campbell, A. and Rydon, H. M. (1953). *J. Chem. Soc.*, 3002
14. Treibs, W. (1947). *Chem. Ber.*, **80,** 56
15. Barton, D. H. R. and de Mayo, P. (1957). *Quart. Rev. Chem. Soc.*, **11,** 189
16. Henderson, G. G., Robertson, J. M. and Kerr, C. A. (1926). *J. Chem. Soc.*, 62
17. Robertson, J. M. and Todd, G. (1953). *Chem. Ind.* (*London*), 437
18. Robertson, J. M. and Todd, G. (1955). *J. Chem. Soc.*, 1254
19. Robertson, J. M. (1936). *J. Chem. Soc.*, 1195
20. Deussen, E. (1926). *J. Prakt. Chem.*, **114,** 63
21. Ramage, G. R., Whitehead, R. and Wilson, B. (1954). *J. Chem. Soc.*, 4341
22. Burkhard, C. A. and Brown, J. F. (1964). *J. Org. Chem.*, **29,** 2235
23. Hawley, D. M., Ferguson, G. and Robertson, J. M. (1968). *J. Chem. Soc.* (*B*), 1255
24. Hawley, D. M., Roberts, J. S., Ferguson, G. and Porte, A. L. (1967). *Chem. Commun.*, 942
25. Hoffmann, A. K. and Henderson, A. T. (1961). *J. Amer. Chem. Soc.*, **83,** 4671
26. McConnell, A. A., Mitchell, S., Porte, A. L., Roberts, J. S. and Thomson, C. (1970). *J. Chem. Soc.* (*B*), 833
27. Robertson, J. M. and Woodward, I. (1937). *J. Chem. Soc.*, 219; (1940). *ibid.*, 36
28. Bijvoet, J. M. (1949). *Proc. Ned. Akad. Wetenschap.*, **52,** 313
29. Roberts, J. S., unpublished results
30. Pauling, L. (1960). *The Nature of the Chemical Bond* (New York: Cornell University Press, Ithaca)
31. Henderson, G. G., McCrone, R. O. O. and Robertson, J. M. (1929). *J. Chem. Soc.*, 1368
32. Bell, J. and Henderson, G. G. (1930). *J. Chem. Soc.*, 1971
33. Ruzicka, L. and Gibson, D. T. (1931). *Helv. Chim. Acta*, **14,** 575
34. Lutz, A. W. and Reid, E. B. (1954). *J. Chem. Soc.*, 2265
35. Doyle, P., Maclean, I. R., Murray, R. D. H., Parker, W. and Raphael, R. A. (1965). *J. Chem. Soc.*, 1344
36. Baines, D., Eck, C. and Parker, W. (1973). *Tetrahedron Lett.*, 3933
37. Rogers, D. and Mazhar-ul-Haque (1963). *Proc. Chem. Soc.*, 371
38. Greenwood, J. M., Qurreshi, I. H. and Sutherland, J. K. (1963). *Proc. Chem. Soc.*, 372
39. Parker, W., Raphael, R. A. and Roberts, J. S. (1969). *J. Chem. Soc.* (*C*), 2634
40. McKillop, T. F. W., Martin, J., Parker, W. and Roberts, J. S. (1967). *Chem. Commun.*, 162
41. Clunie, J. S. and Robertson, J. M. (1960). *Proc. Chem. Soc.*, 82
42. Clunie, J. S. and Robertson, J. M. (1961). *J. Chem. Soc.*, 4382
43. Ferguson, G., Hawley, D. M., McKillop, T. F. W., Martin, J. and Parker, W. (1967). *Chem. Commun.*, 1123
44. Hawley, D. M., Ferguson, G., McKillop, T. F. W. and Robertson, J. M. (1969). *J. Chem. Soc.* (*B*), 599
45. Crane, R. I., Eck, C., Parker, W., Penrose, A. B., McKillop, T. F. W., Hawley, D. M. and Robertson, J. M. (1972). *J. Chem. Soc.* (*Chem. Commun.*), 385
46. Gollnick, K., Schade, G., Cameron, A. F., Hannaway, C., Roberts, J. S. and Robertson, J. M. (1970). *Chem. Commun.*, 248
47. Gollnick, K., Schade, G., Cameron, A. F., Hannaway, C. and Robertson, J. M. (1971). *Chem. Commun.*, 46
48. Schreiner, O. and Kremers, E. (1901). *Pharmacol. Arch.*, **4,** 164
49. Semmoler, F. W. and Mayer, E. W. (1910). *Ber.*, **43,** 3451
50. Corey, E. J., Mitra, R. B. and Uda, H. (1964). *J. Amer. Chem. Soc.*, **86,** 485
51. Eschenmoser, A. and Fürst, A. (1951). *Experientia*, **7,** 290
52. Cameron, A. F., Hannaway, C. and Robertson, J. M. (1973). *J. Chem. Soc.* (*Perkin Trans. II*), 1938
53. Chapman, A. C. (1895). *J. Chem. Soc.*, **67,** 54, 780
54. Halsall, T. G. and Theobold, D. W. (1962). *Quart. Rev. Chem. Soc.*, **16,** 101
55. Parker, W., Roberts, J. S. and Ramage, R. (1967). *Quart. Rev. Chem. Soc.*, **21,** 331
56. Šorm, F., Mleziva, J., Arnold, Z. and Plíva, J. (1949). *Collect. Czech. Chem. Commun.*, **14,** 699
57. Herout, V., Streibl, M., Mleziva, J. and Šorm, F. (1949). *Collect. Czech. Chem. Commun.*, **14,** 716
58. Šorm, F., Streibl, M., Pliva, J. and Herout, V. (1952). *Collect. Czech. Chem. Commun.*, **16,** 639

59. Šorm, F., Streibl, M., Jarolim, V., Novotný, L., Dolejš, L. and Herout, V. (1954). *Collect. Czech. Chem. Commun.*, **19,** 570
60. Clemo, G. R. and Harris, J. O. (1951). *J. Chem. Soc.*, 22; (1952). *ibid.*, 655
61. Harris, J. O. (1953). *J. Chem. Soc.*, 184
62. Fawcett, R. W. and Harris, J. O. (1954). *J. Chem. Soc.*, 2699, 2673
63. Clarke, P. and Ramage, G. R. (1954). *J. Chem. Soc.*, 4345
64. Hildebrand, R. P., Sutherland, M. D. and Waters, O. J. (1959). *Chem. Ind.* (*London*), 489
65. Dev, S. (1960). *Tetrahedron*, **9,** 1
66. Dev, S. (1959). *Tetrahedron Lett.*, 12
67. Hendrickson, J. B. (1959). *Tetrahedron*, **7,** 82
68. Sutherland, M. D. and Waters, O. J. (1961). *Aust. J. Chem.*, **14,** 596
69. McPhail, A. T., Reed, R. I. and Sim, G. A. (1964). *Chem. Ind.* (*London*), 976
70. Hartsuck, J. A. and Paul, I. C. (1964). *Chem. Ind.* (*London*), 977
71. McPhail, A. T. and Sim, G. A. (1966). *J. Chem. Soc.* (*B*), 112
72. Cradwick, M. E., Cradwick, P. D. and Sim, G. A. (1973). *J. Chem. Soc.* (*Perkin Trans. II*), 404
73. Karle, J. and Karle, I. L. (1966). *Acta Crystallogr.*, **21,** 849
74. Asahina, Y. and Tsukamoto, T. (1922). *J. Pharmacol. Soc. Jap.*, **484,** 463
75. Gemmell, K. W., Parker, W., Roberts, J. S. and Sim, G. A. (1964). *J. Amer. Chem. Soc.*, **86,** 1438
76. Gemmell, K. W., Parker, W., Roberts, J. S., Robertson, J. M. and Sim, G. A. (1970). *J. Chem. Soc.* (*B*), 947
77. Nickon, A., McGuire, F. J., Mahajan, J. R., Umizawa, B. and Narang, S. A. (1964). *J. Amer. Chem. Soc.*, **86,** 1437
78. Corey, E. J. and Nozoe, S. (1964). *J. Amer. Chem. Soc.*, **86,** 1652
79. Greenwood, J. M., Sutherland, J. K. and Torre, A. (1965). *Chem. Commun.*, 410
80. Allen, F. H. and Rogers, D. (1968). *J. Chem. Soc.* (*B*), 1047
81. Robertson, J. M. (1954). *Royal Inst. Chem., Lectures, Monographs and Reports, No. 6*
82. Mazhar-ul-Haque and Rogers, D. (1974). *J. Chem. Soc.* (*Perkin Trans. II*), 228

# 4
# Non-bonded Interactions in Organic Molecules

**J. D. DUNITZ, F.R.S. and H. B. BÜRGI**
Swiss Federal Institute of Technology, Zurich

---

## 4.1 INTRODUCTION

Non-bonded interactions—a concept that was saddled with a negative definition and left with it!

The idea that molecules are describable in terms of bonds with characteristic properties is so deeply embedded in chemistry that whenever facts emerged that did not quite fit into this scheme they were promptly attributed to additional factors—conjugation, hyperconjugation, donor–acceptor interactions, electrostatic interactions, steric repulsion, van der Waals' attractions, geometric constraints—that were supposed to influence molecular structure and energy content. Of course, the separation of the total energy of a molecule into various contributions, some associated with bonded and others with non-bonded interactions, has no rigourous justification in quantum mechanics, but this should not deter us from thinking in terms of these concepts. Chemistry, it has been said somewhere, deals with a large number of weak facts, and any concept is useful if it serves as a guide in classifying and comparing the known properties of closely related molecules and in predicting the properties of other related molecules.

Historically, most notions about non-bonded interactions have come from facts about thermochemistry and about structure, and it may be worthwhile to try to clarify some of the ideas involved. The idea that atoms possess characteristic combining power, expressed by the formation of bonds in definite directions in space, emerged from 19th century chemistry long before the discovery of the electron and has coloured chemical thinking ever since. The classical structural formulae of organic chemistry express the idea that a molecule can be regarded as an assemblage of atoms interacting through bonds with neighbouring atoms, and that the number of bonds formed by any kind of atom is characteristic for that atom. The formula is a graph[1,2] that can be described in terms of its adjacency matrix. The adjacency matrices for the molecules of water and of ethane can be written:

*Water*

| | | |
|---|---|---|
| O | 1 | 1 |
| 1 | H | 0 |
| 1 | 0 | H |

*Ethane*

| | | | | | | | |
|---|---|---|---|---|---|---|---|
| H | 0 | 0 | 1 | 0 | 0 | 0 | 0 |
| 0 | H | 0 | 1 | 0 | 0 | 0 | 0 |
| 0 | 0 | H | 1 | 0 | 0 | 0 | 0 |
| 1 | 1 | 1 | C | 1 | 0 | 0 | 0 |
| 0 | 0 | 0 | 1 | C | 1 | 0 | 0 |
| 0 | 0 | 0 | 0 | 1 | H | 0 | 0 |
| 0 | 0 | 0 | 0 | 1 | 0 | H | 0 |
| 0 | 0 | 0 | 0 | 1 | 0 | 0 | H |

where the atom-types are inserted in the otherwise empty diagonal elements. The ethane matrix can obviously be rewritten in the form

$$\begin{matrix} \boldsymbol{A} & 1 \\ 1 & \boldsymbol{A} \end{matrix}$$

expressing the breakdown into two groups that interact only through a C—C bond. If we wished to express the internal energy of these molecules as sums of pairwise interactions the first step would be to replace the 1's (denoting the occurrence of bonds) by numbers denoting the respective bond energies. For many molecules the sum of these bond energies would give the standard enthalpy of formation (from monatomic gases) in quite good approximation, but for other molecules the agreement would be unsatisfactory. The introduction of non-bonded interactions would correspond, in this scheme, to replacing the off-diagonal 0's by appropriate numbers denoting the interaction energies between non-bonded pairs of atoms.

With the development of physical methods for investigating molecular structure it became possible to get more or less precise information about the equilibrium arrangement of atoms, enough to establish in most cases what the organic chemist would call the molecular structure, that is, the constitution and configuration, together with the conformation. Structure analyses that provide this kind of information—more or less routine nowadays—have been most important in the development of 20th century chemistry. They gave some sort of physical reality to the abstractness of structural formulae and they put these formulae on a metrical basis that can be summarised in tables of interatomic distances and valency angles. The distances could be systematised in terms of standard bond lengths, which could in turn be expressed as sums of standard covalent or ionic radii of the atoms involved in the bond, and in terms of typical contact distances between pairs of atoms not bonded to each other and not forced together by other bonds. These contact distances are roughly the same, whether the atoms involved belong to the same molecule or to different molecules packed together in a crystal and they can be expressed as sums of 'non-bonded' or van der Waals' radii, which are typically *ca.* 1 Å larger than corresponding covalent radii and roughly equal to the corresponding ionic radii. In some molecules, atoms are forced by a particular arrangement of bonds into closer contact than that given by their typical non-bonded radii. Such contacts, which are often evident from inspection of molecular models, are usually associated with repulsive interactions between the atoms concerned and can have important consequences on chemical properties (steric hindrance).

Attractive non-bonded interactions are known to be responsible for many characteristics of the liquid and solid state—indeed for the existence of these states. The energy associated with such interactions (van der Waals' attraction) is typically much smaller than that associated with chemical bonding, but the existence of very stable molecular complexes[3] testifies to a range of attractive interactions intermediate between van der Waals' attractions and chemical bonds. For certain kinds of atom pairs there appears to be an almost continuous transition from bonded to non-bonded distances, suggestive of the geometrical changes that have to occur in the course of bond making and breaking, i.e. in the course of chemical reactions[4–7].

Any fundamental theory of these phenomena would have to be based on quantum mechanics, but there are formidable difficulties. In the following section we discuss the concepts and approximations behind various quantum mechanical models that provide some degree of insight into the behaviour of simple molecular systems. The remainder of this article is divided into two parts. In the first, we discuss the structure and parametrisation of molecular force fields based on a complex interplay of approximate theoretical considerations and experimental data, and show how non-bonded interactions can be accommodated within this framework. In the second part, we discuss, by way of example, successes and failures, difficulties and open problems of molecular force-field calculations.

## 4.2 A MINIMUM OF NON-MATHEMATICAL QUANTUM MECHANICS

The properties of a chemical species, such as energy, geometry, electron density, vibrational and electronic transitions, etc. can be calculated in principle by quantum mechanics. For most problems of chemical interest this is impracticable, and in order to get round some formidable difficulties a cascade of approximations has to be introduced into the theory. Experimental results, though often of limited accuracy, are unaffected by these approximations, so that in order to ensure the connection between theory and experiment it becomes necessary to gauge the capabilities of a theoretical model at any level of approximation.

Prime among these approximations is the Born–Oppenheimer approximation, in which the Schrödinger equation is solved only for the electrons, with all nuclei fixed in space. The calculation is repeated for a sufficient number of nuclear arrangements, and the multidimensional energy surface, the Born–Oppenheimer (BO) surface, is thereby mapped out for the species under consideration. The BO surface serves to describe the equilibrium structure and the vibrational characteristics of the species, the only properties we shall be concerned with here. However, for all cases of interest in this article the calculation of the BO surface would still be much too tedious to be useful in practice. The opposite procedure, the reconstruction of the BO surface from experimental data, therefore offers a worthwhile challenge and will be considered in the following chapter.

A further approximation is the assumption that the repulsion between electrons does not depend critically on their instantaneous distribution but can be simulated by considering the repulsion of each electron with the smeared-out averaged density of all other electrons. The corresponding one-electron wave functions $\psi(x_i,y_i,z_i)$ are known as orbitals—atomic orbitals (AOs) in the case of atoms, molecular orbitals (MOs) in the case of polyatomic species. The MOs are often expressed as linear combinations of AOs. The part of the total energy neglected in such Hartree–Fock self-consistent-field calculations (HF–SCF) is called intra- or inter-orbital correlation or intermolecular dispersion energy; it is attractive and at large internuclear distances $r$ is approximately proportional to $r^{-6}$. This reflects the tendency of the HF–SCF approximation to underestimate the extent to which individual electrons avoid one another by virtue of their charge. HF–SCF theory, or any of its more simplified variants, is able to produce analogies to most of the concepts used for describing attractive and repulsive non-bonded interactions except, of course, correlation energy. It thus provides some sort of link between quantum mechanical theory and the various additivity schemes that have been found so useful by chemists—between the valleys, passes and mountains of the BO surface and the less well defined, semi-quantitative chemical concepts which possibly form the strata underlying this landscape.

In the following discussion we shall forget involved theories and methods and use instead a qualitative perturbation theory[8–12] which has become widely used in organic chemistry. In this approach the interaction between two chemical moieties in the same or in different molecules is described in terms of an interaction between their unperturbed filled and empty MOs, as shown in the following examples.

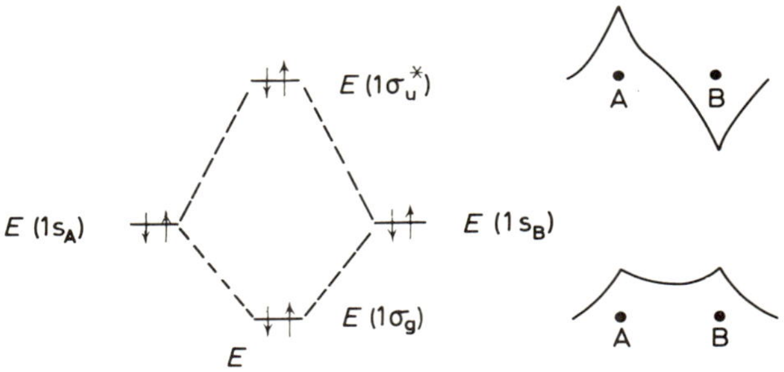


**Figure 4.1** Schematic representation of interaction between 1s orbitals of two helium atoms, leading to net repulsion. Left: orbital energies. Right: wave functions.

Two helium atoms at infinite separation may be described by two doubly occupied AOs, $1s_A$ and $1s_B$, each associated with an energy $E(1s)$. As the two atoms approach the orbitals start 'feeling' each other and a better description of the composite system is given by linear combinations (MOs) of the separate orbitals. The combination with the two 1s orbitals in phase ($1\sigma_g$) corresponds to a de Broglie standing half-wave and is associated with an energy $E(1\sigma_g) < E(1s)$; it takes two electrons. The other combination with the two 1s orbitals out of phase ($1\sigma_u^*$) corresponds to a full de Broglie standing wave with an energy $E(1\sigma_u) > E(1s)$. The respective energies classify $1\sigma_g$ as a bonding orbital and $1\sigma_u^*$ as an antibonding orbital (Figure 4.1). It is well known that 'the antibonding character of $1\sigma_u^*$ dominates the bonding character of $1\sigma_g$ because the presence of a nodal plane necessitates larger coefficients for the atomic orbitals in $1\sigma_u^*$[13]. This is the reason why two helium atoms repel each other at short distances. In general, the interaction between *any* pair of non-orthogonal MOs leads to a pair of bonding–antibonding combinations with a tendency for the antibonding member to dominate[14]. Transferring this simple picture to intramolecular non-bonded repulsions involves some difficulties, but it nevertheless may be considered to portray the essential part of non-bonded repulsions, Pauli repulsions, steric hindrance or other synonymous concepts.

The preceding discussion applies to systems in which the low-lying AOs or MOs are completely filled with electrons. However, certain kinds of functional groups (carbonyl, carboxylic acid, ester, polyene, etc.) have low-lying empty orbitals while others (amino-nitrogen, ether-oxygen, etc.) have high-lying filled orbitals. An interaction between such a pair of orbitals depends on spatial parameters (distance and orientation) and leads to a stabilised bonding MO and a destabilised antibonding one. The electrons available—two in Figure 4.2—reside in the stabilised bonding MO, thereby contributing to a net

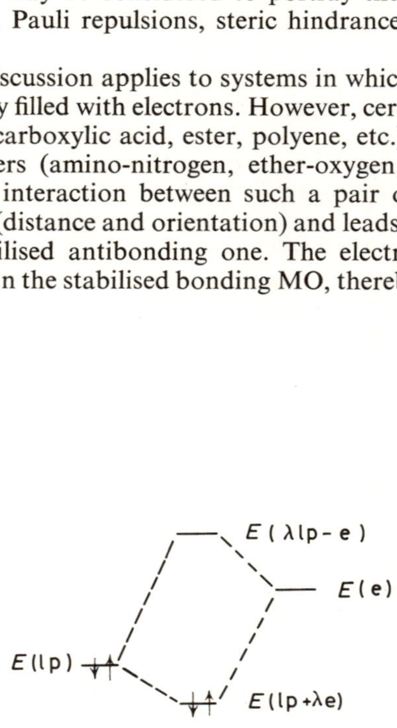


**Figure 4.2** Schematic representation of interaction between high lying filled orbital (1p) and low lying empty orbital (e), leading to net energy gain.

overall stabilisation of the system. This kind of analysis, crude though it may be, provides insight into the nature of donor–acceptor complexes between energetically high-lying lone pair or $\pi$-type donor orbitals and low-lying antibonding $\pi^*$ or $\sigma^*$ orbitals.

These simple ideas are in no way intended as a complete basis for discussing non-bonded interactions. In particular, electrostatic effects due to the detailed distribution of nuclear and electronic charges may often be of comparable magnitude with those described above. However, the ability to handle such qualitative concepts may help to link deviations from standard values of energy and interatomic distance with qualitative features of the BO surface. In particular, mathematical functions that express the essence of relevant parts of this surface can sometimes be derived by qualitative analysis of the shapes and mutual orientations of the orbitals involved.

## 4.3 MOLECULAR FORCE FIELDS

### 4.3.1 Reconstruction of the Born–Oppenheimer (BO) surface

It has been pointed out that for molecules of medium complexity, the calculation of the BO surface by quantum mechanics is impractical and suffers from many approximations. It is, therefore, very desirable to use experimental data to construct the BO surface, and this is sometimes possible, at least in regions of the parameter space close to local or absolute minima.

As a simple example, the energy of a diatomic molecule may be described by a one-dimensional function depending only on the interatomic distance $r$ and showing the general features of Figure 4.3. We are particularly interested in the value of the equilibrium distance $r_e$, the force constant or curvature at the minimum, $k=(d^2V/dr^2)_{r=r_e}$, possibly the anharmonicities, $(d^nV/dr^n)_{r=r_e}$ $(n>2)$, and the dissociation energy, $D_e$. Experimentally, none of these quanti-

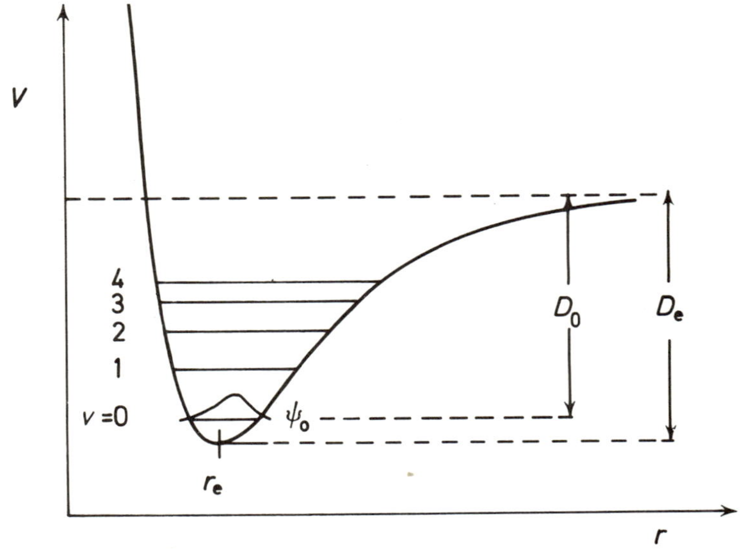


**Figure 4.3** Typical dependence of potential energy on internuclear distance for diatomic molecule.

ties is directly accessible. The value of the observed distance $r_{obs}$ represents some kind of average over the ground state and possibly excited state vibrational wave functions $\Psi_v$. Because $\Psi_v$ is not mirror-symmetric with respect to $r_e$, $r_{obs}$ differs from $r_e$ by an amount that depends on the type of experiment performed. Similarly, we cannot observe $(d^nV/dr^n)_{r=r_e}$ directly; instead we observe energy differences between vibrational levels, which become smaller and smaller with increasing vibrational quantum number $v$. Values measured for the dissociation energy represent the difference between the energy of the atoms at infinite separation and that of the diatomic molecule in its vibrational ground state. All of the above experimentally observable quantities may be converted, however, by complicated but established procedures[15] into the desired quantities, $r_e$, $(d^nV/dr^n)_{r=r_e}$, $D_e$, i.e. into a potential energy curve $V(r)$.

The same basic ideas are applicable to polyatomic molecules. Potential energy is now a function of $3N-6$ independent variables, which are usually chosen as structural parameters, such as distances and angles. Again the BO surface cannot be observed directly but has to be inferred by means of even more complicated algorithms from structural and spectroscopic data[16–20]. As a consequence, information is usually obtainable only about the immediate neighbourhood of the minima. For some tri- and tetra-atomic molecules quite accurate equilibrium structures and force constants up to fourth order [e.g. $(d^4V/dp_i^4)_{p=p_e}$] have been determined[21]. For more complicated systems, equilibrium molecular structures are not available. In their place some kind of averaged structure, as obtained from x-ray diffraction, electron diffraction or microwave spectroscopic data, has to be used. Anharmonicity effects are usually neglected with the hope that they can be absorbed by an appropriate parametrisation based on the uncorrected structural data.

### 4.3.2 Molecular mechanics

A technique, known by the name of molecular mechanics, has been developed to deal with the complexity of forces attributed to polyatomic molecules. Early applications to various problems were made 20–30 years ago[22–27], but it is only with the more recent increased power and availability of modern computers that the method has come into its own[28,29]. Its recent successes have led to its widespread application to many problems of molecular structure. The basic idea is that the potential energy of a diatomic molecule may be simulated by a mechanical model consisting of two masses connected by a spring with the special property that the spring gets harder on compression and softer on extension. The potential energy of diatomic molecules, for example, is given in good approximation by the Morse curve:

$$V(r) = -D_e + D_e[1 - \exp\{-a(r-r_e)\}]^2 \tag{4.1}$$

or

$$V(r) = -D_e[1 - a^2(r-r_e)^2 + a^3(r-r_e)^3 + \ldots] \tag{4.2}$$

by expanding in a Taylor series about the equilibrium $r_e$. The force constant $k$ equals $2a^2D_e$. For the hydrogen molecule: $D_e = 108.9$ kcal mol$^{-1}$; $r_e = 0.74$ Å; $k = 5.2$ mdyn Å$^{-1}$; $a \approx 1.86$ Å$^{-1}$, with the same orders of magnitude applicable to most organic molecules. Polyatomic molecules may be described in similar terms. However, the same set of masses may be connected in various ways by different sets of springs to simulate any given molecule or class of molecules. In practice, therefore, different kinds of coordinate systems and force fields may be used to describe the same molecule.

### 4.3.3 The $H_2O$ molecule in different force fields

As a simple example we show how different force fields can be used to describe regions of the BO surface of the $H_2O$ molecule[30]. The descriptions are valid only in the immediate vicinity of the equilibrium structure, $r(\mathrm{OH})=0.957$ Å, $a(\mathrm{HOH})=104.5°$.

Use of the interatomic distances as variables leads to a general central force field (GCFF)[31]:

$$2V(\mathrm{GCFF})=k_r(\delta r_1^2+\delta r_2^2)+k_q\delta q^2+2k_{rr}\delta r_1\delta r_2+2k_{qr}(\delta r_1+\delta r_2)\delta q \quad (4.3)$$

Bond lengths and valency angles go into (general) valence force fields (GVFF or VFF)[31]:

$$2V(\mathrm{GVFF})=K_r(\delta r_1^2+\delta r_2^2)+H_\alpha r_e^2\delta a^2+2F_{rr}\delta r_1\delta r_2+2F_{r\alpha}r_e(\delta r_1+\delta r_2)\delta a \quad (4.4)$$

The Urey–Bradley force field (UBFF)[32,33] includes terms in $\delta r_1^2$, $\delta r_2^2$, $\delta a^2$ and in $\delta q^2$ and omits cross terms. The potential energy expression has to contain linear terms as well as quadratic ones since $\delta r_1$, $\delta r_2$, $\delta a$ and $\delta q$ are not independent:

$$\begin{aligned}2V(\mathrm{UBFF})=2K'r_e(\delta r_1+\delta r_2)+K(\delta r_1^2+\delta r_2^2)+2H'r_e^2\delta a+\\ Hr_e^2\delta a^2+2F'q_e\delta q+F\delta q^2\end{aligned} \quad (4.5)$$

The first-order constants $H'$, $K'$ and $F'$ are related by the condition that the derivatives of $V$ with respect to any infinitesimal displacement must vanish at equilibrium. Interaction constants in the UBFF analogous to $F_{rr}$ and $F_{r\alpha}$ in a GVFF are hidden in the terms containing $F$ and $F'$ as shown in equation (4.6). The absence of cross terms in the UBFF means that for molecules more complicated than $H_2O$ it has fewer force constants than a GCFF or GVFF. Moreover, the term in $\delta q^2$ can be plausibly identified with a repulsion between non-bonded atoms (Section 4.3.5). Apart from its simplicity, the main virtue of the UBFF is that its constants are reasonably transferrable among molecules of the same type, a property not shared with the GCFF and GVFF or with their simplified variants, CFF and VFF, where cross terms are omitted.

With the help of $q^2=r_1^2+r_2^2-2r_1r_2\cos a$, both the GCFF and the UBFF may be transformed into a GVFF, yielding the relationships given below, where $s=\sin(a/2)$, $c=\cos(a/2)$:

| GVFF | GCFF | UBFF | |
|---|---|---|---|
| $K_r$ | $k_r+k_qs^2+2k_{rq}s$ | $K+Fs^2+F'c^2$ | |
| $H_\alpha$ | $k_qc^2$ | $H+Fc^2-F's^2$ | (4.6) |
| $F_{rr}$ | $k_{rr}+k_qs^2+2k_{rq}s$ | $Fs^2-F'c^2$ | |
| $F_{r\alpha}$ | $k_qsc+k_{rq}c$ | $(F+F')sc$ | |

In order to make full use of possible symmetry, linear combinations of the above coordinates can be chosen such that they are irreducible representations of the point group describing the molecule at equilibrium. For the water molecule ($C_{2v}$ symmetry) these symmetry coordinates (and their irreducible representations) are:

$$\begin{aligned} S_1\,(A_1) &= (\delta r_1+\delta r_2)/\sqrt{2} \\ S_2\,(A_1) &= r_e\delta a \text{ or } \delta q \\ S_3\,(B_1) &= (\delta r_1-\delta r_2)/\sqrt{2} \end{aligned} \qquad (4.7)$$

The potential energy is given by

$$2V(\text{SFF})=F_{11}S_1^2+F_{22}S_2^2+2F_{12}S_1S_2+F_{33}S_3^2 \qquad (4.8)$$

with cross-terms only between coordinates of like symmetry because $V$ must be totally symmetric. The relationships between symmetry and valence force constants are:

| | | | | |
|---|---|---|---|---|
| Symmetry force constants | $F_{11}$ | $F_{12}$ | $F_{22}$ | $F_{33}$ |
| Valence force constants | $K_r+F_{rr}$ | $F_{r\alpha}$ | $H_\alpha$ | $K_r-F_{rr}$ |
| Central force constants | $k_r+k_{rr}$ | $k_{qr}$ | $k_q$ | $k_r-k_{rr}$ |

(4.9)

Table 4.1 summarises numerical values of force constants of $H_2O$ corresponding to the different force fields discussed. Some of the values are widely different. *Force constants appropriate for one force field are not, in general, appropriate for another.*

**Table 4.1 Force constants of water[30] [$\alpha$(HOH=104.5°; $r$(OH)=0.957 Å] in mdyn Å$^{-1}$**

| *SFF* | *GVFF* | *UBFF* | *GCFF* |
|---|---|---|---|
| $F_{11}$=8.3527 | $K_r$ = 8.4536 | $K$ =8.259 | $k_r$ = 9.135 |
| $F_{22}$=0.7605 | $H_\alpha$ = 0.7605 | $H$ =0.979 | $k_q$ = 2.029 |
| $F_{12}$=0.2274 | $F_{r\alpha}$ = 0.2274 | $F'$ =0.395 | $k_{rq}$ =−1.233 |
| $F_{33}$=8.5545 | $F_{rr}$ =−0.1009 | $F$ =0.075 | $k_{rr}$ = 0.580 |

## 4.3.4 Complex molecules: general valence force field (GVFF)

The force fields used in vibrational analysis and molecular mechanics are usually of the UBFF or GVFF type. The latter, in its most complete form, is given by*:

$$\begin{aligned} 2V(\text{GVFF}) = {} & \sum_b K_b(b-b_0)^2+\sum\sum_{b\neq b'} F_{bb'}(b-b_0)(b'-b'_0)+\sum_a H_\alpha(a-a_0)^2+ \\ & \sum\sum_{a\neq a'} F_{\alpha\alpha'}(a-a_0)(a'-a'_0)+\sum H_\omega(\omega-\omega_0)^2+\sum\sum_{\omega\neq\omega'} F_{\omega\omega'}\times \\ & (\omega-\omega_0)(\omega'-\omega'_0)+\sum_b\sum_a F_{b\alpha}(b-b_0)(a-a_0)+\sum_b\sum_\omega F_{b\omega}\times \\ & (b-b_0)(\omega-\omega_0)+\sum_a\sum_\omega F_{\alpha\omega}(a-a_0)(\omega-\omega_0) \end{aligned} \qquad (4.10)$$

* The quantities $r_e^2$ or $r_e$, appearing explicitly in equations (4.4), (4.5) and (4.7), are included in the respective force constants.

The first term allows for bond stretching (two bonded atoms), the third for bond-angle bending (three bonded atoms) and the fifth for out-of-plane bending of groups that are planar at equilibrium or for deviations of torsion angles from reference values (four bonded atoms). The remaining cross-terms allow for coupling between these deformations. Fortunately, many of these cross-terms are negligible, but, unfortunately, it is difficult to know in advance which ones are non-negligible. In general, it is sufficient to consider stretch–stretch and stretch–bend terms between bonds and bond angles at a given atom, and bend–bend terms between angles at the same atom or at two atoms directly bonded to each other. The sums in equation (4.10) usually include all distances and angles, both dependent and independent. The reference distances and angles $b_0$, $a_0$ and $\omega_0$ may be determined experimentally by x-ray diffraction, electron diffraction, microwave spectroscopy or estimated. Force fields of this type are in line with the ideas outlined in the introduction, namely that atoms form characteristic bonds in definite directions in space.

In view of the complexity of the force-field determination it is important to know under which conditions force constants are transferable from one molecule to another. It is generally accepted that this is possible only for closely related molecules with equal bond lengths, equal bond angles and similar conformations[34]! Whereas changes in conformation affect primarily the interaction constants[35] (important for spectra, less important for structure), differences in bond lengths and angles may affect the stretching and bending constants, as shown, for example, by Schachtschneider and Snyder for different kinds of C—C bonds in normal and branched alkanes[36]. In such cases extreme care in transferring force constants is required. For overcrowded molecules, such as the medium-ring cycloalkanes or tri-t-butylmethane, where hydrogen atoms separated by three or more bonds come very close together, equation (4.10), with force constants, reference distances and angles taken over from similar but less crowded molecules, e.g. n-alkanes, would give a very poor representation of the potential energy. Determining a new set of constants for every overcrowded molecule would be tedious and out of line with the chemical strategy of profiting from analogies wherever possible.

### 4.3.5 Non-bonded interactions as perturbations (and other non-quadratic terms)

How is it possible to preserve the conceptional simplicity of equation (4.10) and, at the same time, to transfer force constants from one molecule to another when non-bonded interactions come into play? Whereas in vibrational spectroscopy transferability is often sacrificed, molecular mechanics tends to abandon a purely quadratic force field and introduces non-bonded interactions as a perturbation on a hypothetical unstrained system. This is achieved by adding an extra term

$$V_{\mathrm{nb}}=\sum_{q_{ij}} V_{\mathrm{nb}}(q_{ij}) \tag{4.11}$$

to equation (4.10). The functional form $V_{\mathrm{nb}}(q_{ij})$ is usually taken over from known or assumed *inter*molecular interaction potentials[37]. The most important of these are:

(i) Modifications of the Lennard-Jones potential

$$V_{nb}(q_{ij}) = Dq_{ij}^{-n} - Eq_{ij}^{-m} \tag{4.12}$$

with $n=12$ or 9 and $m=6$ most often used.

(ii) Buckingham potentials

$$V_{nb}(q_{ij}) = A\exp(-Bq_{ij}) - Cq_{ij}^{-6} \tag{4.13}$$

(iii) Any combination of the two, in general form

$$V_{nb}(q_{ij}) = \sum_k A_k q_{ij}^{D_k} \exp(-B_k q_{ij}) \tag{4.14}$$

where $k=1$, 2 or 3, including repulsive (Pauli repulsion, Figure 4.1), attractive (dispersion energy) and Coulombic terms.

Considering $V_{nb}$ as a perturbation superimposed on an otherwise unstrained system necessitates a reinterpretation of the reference values $r_0$, $a_0$ and $\omega_0$. These no longer correspond to the observed or equilibrium structural parameters of the molecule in question, but, rather like the force constants, they pertain to a hypothetical unstrained system in which the non-bonded interactions have been 'switched off'[9]. A simple model[38] illustrates this point: consider a C—H bond of length $r$ perturbed by a colinear atom Y, i.e. C—H · · · Y, with the position of C fixed. At a given distance $R$(C · · · Y) the potential energy of the hydrogen atom as a function of the C—H distance $r$ is given by

$$V = K(r-r_0)^2/2 + V_{nb}(R-r) \tag{4.15}$$

$$= K(r-r_0)^2/2 + V_{nb}(R-r_0) - (r-r_0)F' + (r-r_0)^2 F/2 \tag{4.16}$$

where

$$F' = \left(\frac{dV_{nb}}{d(R-r)}\right)_{r_0} \quad F = \left(\frac{d^2V_{nb}}{d(R-r)^2}\right)_{r_0} \tag{4.17}$$

With $x=r-r_0$, the equilibrium condition is given by

$$\frac{dV}{dx} = 0 = Kx - F' + Fx \tag{4.18}$$

where

$$x_e = F'/(K+F) \tag{4.19}$$

as illustrated in Figure 4.4. The 'switching on' of non-bonded interactions has reduced the C—H equilibrium distance from $r_0$ to $r_0 + x_e$ ($x_e$ has a negative sign in this case).

It follows from equation (4.18) that

$$\left(\frac{d^2V}{dx^2}\right)_{x=x_e} = k(x_e) = K + F \tag{4.20}$$

i.e. the harmonic force constant $K$ of the unperturbed system is also modified by the perturbation.

Extensive experimental information on internal relation barriers has prompted another, minor modification of the GVFF [equation (4.10]. Instead of the quadratic potential the energy associated with deviations of torsion angles from preferred values is expressed as a trigonometric function:

$$V(\phi)=\frac{1}{2}K_\phi(1\pm\cos n\phi) \tag{4.21}$$

where $n$ indicates the rotational symmetry of the barrier (e.g. $n=3$ for ethane) and the plus or minus signs determine whether $V(\phi)$ has a maximum (+) or a minimum (−) at $\phi=0$.

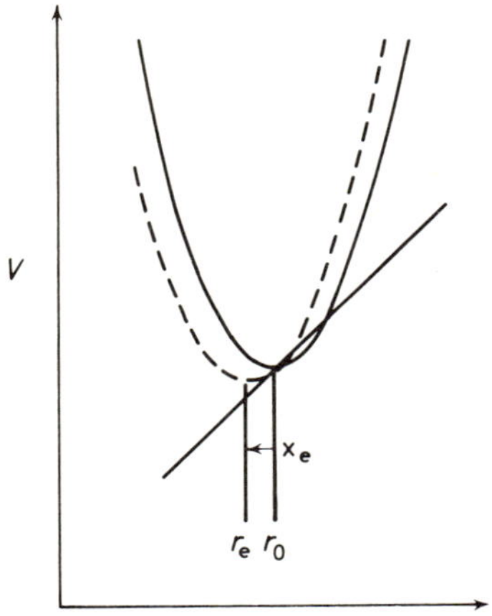


**Figure 4.4** The energy minimum is displaced from $r_0$ to $r_e$ by superimposing the linear potential on the harmonic one.

A valence force field implementing these modifications takes the form:

$$\begin{aligned}2V(\text{VFF})= &\sum K_b(b-b_0)^2+\sum_{b\neq b'}\sum F_{bb'}(b-b_0)(b'-b'_0)\\ &+\sum H_\alpha(a-a_0)^2+\sum_{a\neq a'}\sum F_{\alpha\alpha'}(a-a_0)(a'-a'_0)\\ &+\sum H_\chi\chi^2+\sum_{\chi\neq\chi'}\sum F_{\chi\chi'}\chi\chi'\\ &+\sum_b\sum_\alpha F_{b\alpha}(b-b_0)(a-a_0)\\ &+\sum_\phi K_\phi(1\pm\cos n\phi)\\ &+\sum_{ij}V_{\text{nb}}(q_{ij})\end{aligned} \tag{4.22}$$

where most of the symbols are the same as in equation (4.10); the $\chi$'s stand for out-of-plane bending parameters, the $\phi$'s for torsion angles; $V_{\text{nb}}(q_{ij})$ is summed over all pairs of non-bonded atoms separated by three or more bonds. The experimental torsional barrier $K_\phi$(exp), is not, in general, identical with the barrier $K_\phi$ appearing in equation (4.22); it is the sum of $K_\phi$ and a term describing the change in relevant non-bonded interactions as a function of $\phi$. This situation is analogous to the one described by equation (4.15) and Figure 4.4.

Comparison of equation (4.10) and equation (4.22) shows that some of the cross-terms between $b$, $a$, $\chi$ and $\phi$ are missing in the latter. They have been deleted because the best available parametrisation of this type[39] for alkenes and alkanes has shown them to be either negligibly small or indeterminable within the data set used. This is not necessarily the case, however, for other classes of compound. It should also be pointed out that some anharmonicity is built into equation (4.22) through the last two terms, which are not quadratic.

### 4.3.6 Complex molecules: Urey–Bradley force field (UBFF)

The ideas behind equation (4.15) and Figure 4.4 lead in a natural way to the second important type of force field: the UBFF, as used in the analysis of vibrational spectra, or the modified UBFF (MUBFF), used in molecular mechanics[40]. This force field contains contributions due to bond stretching and angle bending, includes *all* non-bonded interactions, even those between atoms separated by *two* bonds, and omits explicit cross-terms:

$$\begin{aligned} 2V(\text{MUBFF}) = {} & \sum_b K_b(b-b_0)^2 + \sum_a K_\alpha(a-a_0)^2 \\ & + \sum_\phi K_\phi(1 \pm \cos n\phi) \\ & + \sum_{ij} V_{\text{nb}}(q_{ij}) \end{aligned} \tag{4.23}$$

The non-bonded energy contributions $V_{\text{nb}}$ are usually described by the anharmonic functions given in equations (4.12), (4.13) or (4.14). It may be shown, through Taylor-series expansion at equilibrium and transformation to VFF coordinates, i.e. to bond lengths and bond angles, that the linear and quadratic parts of these functions [$F'$ and $F$ of equation (4.17)] partially compensate for the lack of explicit cross-terms in equation (4.23) [compare the discussion of equations (4.5) and (4.6)]. The harmonic part of a MUBFF is nevertheless not equivalent, in general, to a complete GVFF[41].

The anharmonicities in $V_{\text{nb}}$ may be taken up by anharmonic bond-stretching and angle-bending potential functions. The extent to which the anharmonicity of $V_{\text{nb}}$ can compensate for the inadequacy of the harmonic part has not been studied in detail. In any case, it seems unlikely that a force field as simple as that given by equation (4.23) is capable of simulating the detailed shape of the BO surface of actual molecules, even in the neighbourhood of the energy minima.

The force constants and reference values in the MUBFF have to be interpreted in the light of our discussion of the hypothetical C—H · · · Y system [equations (4.19) and (4.20) and Figure 4.4]. The need for caution in taking parameters from one force field and using them in another (already evident from Table 4.1) has to be stressed again. Table 4.2 shows the heavy dependence of C—C stretching force constant and reference length on the type of force field and on the non-bonded potentials $V_{\text{nb}}$.

**Table 4.2**

| *Force field* | $K$(C—C)/mdyn Å$^{-1}$ | $b_0$(C—C)/Å | *Ref.* |
|---|---|---|---|
| MUBFF* | 2.2 | .24 | 40 |
| MUBFF* | 1.542 | .455 | 42 |
| UBFF† | 2.229 | .54 | 36 |
| VFF‡ | 4.423 | .54 | 36 |
| VFF§ | 4.483 | .526 | 39 |

* Equation (4.23); † equation (4.23) with only first and second derivatives of $V_{nb}$ ($F'$,$F$); ‡ equation (4.10); § equation (4.22).

The VFF and MUBFF are both rather simple in their basic structure. The VFF is just the Taylor expansion of the BO surface around an energy minimum, including all quadratic terms and expressed in valence coordinates (bond lengths and bond angles). The MUBFF uses only squared terms and has to make up for the absence of cross-terms by including explicitly all non-bonded interactions, but at the cost of abandoning the conceptual simplicity of an expansion about equilibrium. These force fields are, however, by no means the only ones possible for complex molecules. In fact, the current literature on molecular mechanics contains nearly as many different force fields as publications. Most of them stand somewhere between the extremes of equations (4.10) and (4.23) and differ among themselves by additions or deletions of cross-terms, by the analytical form of $V_{nb}$ or by the extent of anharmonicity built into them. One might appeal to workers in this field to refrain from introducing new energy terms unless the extra flexibility gained thereby is clearly necessary for the problem under consideration.

### 4.3.7 Parametrisation of force fields

The aim of parametrising force fields is to evaluate the constants in equation (4.22) or (4.23) from experimental data. A successful parametrisation makes it possible to summarise the data in compact form, to infer properties of molecules closely related to those used in the parametrisation and, possibly, to extrapolate properties of other molecules, especially overcrowded ones. However, the problem is by no means a straightforward one.

For example, the non-linear $H_2O$ molecule discussed in Section 4.3.3 has only three ($3\times3-6$) degrees of vibrational freedom to determine four quadratic force constants, and this problem of underdetermination becomes more serious the more atoms in the molecule and the lower its symmetry. One way to overcome it is by extensive isotopic substitution, and another, used, for example, by Snyder and Schachtschneider[36], is to attempt to fit the spectra of a large number of similar compounds (e.g. n-alkanes) to a single set of force constants. The strategies used in parametrisation range from the almost naive to the highly sophisticated. We choose one of each kind to show how valency force constants and non-bonded interaction potentials may be evaluated.

For his appealingly simple MUBFF[40] [equation (4.23)], Bartell has used the Urey–Bradley force constants derived for alkanes by Schachtschneider and Snyder[36] together with his own non-bonded potentials[43]. Reference bond angles were chosen as tetrahedral. The reference length $b_0$ (C--H) was adjusted to reproduce the C—H bond length in methane as determined by electron diffraction. The reference length $b_0$ (C—C) was adjusted to give an optimal fit to both the C—C distance and the HCC angle of ethane (electron diffraction). Simple as it is, this force field has been successful in reproducing observed variations in bond lengths and bond angles of molecules as complicated as tri-t-butylmethane[38], although no molecule with comparable strain was used in the parametrisation: $r(C_{tert}—C_{quart})$, 1.611 ± 0.006 Å (1.595 Å, calc); $\langle r(C_{quart}—CH_3)\rangle$, 1.548 ± 0.002 Å (1.552 Å, calc); $\alpha(H—C_{tert}—C_{quart})$, 101.6 ± 0.4° (102.4°, calc); $\alpha(C_{tert}—C_{quart}—CH_3)$, 113.0 ± 0.2° (112.5°, calc); $\alpha(C_{quart}—C—H)$, 114.2 ± 1.0° (112.4°, calc). The force field also accounts for subtle structural details of normal and branched alkanes as well as for approximate energies of isomerisation (branched to n-alkanes)[40].

Lifson and Warshel[42], and later Ermer and Lifson[39], have used an elaborate least-squares procedure to find optimal potential constants from structural, thermodynamic and vibrational data (consistent force field, CFF). The scheme adopted by these authors has three distinct advantages:

(i) Every potential constant is associated with a standard deviation. This allows potential constants not significantly different from zero to be eliminated and those not significantly different from each other to be grouped together.

(ii) The correlation matrix obtained by inversion of the least-squares normal equations makes it possible to distinguish interdependent from independent groups of parameters. For example, if the three reference angles at trigonal carbon are chosen as variable parameters in the least-squares analysis, they are found to be highly correlated. Note that the *reference* angles, in contrast to the *observed* angles around trigonal carbon, need not add up to 360°.

(iii) The algorithm developed by Lifson and Warshel[42] uses a Jacobian matrix with elements $\partial y_i/\partial p_l$ where $y_i$ is an observed quantity and $p_l$ is one of the potential constants to be determined. This matrix measures the degree to which any given potential constant depends on the experimental data and enables one to recognise which observations are most relevant to the potential constant in question.

Since the observational equations pertinent to the least-squares analysis are, in general, highly non-linear, a reasonably good 'trial structure' has to be available at the start. This was taken from spectroscopic work[34]. The observational data used by Ermer and Lifson[39] consisted of 259 vibrational frequencies, 44 structural and 10 thermochemical quantities. The weighting scheme for these diverse types of observation was assigned such that their influence on the force field parameters was reasonably balanced; the final sum of weighted differences squared, $\sum_i[\omega_i(y_{obs}-y_{calc})]^2$, was composed of 75% vibrational, 16% structural and 9% thermochemical contributions. Average absolute differences between observed and calculated bond lengths, bond angles and torsion angles used in the optimisation are 0.003 Å, 0.5° and 1.0°, respectively, between observed and calculated heats of hydrogenation 0.92 kcal mol$^{-1}$, and between observed and calculated frequencies 15.4 cm$^{-1}$.

Other workers have used trial and error methods to determine force-field parameters—with varying degrees of ingenuity, chemical intuition and success. Most of them started from spectroscopic force-fields for potential constants and observed structures for reference values, both of which were subsequently modified to serve the envisaged purpose.

Similar remarks are applicable to the derivation of non-bonded interaction potentials. Reasonably accurate functions are available for rare-gas atoms[37], but their applicability to intramolecular interactions involving other atoms is an *ad hoc* assumption with some success in its favour but with hardly any sound theoretical backing. In view of these uncertainties, estimates of non-bonded potentials should either have the virtues of simplicity and wide generality or else be derived on the basis of a clearly defined physical model or mathematical procedure.

Kitaigorodski[44] has chosen the first possibility. He estimated intermolecular non-bonded equilibrium distances in alkanes to be *ca.* 3.8 Å for C · · · C, 3.15 Å for C · · · H and 2.6 Å for H · · · H, and noticed that non-bonded distances between atoms separated through two bonds are about two-thirds of these values. Using a Buckingham potential [equation (4.13)], non-bonded interaction energies were expressed in units of the energy applicable at two-thirds of the equilibrium non-bonded distances

$$V(z)=V_{2/3}[1/z^6-6\exp\{a(1-z)\}/a]/[1/(2/3)^6-6\exp(a/3)/a] \qquad (4.24)$$

with $V_{2/3}=V(q=2q_0/3)$, $z=q/q_0$ and $a=q_0B$. A single pair of values, $V_{2/3}=$ 3.5 kcal $mol^{-1}$ and $a=13$, used for all three kinds of interaction, accounted qualitatively for a diverse collection of experimental data—thermochemical, mechanical and structural. Williams[45], on the other hand, has derived non-bonded potentials from a least-squares fit to observed heats of sublimation, lattice constants and some structural details of aromatic and non-aromatic hydrocarbon molecules. Similarly, the Lifson group[39,42,46] determined non-bonded potential constants with their least-squares method. Both groups report heavy correlations among non-bonded potential constants and usually have to resort to some kind of constrained refinement procedure. These difficulties may be partly due to present inaccuracies in hydrogen positions in crystals and overcrowded molecules. Moreover, the use of spherically symmetrical functions to represent non-bonded potentials is a drastic simplification. For interactions between atoms belonging to polar groups or with low coordination numbers, especially between hydrogens (with no core electrons), it is probably an oversimplification. Improvements in non-bonded potentials may come from the use of analytical functions with other than spherical symmetry and from more extensive experimental investigations of heavily overcrowded compounds.

For unsaturated molecules, the simple approaches of equation (4.22) or (4.23) have been combined with approximate quantum mechanical theories of delocalised $\pi$-electron systems[47,48]. Although these force fields lead to quantitative agreement between calculated and experimental data, they have lost transparency without giving convincing evidence that 'pure' molecular mechanics are inappropriate for the unsaturated molecules.

Recent progress in the parametrisation of molecular force fields should not blind us to the intrinsic weakness of the approach, which is essentially one of curve-fitting based on mechanical models and aided by all sorts of chemical intuition. However, the attempt to reconstruct the BO surface in terms of an additive scheme is in line with the general experience that the chemical and some physical properties of molecules can be understood in terms of bond properties. The usual quantum mechanical description of molecules in terms of delocalised MOs can be transformed to one in terms of localised bonds[49], but there is so far no rigorous justification for the success of molecular mechanics. Some of the problems have been discussed elsewhere[50].

### 4.3.8 Energy minimisation

Once the analytical form of a model molecular force field is set up and its parameters determined, features of the BO surface, in as many dimensions as necessary, can be obtained by numerical calculation. In general, we are interested mainly in the properties of the equilibrium structure. The two main methods for locating the corresponding energy minimum, starting from a reasonable trial structure, are: (i) the method of steepest descent; (ii) a quadratic extrapolation procedure (Newton–Raphson method)[51]. These methods are not always the most efficient ones to be used on a digital computer but they rest directly on the condition of mechanical equilibrium. The potential energy is expanded in a Taylor series about a point corresponding to an initial trial structure ($\boldsymbol{x}_0$) of the molecule whose energy is to be minimised:

$$V(\boldsymbol{x})=V(\boldsymbol{x}_0)+\sum_i\left(\frac{\partial V}{\partial x_i}\right)_{x_0}\Delta x_i+\frac{1}{2}\sum_i\sum_j\left(\frac{\partial^2 V}{\partial x_i \partial x_j}\right)_{x_0}\Delta x_i \Delta x_j \qquad (4.25)$$

The $x_i$s are the $3N-6$ independent coordinates. If $\boldsymbol{x}_0$ is not equal to $\boldsymbol{x}_e$ (equilibrium) the first derivatives in equation (4.25) are different from zero and the path of steepest descent is along the negative gradient of $V$. Each coordinate is changed by an amount proportional to its partial derivative, and the calculation is repeated until the equilibrium condition

$$\frac{\partial V}{\partial x_i}=0, i=(1, 2, \ldots . 3N-6) \qquad (4.26)$$

is reached. In practice, as equilibrium is approached, successive iterations show oscillatory behaviour, and convergence to the equilibrium state becomes painfully slow. At this stage it is often advantageous to go over to the quadratic extrapolation procedure and calculate the matrix of second derivatives

$$F_{ij}=\frac{\partial^2 V}{\partial x_i \partial x_j} \qquad (4.27)$$

The coordinate shifts required to reach equilibrium are now given by

$$\Delta \boldsymbol{x}=-\boldsymbol{F}^{-1} \text{ grad } V \qquad (4.28)$$

The increase in computational time required to calculate the matrix elements $F_{ij}$ will often be more than compensated by the more efficient convergence. In any case, the matrix elements $F_{ij}$ calculated for the equilibrium structure $\boldsymbol{x}_e$ are of interest since they correspond to the mass-independent generalised force constants of the system. The force constants corresponding to the normal vibrations of the molecule are just the eigenvalues of the matrix with elements ($G_{ij}$ $F_{ij}$) where $\boldsymbol{G}$ is the matrix of reduced masses expressed in the appropriate coordinate system

$$|\boldsymbol{G}\cdot\boldsymbol{F}-\boldsymbol{\Lambda}|=0 \tag{4.29}$$

We have skipped over a number of difficulties. If Cartesian coordinates are used the vector $\boldsymbol{x}$ has $3N$ components, not $3N-6$, and the matrix $\boldsymbol{F}$ is singular with six eigenvalues zero. If internal coordinates are used the calculation of the first and second derivatives can be very cumbersome. Moreover, if the trial model contains symmetry elements, then certain first derivatives are necessarily zero and some diagonal second derivatives may be negative, corresponding to energy maxima situated between multiple energy minima. Trial models corresponding to maxima or saddle-points in the potential energy surface, and hence yielding zero first derivatives and negative diagonal second derivatives, may always be recognised from negative eigenvalues obtained on solving equation (4.29). Such eigenvalues correspond to the negative curvature of potential energy maxima.

Information about other minimisation procedures using search techniques is given elsewhere[50].

## 4.4 APPLICATIONS OF FORCE FIELD CALCULATIONS

### 4.4.1 Areas of application

Over the past decade molecular mechanics has developed from a rather crude tool yielding results of dubious relability into a reasonably dependable method of investigating features of the BO surface of wide classes of molecules. It is basically an empirical, inductive method that begins with the attempt to systematise various kinds of experimental data in terms of an assumed force field containing a limited number of parameters, and proceeds, by suitable interpolations or extrapolations, to predict further properties, often those of new and unobserved systems, sometimes even of unobservable ones, e.g. transition states. In favourable cases the reliability of such predictions is high and their accuracy rivals that attainable by well established experimental methods for determining equilibrium structures, vibrational spectra and thermodynamical properties.

In a recent paper, Engler *et al.*[52] furnish a detailed bibliography listing the following areas where molecular mechanics techniques have been applied: (i) simple and cyclic alkanes; (ii) polycycloalkanes; (iii) torsional barriers in substituted ethanes; (iv) functional groups (olefins, carbonyls, halogens, oxygen and nitrogen containing compounds); (v) aromatic compounds; (vi) silanes, Group IV organometallics, transition metal complexes; (vii) polymer and polypeptide conformations; (viii) radical and carbonium ion reactivity (solvolytic reactivity); (ix) reactivity calculations, transition state evaluations; (x) conformational interconversions, potential energy surfaces. A few additional applications could be mentioned to augment the list: (xi) force constants and interpretation of vibrational spectra; (xii) predictions of crystal structures; (xiii) lattice dynamics, molecular motions in crystals; (xiv) effect of crystal packing forces on molecular structure.

The force fields used in these applications range from the ridiculously simple to the most elaborate—it is not always easy to know in advance, or even sometimes in retrospect, whether a given force field is really adequate for the type of problem it may be called on to deal with. A long run of calculations that gives excellent agreement with observed properties of a large number of molecules may be followed by one that gives catastrophic disagreement. In a comparison[52] of predicted heats of formation of 80 alkanes of diverse types using two different force fields, each parametrised for a set of 40 alkanes with known heats of formation, the agreement between the two series of calculations was within 2.0 kcal $mol^{-1}$ for 53 of the compounds included. However, the largest disagreement amounted to 45.5 kcal $mol^{-1}$! In using such calculations it is just as well to be on guard against undue optimism. However, experimental methods have their own limitations. To determine the properties of a molecule experimentally, we need to have the compound in our hands, a limitation that is of no concern in molecular mechanics.

In general, calculated equilibrium structures (bond lengths, bond angles, torsion angles) are relatively insensitive to the quality of the force field employed; force fields of different types and with quite different kinds of parametrisation often lead to structures that can hardly be distinguished from one another or from the observed ones[52]. This may not be the case when highly polar groups are present, since calculated interactions between such groups are very sensitive to the multipole distribution and the 'effective dielectric constant' assumed. The calculation of energy differences between rival conformations places higher demands on the force field. In order to compare calculated energy differences with experimental quantities (usually strain enthalpies or enthalpies of formation in the gaseous state) it is necessary that the calculated quantities contain the corresponding enthalpies of vibration, rotation and translation. These can be included by making empirical corrections (bond or group contributions) or, more in line with the spirit of molecular mechanics, by calculating the frequencies of the normal vibration modes[42,120] and inserting them into the appropriate statistical mechanical expression

$$H_v = RT^2 \frac{\partial \ln f}{\partial T} = RT \sum_{a=1}^{3n-6} x_\alpha \left( \frac{1}{2} + \frac{1}{\exp x_\alpha - 1} \right) \tag{4.30}$$

with $x_\alpha = h\nu_\alpha / kT$. This requires the evaluation of the frequencies from the force constants of the $3n-6$ normal vibrations, which depend on the second derivatives $\partial^2 V / \partial p_i^2$ at equilibrium. Thus, for complete agreement, we need to know not only the position of the energy minimum but also its depth and curvature. A force field that reproduces correctly the equilibrium structure may give very poor agreement between observed and calculated energy quantities or frequencies.

### 4.4.2 Medium-ring cycloalkanes

Several of the above points are well illustrated by examples taken from calculations of medium-ring cycloalkanes, a class of molecules that has provided a favourite playground for exercises in molecular mechanics. There are several reasons for this popularity. First, the molecules are formed from one kind of building block, the —$CH_2$— group. Secondly, extensive thermochemical[53] and infrared spectroscopic data[54,55] are available for these compounds and their derivatives. Thirdly, concurrent with the development of the force-field approach the stable conformations of nearly all these molecules were gradually established by x-ray analysis of suitable derivatives[56]. N.m.r. studies have also been informative in some cases[57].

#### 4.4.2.1 Cyclodecane

The evolution in our knowledge about the cyclodecane system illustrates the interplay between experimental investigation and molecular mechanics calculations. Molecular models (a simple, cheap, but sometimes misleading version of molecular mechanics) suggest that, in contrast to cyclohexane with its relatively rigid chair form and its single family of flexible twist forms, the cyclodecane ring has an enormous variety of highly flexible conformations. Closer inspection reveals that all these conformations are characterised by unfavourable features, such as unacceptably short transannular H · · · H distances or eclipsed torsion angles. In the absence of quantitative criteria it was impossible to decide which, if any, of the apparent multiplicity of con-

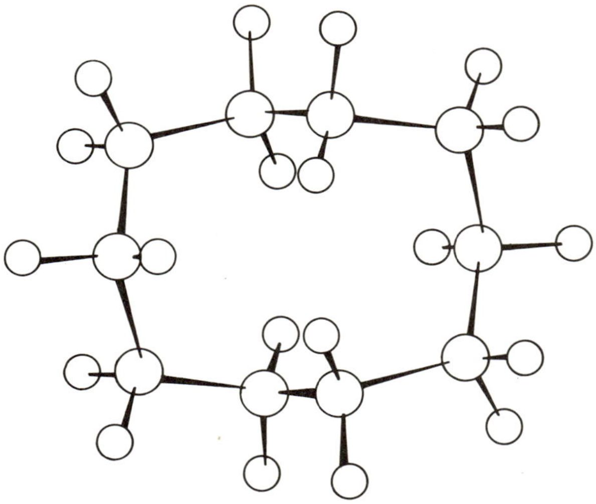

**Figure 4.5** Minimum energy conformation of cyclodecane ($C_{2h}$ symmetry).

formations represented the energetically stable one, a state of affairs described by Prelog[58] in 1956 as 'all the more regrettable because the experimental data that could provide information about the geometry . . . are almost completely lacking'.

By 1961, the thermochemical strain energies (excess combustion enthalpies) of all the medium-ring cycloalkanes had been determined by the labours of Coops and his colleagues[53]. The 9- and 10-membered rings showed the largest strain.

For cyclodecane, the existence of a preferred conformation had also been inferred by 1961 from results of x-ray analyses of three crystalline derivatives of 1,6-diamine dihydrochlorides. The occurrence of a single, common conformation of the 10-membered ring, in spite of radical differences in structure and packing, was interpreted[59] to mean that this conformation (Figure 4.5) was appreciably lower in energy than its competitors. Early force-field calculations by Hendrickson[60], by Wiberg[29] and by Bixon and Lifson[61] appeared to confirm that the x-ray conformation was several kcal $mol^{-1}$ lower in energy than other conformations considered and gave skeletal bond angles and torsion angles in good agreement with the observed ones. However, it is by no means easy to locate all the minima in the potential energy hypersurface and to find those that correspond to low-energy conformations.

In principle it should be possible to find these conformations experimentally by studying molecules in which the normally preferred conformation is destabilised by introducing substituents at suitable positions. For the cyclodecane conformation shown in Figure 4.5, this destabilisation can be accomplished by forcing a substituent to occupy one of the intra-annular positions, as would necessarily be the case in a 1,1,2,2- or 1,1,5,5-tetrasubstituted derivative. An x-ray analysis of 1,1,5,5-tetramethylcyclodecane-8-carboxylic acid[62] led to a ring conformation with torsion angles quite different from those of the preferred conformation; however, some of the observed C—C bond

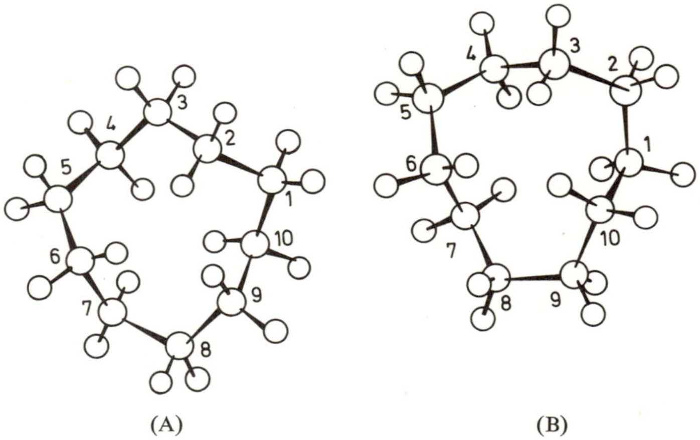


(A) (B)

**Figure 4.6** Two low-energy conformations of cyclodecane ($C_2$ symmetry).

lengths were abnormally short (by *ca*. 0.2 Å!), some of the observed C—C—C angles were abnormally wide (up to 135°!) and some of the temperature factors were strongly anisotropic with *B*-values in certain directions as high as 20 $Å^2$. Since there was no reason to doubt the overall accuracy of the analysis ($R$= 0.087 for about 1000 diffractometer-measured reflections), the possibility of a disordered structure had to be considered—the possibility that the averaged unit cell corresponds to a superposition of molecules with not quite identical conformations. When the 'observed' conformation, with all its obviously incorrect features, was used as starting point for a series of energy-minimisation calculations[63] with the Bixon–Lifson force field, convergence occurred rapidly to a conformation that was quite different from any of those considered in the earlier calculations[29,60,61]. This conformation (A) is shown (without substituents) in Figure 4.6 (left); it has a twofold axis passing through a pair of opposite atoms and its calculated energy was only 2.1 kcal $mol^{-1}$ higher than that of the preferred conformation (Figure 4.5)—much closer than any of the other conformations considered. Small random variations were then made in the parameters of this new conformation to provide a further series of starting models for energy-minimisation calculations, in the course of which yet another new conformation (B) was discovered (Figure 4.6, right); this conformation has a twofold axis passing through a pair of opposite bonds, and its energy was calculated to be 3.1 kcal $mol^{-1}$ higher than that of the preferred conformation, 1.0 kcal $mol^{-1}$ higher than that of A. The two conformations A and B look quite different at first sight but the relative positions of the eight atoms in the sequence 3 to 10 are actually quite similar—and also similar to those of a corresponding sequence of atoms in the observed structure. It could be shown that the x-ray results, including magnitudes and directions of the thermal vibration ellipsoids, are fully consistent with a 4 : 1 mixture of the two new conformations A and B with normal bond distances and angles.

The preferred cyclodecane conformation of Figure 4.5 has subsequently been observed in several other derivatives, most recently in cyclodecane-1,6-*trans*-diol[64] and in a 2 : 1 molecular complex of the *trans*- and *cis*-diols[65]. A neutron-diffraction analysis[64], carried out for the 1,6-*trans*-diol, has yielded experimental verification that the transannular H · · · H non-bonded distances are extremely short (*ca.* 1.95 Å) and that the methylene groups involved show quite large distortions from the local $C_{2v}$ symmetry sometimes assumed for them. In every case the sense of the distortion is such as to increase the short

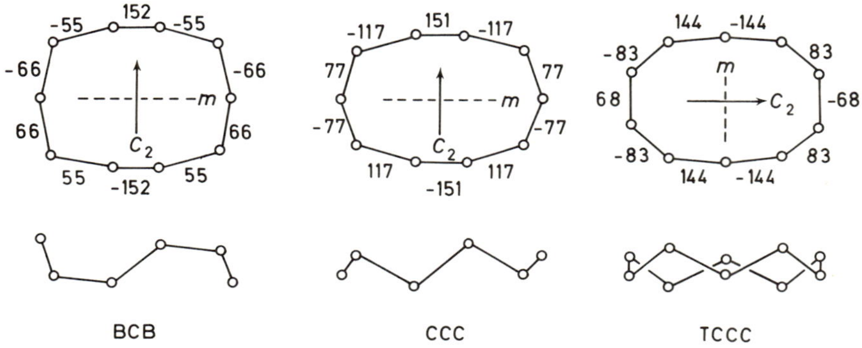


**Figure 4.7** Schematic representations of BCB (minimum energy), CCC and TCCC conformations of cyclodecane, showing approximate torsion angles. All have $C_{2h}$ symmetry or very nearly.

transannular H · · · H distances. Force-field calculations in which the methylene groups are allowed to relax from this symmetry reproduce the observed distortions satisfactorily[64].

It has recently been suggested[52] that if symmetry constraints on the carbon skeleton are relaxed in force-field calculations, the cyclodecane conformation shown in Figure 4.5 is not the most stable one. A conformation described as a twisted chair–chair–chair (TCCC) structure was calculated to be 0.6–1.7 kcal mol$^{-1}$ more stable than the normally preferred boat–chair–boat (BCB) one. The torsion angles cited for the TCCC conformation show that it has just the same symmetry ($C_{2h}$) as CCC and BCB, though with a different orientation of the symmetry elements. The three conformations in question are shown schematically in Figure 4.7. The BCB conformation has been observed in at least eight different crystal environments, whereas the evidence for the TCCC comes only from force-field calculations. A recent electron-diffraction study[66] suggests that gaseous cyclodecane at 130 °C consists of a mixture of BCB (50%), A (35%), B (8%) with *ca.* 8% of yet another conformation BCC. I.r. and Raman spectra[54] of liquid and solid cyclodecane show that the liquid consists essentially of a single conformation, the same one as in the solid. Lifson and Warshel[42] obtained good agreement between the observed frequencies and those calculated for the BCB conformation in the gas phase.

Although BCB seems to be identified as the preferred cyclodecane conformation, the full range of subsidiary stable conformations and the network of interconnecting minimum energy pathways have by no means been fully explored.

### 4.4.2.2 Cycloheptane

The twist-chair (skew-chair) form (Figure 4.8) with approximate $C_2$ symmetry appears to be 3–5 kcal mol$^{-1}$ lower in energy than competitive forms[28,52,61,67]. The crystal structure of 1-aminocycloheptane-1-carboxylic acid hydrobromide

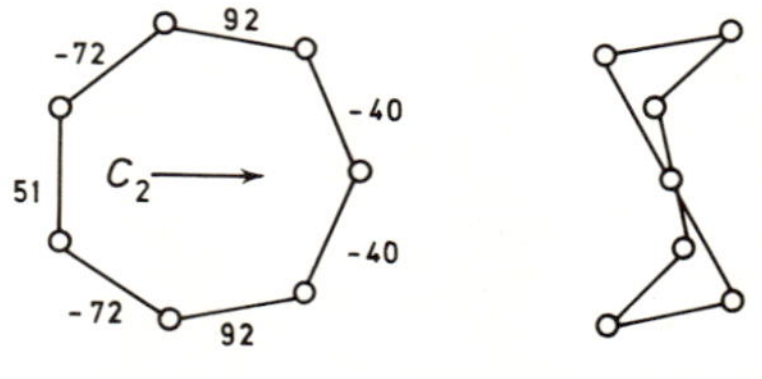


**Figure 4.8** Schematic representation of twist-chair conformation of cycloheptane, showing approximate torsion angles (deg) and symmetry.

monohydrate[68] is disordered and appears to contain a mixture of 'dominant' and 'alternative' conformations of the 7-membered ring, both similar to the calculated minimum energy form and to the conformation found in crystals of dimeric cycloheptanone peroxide[69].

#### 4.4.2.3 *Cyclo-octane*

Crystal structure studies on simple derivatives indicate the boat–chair form (Figure 4.9) as the preferred one[70–72]. N.m.r. studies of deuterated[73] and fluorinated[74] cyclo-octanes point in the same direction. The early force-field calculations by Hendrickson[60] and by Bixon and Lifson[61] suggested that the boat–chair form is slightly favoured (by 1–2 kcal $mol^{-1}$) over the crown form (Figure 4.9), which had often been assumed to be the stable one, although for no good reason. The clarity of the picture seems to be obscured again by more

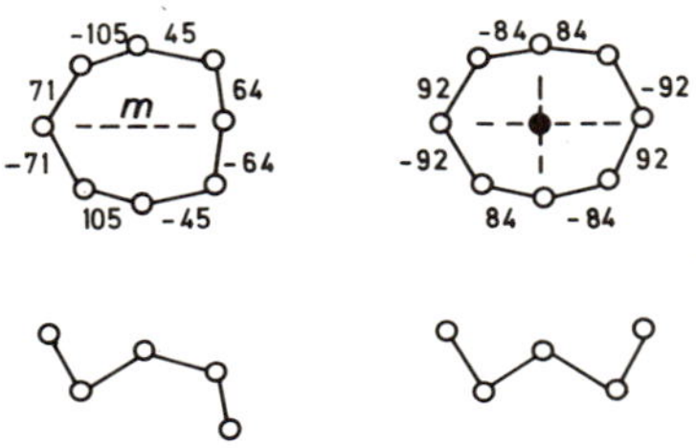


**Figure 4.9** Schematic representation of boat–chair (BC) and crown conformations of cyclo-octane, showing approximate torsion angles (deg) and symmetry.

recent calculations; whereas according to one extensively tested force field[67] the boat–chair form is 1.8 kcal $mol^{-1}$ more stable than its lowest competitor, according to another[52] the energies of three forms (boat–chair, twist–chair–chair and crown) are within 0.2 kcal $mol^{-1}$ of one another. The reason for this discrepancy in the calculated energies has still to be clarified.

**Figure 4.10** Schematic representation of two stable conformations of cyclononane, showing approximate torsion angles (deg) and symmetry.

#### 4.4.2.4 Cyclononane

The crystal structure of cyclononylamine hydrobromide[75] contains two independent molecules with conformations corresponding to different deformations of an idealised form (Figure 4.10) with $C_2$ symmetry. Hendrickson[60], and later Bixon and Lifson[61], found that another conformation, with $D_3$ symmetry (Figure 4.10), was preferred. The $D_3$ conformation actually occurs in crystals of cyclononanol-1-dimethylphosphonate[76], where some of the atoms show extremely large and anisotropic thermal motion tensors, suggesting the possibility of disorder. The observed conformations seem to be sensitive to crystal packing forces, the calculated ones to slight differences in the force field assumed[52], suggesting that cyclononane does not have a characteristic stable conformation but that the $C_2$ and $D_3$ forms correspond to two out of several rather broad, ill-defined minima of roughly equal energy.

#### 4.4.2.5 Cycloundecane

No experimental information is available. A conformation containing a two-fold axis has been proposed[61] and it may be the most stable one[52].

#### 4.4.2.6 Cyclododecane

A conformation (Figure 4.11) with $D_2$ (approximate $D_4$) symmetry was inferred from the results of an x-ray analysis of the hydrocarbon itself[77]—the crystal structure is disordered. A very similar conformation has been found in the structure of 1-hydroxycyclododecanedimethylphosphonate[78]. Calculations

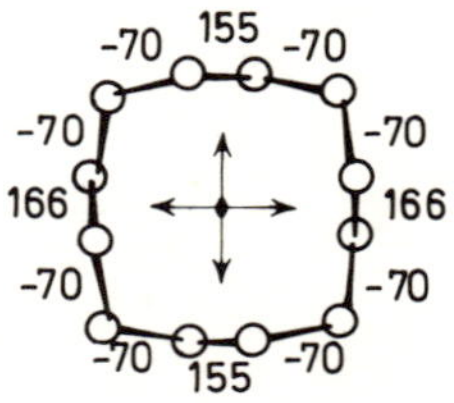


**Figure 4.11** Stable conformation of cyclododecane, showing approximate torsion angles (deg) and symmetry.

with different force fields[42,52,61,67] yield almost identical geometric parameters, in good agreement with the observed ones, but differ with respect to the strain energy, the experimental value of which is also in doubt because of uncertainty in the heat of sublimation of the solid hydrocarbon. The value given in the original study[53,79] for the excess combustion enthalpy was 3.4 $\pm$ 1.2 kcal mol$^{-1}$, but later estimates range as high as 8.5 kcal mol$^{-1}$. Calculated values range from 2.1 to 8.7 kcal mol$^{-1}$. Although recent arguments have been given for the higher value[52,67], the consistent force field of Lifson and Warshel[42] leads to the lower one and to the conclusion that cyclododecane is more or less strain-free, the excess enthalpy at 25 °C being almost entirely attributable to the vibrational–rotational–translational enthalpy $\Delta H_{vrt}$. This may be a case where explicit evaluation of this term from the calculated vibrational frequencies is preferable to burying it in group or bond increments of a force-field parametrisation, and it illustrates how different force fields may yield closely similar geometries but widely different energies for a given conformation.

### 4.4.3 Small-ring cycloalkanes

#### *4.4.3.1 Cyclohexane*

A force field that does not lead to a minimum energy conformation of $D_{3d}$ symmetry (chair form) with CCC angle of 111.0–111.5° can be rejected. Although it digresses from our main theme, it is worth mentioning that a great deal of information is available about equilibration of axial and equatorial substituents on the cyclohexane ring[80]. The corresponding free energy differences are often called 'conformational energies' and may represent a useful source of data for parametrising force fields for a variety of chemical groups.

#### *4.4.3.2 Cyclopentane*

In spite of its apparent simplicity, the cyclopentane molecule has an important place in the development of concepts concerning conformation, because it provided the first example of the phenomenon of pseudorotation[81,82]. A planar pentagonal molecule has two degenerate out-of-plane deformation coordinates that may be combined to give the out-of-plane displacement of the $j$th atom as

$$z_j = \left(\frac{2}{5}\right)^{\frac{1}{2}} q \cos\left(\frac{4\pi j}{5} + \phi\right) \quad j = 1\text{–}5 \tag{4.31}$$

Any out-of-plane deformation is then described in terms of the two parameters, $q$ the amplitude of the deformation, and $\phi$ its phase angle. The term pseudorotation implies that the cyclic coordinate $\phi$ runs along an essentially equipotential path. The equipotential path seems to be due more to the geometric constraints that apply to a 5-membered ring than to the exact details of force fields. For example, the sum of the bond-angle deviations from 108° and the sum of the squares of the torsion angles are both independent of the phase angle and depend only on the puckering amplitude[83]. This is also very nearly the case for the sum of the squares of the bond-angle deviations from the tetrahedral angle[84]. If the path were strictly equipotential, then the matrix of second derivatives $\boldsymbol{F}$ [equation (4.27)] would have seven instead of the normal six zero eigenvalues. Although calculations based on significantly different force fields yield an equipotential or nearly equipotential pseudorotational path, they differ slightly with respect to the calculated puckering amplitude $q$, which has been estimated by various methods to lie in the range 0.43–0.49 Å; a value in the lower part of this range seems to be indicated by the most recent investigations[84].

Alternatively, the out-of-plane deformation coordinates can be taken as the torsion angles $\omega_i$. The two degenerate out-of-plane deformations of the planar pentagon can then be combined to give

$$\omega_i = \left(\frac{2}{5}\right)^{\frac{1}{2}} \omega_0 \cos\left(\frac{4\pi i}{5} + \phi'\right) \quad i = 1\text{–}5 \qquad (4.32)$$

Since the bond lengths in cyclopentane can be assumed to be virtually independent of conformation, the range of validity of equation (4.31) is restricted to infinitesimal displacements from planarity. In contrast to this, equation (4.32) has been found to hold very well in practice, even for highly puckered rings[42,85]. The maximum torsion angle $\omega_0$ attained in the pseudorotational itinerary of cyclopentane is *ca.* 42° [84]. Somewhat larger values, up to *ca.* 50° [85], occur in ring D of steroids, which is *trans*-fused to the chair form of a 6-membered ring and hence slightly more puckered than cyclopentane.

Non-planar 5-membered rings described in the crystal structure literature as being in the envelope ($C_s$) or twist forms ($C_2$) with special values of $\phi$ or $\phi'$ often turn out on closer inspection to be neither one nor the other, but intermediate between the two with arbitrary values of $\phi$ or $\phi'$.

The equilateral pentagon has the property that if its five angles are equal to one another, then they must equal 108°, i.e. the ring must be planar[86]. All other equilateral, equiangular polygons may be puckered, but not the pentagon. Thus, because the standard bond angle at trivalent As is *ca.* 90°, the 5-membered ring of arsenomethane, $(AsCH_3)_5$, is highly puckered and so the As—As—As angles are markedly different from one another[87,88]. This is an example of how a constraint on one kind of geometric parameter, here a bond angle, can lead to structural features that do not require a special explanation in terms of intramolecular forces.

#### 4.4.3.3 *Cyclobutane and cyclopropane*

The force fields developed for alkanes and larger cycloalkanes cannot be expected to be applicable to very small ring systems without modification. Whereas the typical chemical and spectroscopic properties of the larger cycloalkanes are approximately those of normal alkanes with their ends tied together, this is far from the case for cyclopropane or (to a lesser extent) cyclobutane. It is not surprising that force fields which reproduce the structural parameters and energies of larger ring systems fail for small, highly strained molecules.

### 4.4.4 Cyclo-olefins and lactams

Thanks to Ermer and Lifson[39], a consistent force field of rather high quality is available for non-conjugated olefins, but experimental structural data for cyclo-olefins that might provide a basis for verification and refinement of this force field are sparse and inaccurate. Not that this area has failed to attract the attention of crystallographers, but simple derivatives of cycloalkenes are often low melting and difficult to crystallise. Several x-ray studies[56] have been carried out for adducts of cyclo-olefins with complexing agents, such as $AgNO_3$, but the presence of the heavy metal atoms introduces undesired perturbations of the double-bond system and also has an adverse effect on the attainable accuracy. Because of the sparsity of experimental information, force-field calculations provide the best, and in some cases the only, descriptions of the stable conformations of medium-ring olefins.

However, a series of crystal-structure studies[89] has been carried out for the medium-ring lactams and their adducts with HCl, and the structural parameters found may be compared with those calculated for the corresponding cyclo-olefins.

Caprolactam has a nearly symmetric chair conformation with bond angles and torsion angles close to those calculated[39] for cycloheptene. The similarity to cycloheptene is even closer for caprolactam hydrochloride, where protonation of the carbonyl group should increase the double-bond character of the central C—N bond. Good agreement is also obtained between observed actam conformations and calculated cyclo-olefin ones for enantholactam

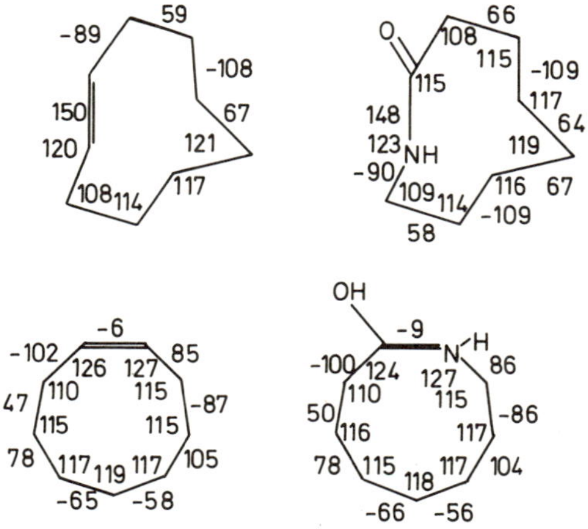


**Figure 4.12** Bond angles and torsion angles calculated for *trans- and cis*-cyclononene compared with observed values in unprotonated and protonated caprylolactam.

hydrochloride (with *cis*-cyclo-octene), caprylolactam (with *trans*-cyclononene), caprylolactam hydrochloride (with *cis*-cyclononene), pelargolactam hemihydrochloride (with one of six roughly equienergetic conformations of *trans*-cyclodecene) and caprinolactam hemihydrochloride (with *trans*-cycloundecene). Pelargolactam itself has a disordered crystal structure (that may contain two of the six roughly equi-energetic conformations of *trans*-cyclodecene). Figure 4.12 shows the calculated parameters for *trans*- and *cis*-cyclononene compared with the experimental ones for caprylolactam and its hydrochloride. The level of agreement is typical for the other cases.

It seems too much to expect that a force field parametrised to describe the behaviour of olefins should be transferrable to amides. The amide group has a far lower rotation barrier than the C=C double bond and also a much lower resistance to out-of-plane bending at one of its ends (the N—H group), apart from other differences in equilibrium parameters and flexibility. Nevertheless, it does resemble the C=C group in that it is a fairly rigid planar system compared with the highly flexible polymethylene segments, and this resemblance seems sufficient to produce the striking similarity in conformation.

The structural data for amides and lactams suggest that the energy surface close to the minimum energy structure is strongly anharmonic. For non-planar amides, the out-of-plane deformation is a mixture of twisting and out-of-plane bending at the N—H group (Figure 4.13)[91], as might be expected from simple

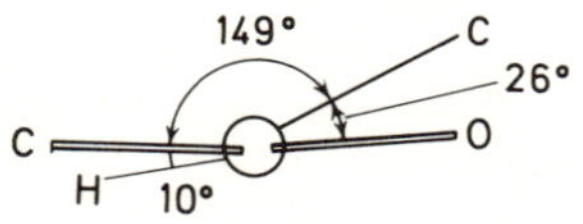


**Figure 4.13** Projection down the C—N bond of a secondary amide showing twisting and out-of-plane bending deformations. The twist angle is the average of the two *syn* torsion angles, i.e. 18°. The out-of-plane bend at N is 21° (180° − 149° − 10°), at C it is 5° (180° − 149° − 26°).

considerations—for a twist angle of 90° with no conjugative stabilisation a pyramidal N—H group should be 5–10 kcal mol$^{-1}$ lower in energy than a planar one.

For caprolactam the experimental torsion angles are closer to the ones calculated for cycloheptene[39] than to ones calculated for caprolactam itself[90], using a force field parametrised by least-squares fitting to the vibrational spectra and conformations of linear amides. This force field apparently fails to give a good description of the out-of-plane deformations of the amide group, which are relatively unimportant in linear amides but highly important in medium-ring lactams.

### 4.4.5 Calculation of transition states for conformational changes

In principle, molecular mechanics can be used to calculate minimum energy paths for interconversion of conformational isomers. A minimum energy path between two minima on the BO hypersurface is one along which the energy is a minimum with respect to all directions normal to the path. The minimum energy path connecting two energy minima must run over a pass or saddle-point in the BO surface, i.e. a point at which all but one of the $3N-6$ eigenvalues of the $\boldsymbol{F}$ matrix are positive, the remaining one negative. This point represents the turning point along the minimum energy path and can be identified with the transition state. Its energy (relative to that of the stable species) corresponds to the activation energy for the isomerisation process, its coordinates to the geometry of the activated complex.

In practice several difficulties are encountered. First of all, straightforward minimisation techniques cannot be applied since the points along the minimum energy path do not correspond to true energy minima. In some cases this problem can be circumvented by assuming certain kinds of constraints, e.g. symmetry restrictions, in the transition state, which would then correspond to an energy minimum in some section through the BO hypersurface. A second difficulty is that force fields that have been parametrised from the properties of stable species may not be transferrable to species that are far

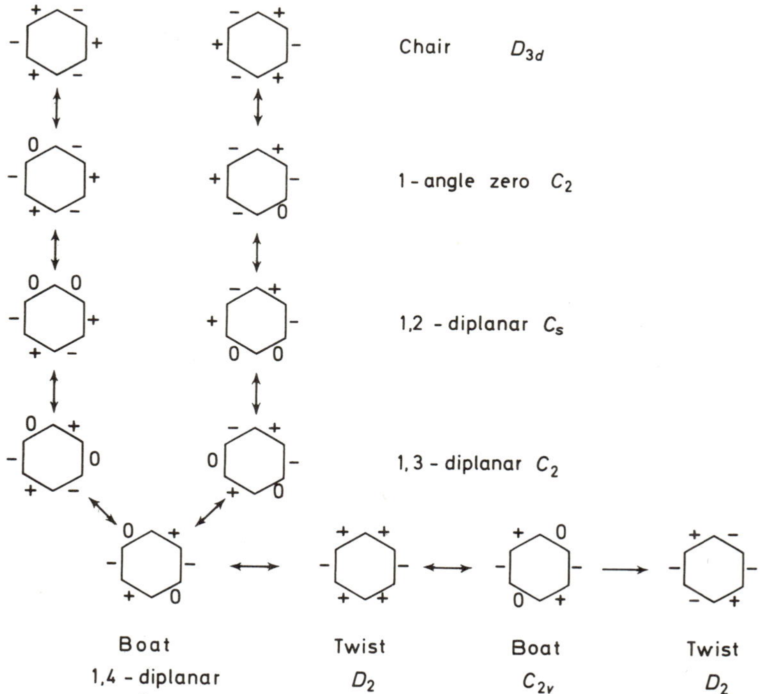


**Figure 4.14** Pattern of signs of torsion angles for inversion of chair and twist conformations of cyclohexane.

from equilibrium. For conformational isomerisations with low activation energies this may not be so serious. Lastly, since transition states are by their nature unobservable species, there are no structural nor spectroscopic data to check the results of calculations. Nevertheless, some useful insight has been obtained into the interconversion paths for certain conformational changes, such as the inversion of the cyclohexane chair conformation.

It is clear that during the course of the inversion itinerary all six torsion angles have to change sign and hence that each of them has to pass through zero at some point of the reaction path. If any three torsion angles are simultaneously zero then the six ring atoms lie in a plane (the other three torsion angles must also be zero) with C—C—C angles of 120° and eclipsed C—C bonds. This is a high energy form that can safely be eliminated as a possible transition state. A path that avoids passing through this form is shown schematically in Figure 4.13. The boat and twist forms may be generated from the regular hexagon with the help of the degenerate pair of $E_{2u}$ out-of-plane deformations. For infinitesimal out-of-plane displacements the family of boat–twist forms are interconvertible along a strictly equipotential surface (pseudorotation) and even for larger displacements the interconversion path should be nearly equipotential. Similarly, it can be shown that the 1-angle-zero and 1,2-diplanar forms (Figure 4.14) also lie on a nearly equipotential surface. The energy variation along the path can only be found from calculations. Since Hendrickson's early work[28] it has generally been accepted that the potential energy maximum (transition state) lies somewhere near the 1-angle-zero or 1,2-diplanar forms with the twist–boat family as intermediate conformations.

Results of a recent set of calculations[92] are shown in Table 4.3. The two

**Table 4.3 Calculated values[92] of energy and thermodynamic quantities (at 25 °C) for various conformations of cyclohexane. $V$ and $\Delta H_f$ are in kcal mol$^{-1}$, $S°$ in cal mol$^{-1}$ deg$^{-1}$. The species $T_1$ has one torsion angle of zero (four consecutive atoms in a plane), $T_2$ has two torsion angles of zero (five atoms in a plane). Slightly different values are obtained with a modified force field**

| | $V$ | $\Delta H_f$ | $S°$ |
|---|---|---|---|
| Chair ($D_{3d}$) | 3.7 | −29.7 | 71.6 |
| Twist ($D_2$, $d,l$ pair) | 9.4 | −24.1 | 75.7 |
| Boat ($C_{2v}$) | 10.0 | −24.0 | 72.1 |
| Activated complex $T_1(C_2)$ | 13.2 | −20.7 | 73.6 |
| Activated complex $T_2(C_2)$ | 13.7 | −20.3 | 75.8 |

models $T_1$ and $T_2$ for the transition state have very similar potential energies ($V$) and enthalpies ($\Delta H_f$). It would appear that $T_1$ is a true saddle point, with $T_2$ slightly displaced to the side of the saddle, although other calculations[93] suggest that there is a continuous family of pseudorotating transition states on an equipotential curve of the energy surface. Note that although $T_1$ lies below $T_2$ on the energy surface, its free energy ($\Delta H - T\Delta S$) would appear to be somewhat higher at 25 °C. Similarly, the twist and boat forms (intermediates along the inversion pathway) are also very nearly equipotential with the former slightly favoured at 25°C because of its larger entropy.

### 4.4.6 Coordination compounds

Another area for application of force-field calculations is that of chelate complexes where polyfunctional organic ligands are coordinated to metal ions. Complexes of $Co^{III}$ with chelating polyamine ligands have been the subject of several studies. These compounds are chemically inert, i.e. ligand substitution reactions require a relatively large activation energy, which makes it possible to separate stereoisomers and investigate their properties in much the same way as for organic compounds.

A favourite problem of this kind is the calculation of the minimum energy conformation of tris-(1,3-diaminopropane)cobalt(III), $[Co(tn)_3]^{3+}$. The possible conformations, three of which show at least $C_3$ symmetry, may be classified according to a proposal made by Corey and Bailar[94] in 1959: (i) the three 6-membered chelate rings are all in the chair conformation (*chair*$_3$); (ii) the chelate rings show a twist conformation with the plane of the three carbon atoms approximately parallel to the threefold axis (*lel*$_3$); (iii) twist conformation of the chelate ring with the plane of the three carbon atoms rather oblique to the threefold axis (*ob*$_3$); (iv) mixed conformations. By considering only the interannular non-bonded H · · · H contacts, which are different in each case, Woldbye[95] suggested that *ob*$_3$ is unfavourable and that *lel*$_3$ is the most stable. However, later calculations by various authors[96–98] using more complete force fields give conflicting results (Table 4.4) concerning the relative

**Table 4.4 Calculated energy differences (kcal mol$^{-1}$) for three conformations of tris-(1,3-diaminopropane)cobalt(III), $[Co(tn)_3]^{3+}$, with various force fields**

| *chair*$_3$ | *lel*$_3$ | *ob*$_3$ | *Ref.* |
|---|---|---|---|
| 0 | 0.60 | — | 96 |
| 0.77 | 0 | 5.83 | 97 |
| 1.66 | 0 | 4.70 | 98 |
| 0 | 0.72 | 2.04 | 98 |

stability of *chair*$_3$ and *lel*$_3$. All calculations concur that *ob*$_3$ is energetically unfavourable by 2–6 kcal mol$^{-1}$ and that in solution a mixture of *chair*$_3$, *lel*$_3$ and possibly mixed conformations is present. The crystal structure of $Co(tn)_3$ halogenides shows the *chair*$_3$ conformation[99].

The force constants used in the above calculations were derived from i.r. data on crystalline $Co(NH_3)_6X_3$ (X=Cl, Br, I)[121], gaseous hydrocarbons and alkylamines. They were used together with various kinds of non-bonded potentials. The experimental data available for most classes of coordination compounds are sparse and fragmentary so that parametrisation of force fields has not yet been carried through in the systematic way described in Section 4.3.7 for hydrocarbons. Results obtainable from the available force fields should be regarded with reserve.

It is also worth recalling that the nature of the environment may be quite decisive for the conformation, indeed for the very existence of these species, whereas the calculations refer to hypothetical isolated ions. The influence of the environment (polar solvent or crystal) has not been studied in detail, but Snow[100] has shown that intermolecular contacts may influence the conformation of chelate rings.

### 4.4.7 Applications to the solid state

Force-field calculations are being applied with increasing success to problems of structure, energy and flexibility of molecular and ionic crystals. We refer below only to a few selected examples.

#### *4.4.7.1 The phase problem*

Early investigations by Rabinovich and Schmidt[101] were directed towards a solution of the phase problem for molecular crystals with the help of information about cell constants, space group, non-bonded contact distances and the approximate structure of the molecule, regarded for this purpose as being rigid.

In this kind of approach, the position and orientation of the molecule are varied within the constraints imposed by crystallographic symmetry; most arrangements lead to unreasonably short intermolecular contacts and can be rejected. For the few remaining arrangements, structure factors are calculated and compared with observed values. Several structures have been solved in this way.

Giglio and co-workers[102] have applied these ideas to crystal structures of various types of compound, including those built from polar or ionic species. For these examples most of the positional parameters are fixed by space group symmetry; the few remaining ones are determined using van der Waals', hydrogen-bonding, ion–dipole and Coulombic potentials. The energy is thus decomposed into a set of large contributions with opposite signs and it shows only a very limited number of deep minima. One of these always corresponds closely to the experimental structure.

#### *4.4.7.2 Influence of crystal packing on molecular structure*

The effects of crystal packing on molecular structure are often tacitly assumed to be small and unimportant. There are, however, some cases where they are known to be considerable.

An example is provided by the crystal and molecular structure of cyclodecane-1,6-*trans*-diol[64] discussed in Section 4.4.2.1. The two independent molecules in the asymmetric unit of this crystal have slightly different ring conformations, both of the general type shown in Figure 4.5 but deviating by more than the usual amount from the idealised $C_{2h}$ symmetry. The deviations observed in one of the two independent molecules amount to 16° between pairs of 'equivalent' torsion angles. Force-field calculations for cyclodecane in which $C_{2h}$ symmetry is relaxed show at most small deviations of dubious significance. The observed distortions are probably caused by packing effects

—there is an intermolecular H · · · H distance of only 2.13 Å (neutron diffraction measurement) that must be associated with a strong repulsion between the hydrogen atoms involved. The molecules apparently yield to this repulsion by a shearing motion of the two longer sides of the 10-membered ring to produce the observed distortions.

Another example is caprylolactam (Section 4.4.4); the amide group is *trans* in the crystal[91], whereas the solution consists of an equilibrium mixture of molecules with *cis* and *trans* amide groups, as shown by i.r. and other investigations[103]. Hydrogen bonding leads to dimers of the *cis* form and to infinite chains of the *trans* form. The free energy of a single hydrogen bond in an infinite chain has been estimated to be 1–2 kcal $mol^{-1}$ lower than in a dimer[104]. This would be sufficient to explain the occurrence of the *trans* form in the crystal.

Biphenyl in the gas phase[105,106] shows an angle of 40–45° between the two phenyl rings, whereas it is planar in the crystal[107,108]. The non-planarity of the isolated molecule can be regarded as the result of a balance between two effects. As the angle between the phenyl rings increases from zero, $\pi$-electron energy becomes less favourable but non-bonded repulsions between *ortho* hydrogens are reduced. Calculations with various crystal packing models (cell constants fixed but space group lowered from $P2_1/a$ to $P2_1$) show that the increase in molecular energy on going to planarity is adequately compensated by achieving relatively favourable intermolecular non-bonded contacts[109].

Kurittu and Pawley [110] adopt a different approach to discuss the subtle differences between the structures of elemental sulphur, $S_8$, in the gas phase and the crystal. The site symmetry in crystals of orthorhombic sulphur is 2, whereas $\bar{8}$ symmetry can be assumed for the isolated molecule. At equilibrium, the crystal forces that tend to deform the molecule from $\bar{8}$ symmetry are exactly balanced by the internal restoring forces. The deforming forces were estimated from a Buckingham potential [equation (4.13)], the restoring ones from force constants derived from infrared spectroscopic data[111]. Calculated deviations from $\bar{8}$ symmetry correspond closely to the experimental ones. The atomic displacements are maximally 0.03 Å, and the total energy associated with the molecular distortion is only about 0.08 kcal $mol^{-1}$.

### *4.4.7.3 Lattice energies and cell constants*

The lattice energy of a hypothetical, vibrationless crystal at 0 K may be thought of as a sum of positive and negative terms, arising from intramolecular deformation and intermolecular repulsion energies on the one hand, and of intermolecular attraction energy (dispersion) on the other. The deformation contributions are usually neglected; they are smaller than the uncertainties implicit in the calculations. Intermolecular interactions are approximated as sums of pairwise contributions, repulsive at short distances, attractive at longer ones, as indicated by equations (4.11)–(4.14). The sum of attractive energies clearly depends on the cutoff limit in the summations. With a limit of 5–6 Å, Williams finds that calculated lattice energies for hydrocarbons (uncorrected for vibrational effects) account for *ca.* 80% of the experimental sublimation enthalpy[45]. Warshel and Lifson[46] calculate the sublimation enthalpies of n-hexane and n-octane with a model in which vibrational contributions are included and found to be *ca.* 10% of the total. This 10% is roughly equally distributed among molecular and lattice vibrations. In order to obtain agreement with experiment, summation limits had to be extended to up to 80 Å.

An alternative approximation is to place the polarisability centres in the bonds and thus to represent the attractive energy as a sum of pairwise interactions between bonds rather than between atoms. With this approach Rae and Mason[112,113] obtain lattice energies for benzene and *sym*-triazine within 10–20% of experimental heats of sublimation. They estimate that the attraction centre of a C—H bond is *ca.* 0.3 Å from the hydrogen nucleus. Since the repulsion centres are maintained at the nuclei, the shift of the attraction centres away from the nuclei leads to an anisotropy in H · · · H interaction energies and hence to a further degree of flexibility that seems to be required[45,46,114].

Lattice energy calculations are usually accompanied by a calculation of the cell constants to check the balance between attractive and repulsive terms in the empirical potential functions. Differences between observed and calculated cell constants amount to *ca.* 5% of the experimental values and are thus many times larger than the standard deviations of the measurements. One reason for the discrepancies is that potentials of the type of equation (4.12) and (4.13) cannot be fitted to reproduce the interaction energies at both long and short distances[37], as would be required for lattice energy calculations. Another factor is the neglect or inadequacy of anisotropic terms in non-bonded potential functions, and still another the neglect of simultaneous interactions between more than two attraction or repulsion centres[37].

### *4.4.7.4 Dynamics of rigid molecules in the crystal lattiec*

The results of modern crystal-structure analyses usually include not only atomic coordinates but also a set of parameters describing the anisotropic thermal motion of individual atoms. In many cases the internal molecular vibrations have much lower amplitudes than the vibrations of the molecules in the crystal lattice. The internal motions can then be neglected and the observed thermal motion parameters interpreted exclusively in terms of rigid-body motion. The usual practice involves a least-squares fit of translational and librational tensors $\boldsymbol{T}$ and $\boldsymbol{L}$ and of an additional tensor $\boldsymbol{S}$ (that allows for the mean quadratic correlation of translation and libration) to the vibration tensors $\boldsymbol{U}$ of the individual atom[115]. The $\boldsymbol{T}$, $\boldsymbol{L}$ and $\boldsymbol{S}$ tensors represent averages over the lattice vibrations and thus contain information about intermolecular interactions in the crystal. Alternatively, from assumptions about the nature of these interactions, the tensors may be calculated and compared with the observed tensors to test the validity of the assumptions made[116,117].

In the Einstein model it is assumed that each molecule vibrates independently of all the others; the corresponding potential is obtained by summing all relevant intermolecular interactions assuming one molecule to be displaced, while all neighbouring molecules remain fixed. Once the potential for any kind of displacement has been calculated, the corresponding mean square amplitude of vibration $\langle\phi^2\rangle$ can be estimated in various ways—it is roughly equal to $kT/(\partial^2 V/\partial\phi^2)_{\phi=0}$ for the displacement parameter $\phi$. Shmueli and Goldberg report good correlation between eigenvalues of $\boldsymbol{L}$ tensors and mean square amplitudes derived from the Einstein model for several molecular crystals[118]. Cerrini and Pawley[117] find that potentials for rigid-body motions derived with this model are, in general, too hard. They also find poor agreement regarding the $\boldsymbol{S}$ tensor and off-diagonal components of $\boldsymbol{T}$ and $\boldsymbol{L}$.

Significantly better agreement is obtained (for pyrene) with the Born–von Kármán model, in which the potential is set up in terms of displacements of all the molecules[117]. The mutual displacements are conveniently expressed in terms of symmetry coordinates determined by the space group [see equation (4.7) for the analogous treatment of a simple point group]. Other calculations[119] on the same basis for anthracene give excellent agreement with observed Raman frequencies; however, the calculated values of diagonal elements of the $\boldsymbol{U}$ tensors for the individual atoms are consistently larger than the experimental values. The authors raise the possibility that the lattice-dynamical results are better than the experimental ones!

## 4.5 SUMMARY

In spite of the difficulty, mentioned in the introduction, of providing a rigourous definition of non-bonded interactions, mathematical models that give qualitative and quantitative expression to this concept have been developed in a form suitable for carrying out calculations. We have tried to show some of the ways in which these models may be used to systematise properties, structures, energies and flexibilities, of classes of known molecular systems and to predict properties of unknown ones.

### Acknowledgements

We are grateful to L. S. Bartell and O. Ermer for many discussions and critical comments.

### References

1. Rouvray, D. H. (1974). *Chem. Brit.*, **10,** 11
2. Balaban, A. T. (1974). *Chemical Applications of Graph Theory* (London: Academic Press)
3. Pfeiffer, P. (1922). *Organische Molekülverbindungen* (Stuttgart: Enke Verlag)
4. Hassel, O. and Romming, C. (1962). *Quart. Rev. Chem. Soc.*, **16,** 1
5. Bent, H. A. (1968). *Chem. Rev.*, **68,** 587
6. Bürgi, H. B. (1973). *Inorg. Chem.*, **12,** 2321
7. Bürgi, H. B., Dunitz, J. D. and Shefter, E. (1973). *J. Amer. Chem. Soc.*, **95,** 5065
8. Bader, R. F. W. (1962). *Can. J. Chem.*, **40,** 1164
9. Bartell, L. S. (1968). *J. Chem. Educ.*, **45,** 754
10. Pearson, R. G. (1973). *Topics in Current Chem.*, **41,** 75
11. Hoffmann, R. (1971). *Accounts Chem. Res.*, **4,** 1
12. Jorgensen, W. L. and Salem, L. (1973). *The Organic Chemists Book of Orbitals* (New York: Academic Press)

13. Lowe, J. P. (1973). *Science*, **179,** 527
14. See however: Kutzelnigg, W. (1973). *Angew. Chem. Int. Ed. Engl.*, **12,** 546
15. Herzberg, G. (1950). *Molecular Spectra and Molecular Structure* I. *Spectra of Diatomic Molecules* (Princeton, N.J.: van Nostrand)
16. Morino, Y., Kuchitsu, K. and Oka, T. (1962). *J. Chem. Phys.*, **36,** 1108
17. Toyama, M., Oka, T. and Morino, Y. (1964). *J. Molec. Spectrosc.*, **13,** 193
18. Herschbach, D. R. and Laurie, V. W. (1962). *J. Chem. Phys.*, **37,** 1668
19. Laurie, V. W. and Herschbach, D. R. (1962). *J. Chem. Phys.*, **37,** 1687
20. Kuchitsu, K. and Bartell, L. S. (1961). *J. Chem. Phys.*, **35,** 1945
21. Morino, Y. (1969). *Pure Appl. Chem.*, **18,** 323
22. Hill, T. L. (1946). *J. Chem. Phys.*, **14,** 465
23. Dostrovsky, I., Hughes, E. D. and Ingold, C. K. (1946). *J. Chem. Soc.*, 173
24. Westheimer, F. H. and Mayer, J. E. (1946). *J. Chem. Phys.*, **14,** 733
25. Pitzer, K. S. (1945). *Science*, **101,** 672
26. Hazebroek, P. and Oosterhoff, L. J. (1951). *Discuss. Faraday Soc.*, **10,** 87
27. Dunitz, J. D. and Schomaker, V. (1952). *J. Chem. Phys.*, **20,** 1703
28. Hendrickson, J. B. (1961). *J. Amer. Chem. Soc.*, **83,** 4537
29. Wiberg, K. B. (1965). *J. Amer. Chem. Soc.*, **87,** 1070
30. Benedict, W. S., Gailar, N. and Plyler, E. K. (1956). *J. Chem. Phys.*, **24,** 1139
31. See for example: Herzberg, G. (1945). *Molecular Spectra and Molecular Structure,* II. *Infrared and Raman spectra of Polyatomic Molecules* (Princeton, N.J.: van Nostrand)
32. Urey, H. C. and Bradley, C. A. (1931). *Phys. Rev.*, **38,** 1969
33. (a) Simanouti, T. (1949). *J. Chem. Phys.*, **17,** 245
    (b) Shimanouchi, T. (1963). *Pure Appl. Chem.*, **7,** 131
34. Califano, S. (1969). *Pure Appl. Chem.*, **18,** 353
35. Warshel, A. and Lifson, S. (1969). *Chem. Phys. Lett.*, **4,** 255
36. Schachtschneider, J. H. and Snyder, R. G. (1963). *Spectrochim. Acta*, **19,** 117
37. For a review see: Maitland, G. C. and Smith, E. B. (1973). *Chem. Soc. Rev.*, **2,** 181
38. Bartell, L. S. and Bürgi, H. B. (1972). *J. Amer. Chem. Soc.*, **94,** 5239
39. Ermer, O. and Lifson, S. (1973). *J. Amer. Chem. Soc.*, **95,** 4121
40. Jacob, E. J., Thompson, H. B. and Bartell, L. S. (1967). *J. Chem. Phys.*, **47,** 3736
41. Bartell, L. S. and Kuchitsu, K. (1962). *J. Chem. Phys.*, **37,** 691
42. Lifson, S. and Warshel, A. (1968). *J. Chem. Phys.*, **49,** 5116
43. Bartell, L. S. (1960). *J. Chem. Phys.*, **32,** 827
44. Kitaigorodski, A. I. (1961). *Tetrahedron*, **14,** 230
45. Williams, D. E. (1967). *J. Chem. Phys.*, **47,** 4680
46. Warshel, A. and Lifson, S. (1970). *J. Chem. Phys.*, **53,** 582
47. Warshel, A. and Karplus, M. (1972). *J. Amer. Chem. Soc.*, **94,** 5612
48. Allinger, N. L. and Sprague, J. T. (1972). *J. Amer. Chem. Soc.*, **94,** 5734
49. England, W., Salmon, L. S. and Ruedenberg, K. (1971). *Topics in Current Chem.*, **23,** 31
50. Williams, J. E., Stang, P. J. and Schleyer, P. v. R. (1968). *Ann. Rev. Phys. Chem.*, **19,** 531
51. Whittaker, E. T. and Robinson, G. (1937). *The Calculus of Observations*, 2nd edn. (London and Glasgow: Blackie)
52. Engler, E. M., Andose, J. D. and Schleyer, P. v. R. (1973). *J. Amer. Chem. Soc.*, **95,** 8005
53. Coops, J., van Kamp, H., Lambregts, W. A., Visser, B. J. and Dekker, H. (1960). *Rec. Trav. Chim. Pays-Bas*, **79,** 1226
54. Billeter, E. and Günthard, H. H. (1958). *Helv. Chim. Acta*, **41,** 338, 686
55. Chiurdoglu, G., Doehard, T. and Türsch (1960). *Bull. Soc. Chim. Fr.*, 1322
56. For a review see Dunitz, J. D. (1968). *Perspectives in Structural Chemistry*, Vol. 2, 1 (J. D. Dunitz and J. A. Ibers, editors) (New York: Wiley)
57. See Anet, F. A. L. (1971). *Conformational Analysis, Scope and Present Limitations*, 15 (G. Chiurdoglu, editor) (New York: Academic Press)
58. Prelog, V. (1956). *Perspectives in Organic Chemistry*, 98 (Sir Alexander Todd, editor) (New York, London: Interscience)
59. Huber-Buser, E., Dunitz, J. D. and Venkatesan, K. (1961). *Proc. Chem. Soc.*, 463

60. Hendrickson, J. B. (1964). *J. Amer. Chem. Soc.*, **86,** 4854
61. Bixon, M. and Lifson, S. (1967). *Tetrahedron*, **23,** 769
62. Dunitz, J. D. and Eser, H. (1967). *Helv. Chim. Acta*, **50,** 1565
63. Dunitz, J. D., Eser, H., Bixon, M. and Lifson, S. (1967). *Helv. Chim. Acta*, **50,** 1572
64. Ermer, O., Dunitz, J. D. and Bernal, I. (1973). *Acta Crystallogr.*, **B29,** 2278
65. Ermer, O. and Dunitz, J. D., unpublished work
66. Hilderbrandt, R. L., Wieser, J. D. and Montgomery, L. K. (1973). *J. Amer. Chem. Soc.*, **95,** 8598
67. Allinger, N. L., Tribble, M. T., Miller, M. A. and Wertz, D. H. (1971). *J. Amer. Chem. Soc.*, **93,** 1637
68. Chacko, K. K., Srinivasan, R. and Zand, R. (1971). *J. Cryst. Molec. Struct.*, **1,** 213
69. Groth, P. (1967). *Acta Chem. Scand.*, **21,** 2631
70. Dobler, M., Dunitz, J. D. and Mugnoli, A. (1966). *Helv. Chim. Acta*, **49,** 2492
71. Bürgi, H. B. and Dunitz, J. D. (1968). *Helv. Chim. Acta*, **51,** 1514
72. Srikrishnan, T., Srinivasan, R. and Zand, R. (1971). *J. Cryst. Molec. Struct.*, **1,** 199
73. Anet, F. A. L. and St. Jacques, M. (1966). *J. Amer. Chem. Soc.*, **88,** 2585, 2586
74. Anderson, J. E., Glazer, E. S., Griffith, D. L., Knorr, R. and Roberts, J. D. (1969). *J. Amer. Chem. Soc.*, **91,** 1386
75. Bryan, R. F. and Dunitz, J. D. (1960). *Helv. Chim. Acta*, **43,** 3
76. Samuel, G. and Weiss, R. (1969). *Tetrahedron Lett.*, 3529
77. Dunitz, J. D. and Shearer, H. M. M. (1960). *Helv. Chim. Acta*, **43,** 18
78. Samuel, G. and Weiss, R. (1969). *Tetrahedron Lett.*, 2803
79. Van Kamp, H. (1957). *Dissertation*, Vrije Universiteit te Amsterdam, 45
80. Hirsch, J. A. (1967). *Topics in Stereochemistry*, Vol. 1, 199 (N. L. Allinger and E. L. Eliel, editor) (New York, London: Interscience)
81. Kilpatrick, J. E., Pitzer, K. S. and Spitzer, R. (1947). *J. Amer. Chem. Soc.*, **69,** 2483
82. Pitzer, K. S. and Donath, W. E. (1959). *J. Amer. Chem. Soc.*, **81,** 3213
83. Dunitz, J. D. (1972). *Tetrahedron*, **28,** 5459
84. Adams, W. J., Geise, H. J. and Bartell, L. S. (1970). *J. Amer. Chem. Soc.*, **92,** 5013
85. Altona, C., Geise, H. J. and Romers, C. (1968). *Tetrahedron*, **24,** 13
86. Dunitz, J. D. and Waser, J. (1972). *Elemente der Mathematik*, **27,** 25
87. Waser, J. and Schomaker, V. (1945). *J. Amer. Chem. Soc.*, **67,** 2014
88. Burns, J. H. and Waser, J. (1957). *J. Amer. Chem. Soc.*, **79,** 859
89. Winkler, F. K. (1973). *Dissertation*, E. T. H. Zürich, Nr. 5171
90. Warshel, A., Levitt, M. and Lifson, S. (1970). *J. Molec. Spectrosc.*, **33,** 84
91. Winkler, F. K. and Dunitz, J. D. (1971). *J. Molec. Biol.*, **59,** 169
92. Wiberg, K. B. and Boyd, R. H. (1972). *J. Amer. Chem. Soc.*, **94,** 8426
93. Pickett, H. M. and Strauss, H. L. (1970). *J. Amer. Chem. Soc.*, **92,** 7281
94. Corey, E. J. and Bailar, J. C. (1959). *J. Amer. Chem. Soc.*, **81,** 2620
95. Woldbye, F. (1967). *Proc. Roy. Soc.* (*London*), **A297,** 79
96. Gollogly, J. R. and Hawkins, C. J. (1972). *Inorg. Chem.*, **11,** 156
97. Geue, R. J. and Snow, M. R. (1971). *J. Chem. Soc. A*, 2981
98. Niketić, S. R. and Woldbye, F. (1973). *Acta Chem. Scand.*, **27,** 621
99. Nomura, T., Marumo, F. and Saito, Y. (1969). *Bull. Chem. Soc. Jap.*, **42,** 1016
100. Snow, M. R. (1970). *J. Amer. Chem. Soc.*, **92,** 3610
101. Rabinovich, D. and Schmidt, G. M. J. (1966). *Nature* (*London*), **211,** 1391
102. Dosi, C., Giglio, E., Pavel, V. and Quagliata, C. (1973). *Acta Crystallogr.*, **A29,** 644
103. Huisgen, R., Brade, H., Walz, H. and Glogger, I. (1957). *Chem. Ber.*, **90,** 1437
104. Graham, L. L. and Chang, C. Y. (1971). *J. Phys. Chem.*, **75,** 776
105. Bastiansen, O. (1949). *Acta Chem. Scand.*, **3,** 408
106. Almenningen, A. and Bastiansen, O. (1958). *Kgl. Norske Vidensk. Selsk. Skv.*, No. 4
107. Trotter, J. (1961). *Acta Crystallogr.*, **14,** 1135
108. Hargreaves, A. and Rizvi, S. H. (1962). *Acta Crystallogr.*, **15,** 365
109. Casalone, G., Mariani, C., Mugnoli, A. and Simonetta, M. (1968). *Molec. Phys.*, **15,** 339
110. Kurittu, J. and Pawley, G. S. (1973). *Acta Crystallogr.*, **A29,** 615
111. Cyvin, S. J. (1970). *Acta Chem. Scand.*, **24,** 3259

112. Rae, A. I. M. and Mason, R. (1968). *Proc. Roy. Soc. (London)*, **A304,** 487
113. Mason, R. and Rae, A. I. M. (1968). *Proc. Roy. Soc. (London)*, **A304,** 501
114. For a review see: Mason, R. (1970). *Perspectives in Structural Chemistry*, Vol. 3, 59 (J. D. Dunitz and J. A. Ibers, editors) (New York: Wiley)
115. Schomaker, V. and Trueblood, K. N. (1968). *Acta Crystallogr.*, **B24,** 63
116. Scheringer, C. (1973). *Acta Crystallogr.*, **A29,** 554
117. Cerrini, S. and Pawley, G. S. (1973). *Acta Crystallogr.*, **A29,** 660
118. Shmueli, U. and Goldberg, I. (1973). *Acta Crystallogr.*, **B29,** 2466
119. Filippini, G., Gramaccioli, C. M., Simonetta, M. and Suffritti, G. B. (1973). *Proc. 1st Europ. Cryst. Meeting*, Bordeaux
120. Boyd, R. H. (1968). *J. Chem. Phys.*, **49,** 2574
121. Nakagawa, I. and Shimanouchi, T. (1966). *Spectrochim. Acta*, **22,** 1707

# 5
# Structural Organotransition Metal Chemistry

**R. MASON, F.R.S.**
University of Sussex

**and**

**D. M. P. MINGOS**
Queen Mary College, University of London

---

---

## 5.1 INTRODUCTION

Organotransition metal chemistry has been, for the past twenty years, the subject of very rapid development. The synthesis of sandwich compounds was virtually in step with a revolution in chemical instrumentation which allowed their rapid characterisation. The continuing interest in the subject owes much to the development of rational syntheses and also to the particular impact of x-ray diffraction methods, which have provided unequivocal and accurate characterisations of molecular geometries and hence the basis for theories of the chemical bond in these molecules. The latter is of obvious importance in connection with attempts to understand the reactivities of coordinated ligands and, closely related to that point, the mechanisms of homogeneous and heterogeneous catalysis by transition metal complexes. We shall not concern ourselves here with reactions of coordinated ligands but choose to simply illustrate some recent structural results and comment on their general implications. The review divides naturally into complexes containing one, two and three or more metal atoms, the latter generally falling into the class of molecules known as cluster complexes[1] where several transition metal atoms are bonded together in a polyhedral or polygonal array together with ligands such as carbon monoxide, acetylene and tertiary phosphines. For previous surveys on organometallic, halide, oxo, chalcogenide and carboxylate cluster compounds the reader is referred to Refs. 2–10.

## 5.2 MONONUCLEAR ORGANOMETALLIC COMPLEXES CONTAINING SIMPLE UNSATURATED LIGANDS

### 5.2.1 Olefin (two electron donor) complexes

Although olefin–transition metal complexes date from Zeise's discovery of the salt $K[PtCl_3(C_2H_4)]$, accurate structural data have only become available, for the most part, during the past five years or so. Representative data, which can be seen to mostly refer to complexes of $d^8$ and $d^{10}$ metal ions, are collected in Table 5.1.

Three generalisations seem justified from these data. Firstly, coordination leads to an increase in the carbon–carbon bond length of the olefin {*cf.* the uncoordinated olefin bond lengths of 1.344 (10) Å ($C_2H_4$), 1.339 (8) Å [$C_2(CN)_4$], and 1.32 (1) Å ($C_2F_4$)}. The bond length increase could originate,

**Table 5.1 Molecular geometries of metal–alkene complexes**

| *Molecule* | | C=C/Å | M—C/Å |
|---|---|---|---|
| $[(C_2H_4)PtCl_3]H_2O$ [11] | $Pt^{II}$ | 1.37(3) | 2.13(2) |
| | | 1.354(15) | 2.14(1) |
| $(Me_2H_2C_2)PtCl_2$(phenylethylamine) [12] | | 1.36(6) | 2.17(5) |
| PtCl(OMe)(dicyclopentadiene) [13] | | 1.37 | 2.09 |
| $[(C_4H_{12}N_2)PtCl_3]^+$ [14] | | 1.416(12) | 2.13(1) |
| $(C_{10}H_{14}O_3)PtCl_2$ [15] | | 1.42(3) | 2.09(2), 2.33(2) |
| $(C_2H_4)(C_2F_4)Rh(C_5H_5)$ [16] | $Rh^I$ | 1.358(9) ($C_2H_4$) | 2.167(2) |
| | | 1.405(7) ($C_2F_4$) | 2.024(2) |
| $(C_2H_4)(C_2F_4)Rh(C_5H_7O_2)$ [17] | | 1.42(2) ($C_2H_4$) | 2.19(1) |
| | | 1.40(2) ($C_2F_4$) | 2.01(1) |
| $(C_2H_4)_2Rh(C_5H_7O_2)$ [18] | | 1.41(3) | 2.14(1) |
| $(C_2F_4)Rh(Ph_3P)_2Cl$ [19] | | 1.41(2) | 2.01(2) |
| $\{(o\text{-vinylPh})_3P\}RhBr$ [20] | | 1.47(3) | 2.23(2) |
| $\{H_2(CN)_2C_2\}IrH(CO)(Ph_3P)_2$ [21] | $Ir^I$ | 1.431(20) | 2.110(9) |
| $\{(CN)_4C_2\}IrBr(CO)(Ph_3P)_2$ [22] | | 1.506(15) | 2.146(11) |
| $\{(CN)_4C_2\}Ir(C_6N_4H)(CO)(Ph_3P)_2$ [23] | | 1.526(12) | 2.166(15) |
| $(C_8H_{12})$(diphos)$IrCH_3$ [24] | | 1.459(21) | 2.011(9) (equatorial) |
| | | 1.374(21) | 2.106(9) (axial) |
| $(C_8H_{12})\{Ph_2P(CH_2)_3PPh_2\}IrCH_3$ [25] | | 1.475(25) | 2.033(12) (equatorial) |
| | | 1.394(25) | 2.127(12) (axial) |
| $(C_2H_3CN)Fe(CO)_4$ [26] | $Fe^0$ | 1.40(2) | 2.10(2) |
| $(C_2H_2CO_2H)Fe(CO)_4$ [27] | | 1.40(5) | 2.10(2) |
| $\{o\text{-}CH_2{=}CH(C_6H_4)PPh_2\}_2Fe(CO)_2$ [28] | | 1.455(8) | 2.10(1) |
| $(C_2H_4)Ni\{(C_6H_{11})_3P\}_2$ [29] | $Ni^0$ | 1.401(14) | 2.01(1) |
| $(C_2H_4)Ni(Ph_3P)_2$ [30] | | 1.41(1) | 1.93(1) |
| $(C_2H_4)Ni\{(o\text{-}MeC_6H_4O)_3P\}_2$ [31] | | 1.43(1) | 1.99(1) |
| $(C_2H_3CN)Ni\{(o\text{-}MeC_6H_4O)_3P\}_2$ [32] | | 1.46(2) | 2.016(10) |
| $\{C_2(CN)_4\}Ni(C_4H_9NC)_2$ [33] | | 1.476(5) | 1.954(4) |
| $(C_2F_4)Ni(Me_3C(Me)C(CH_2PPh_2)_3$ [34] | | 1.37(3) | 1.86(2) |
| $\{Cl_2C_2(CN)_2\}Pt(Ph_3P)_2$ [35] | $Pt^0$ | 1.42(3) | 2.00(2), 2.10(2) |
| $(C_2H_4)Pt(Ph_3P)_2$ [36] | | 1.43 | 1.99(1) |
| $\{C_2(CN)_4\}Pt(Ph_3P)_2$ [37] | | 1.49(5) | 2.10(3) |
| $(C_2Cl_4)Pt(Ph_3P)_2$ [38] | | 1.62(3) | 2.03(3) |

within the simple Dewar–Chatt[39] bonding model for metal–olefin complexes, through either σ-bond (ligand to metal) charge transfer from the filled $2p_\pi$ olefin orbital or from π-bond (metal to ligand) electron transfer into the ligand $2p_{\pi*}$ orbital; moreover, one would expect the relative importance of these processes to reflect the formal oxidation state of the metal with the ligand acting as a net Lewis base in, say, a platinum(II) complex and as an overall Lewis acid in a zerovalent complex. The mean $Pt^{II}$—C(olefin) bond length is approximately 0.1 Å larger than the $Pt^0$—C(olefin) value and, since the σ-covalent radius of platinum(II) must be somewhat smaller than that of platinum(0), it carries the implication of substantial back-donation from platinum(0). The general pattern is for the carbon–carbon bond lengths in the zerovalent complexes to be also greater but the significance of this result is low. The strong π-basicity of zerovalent metal fragments is also reflected in ESCA metal binding energies[40]. The platinum complexes are also useful in demonstrating asymmetric olefin bonding: Mason and Robertson[15] commented on the difference of 0.2 Å between the metal–carbon bond lengths in $PtCl_2(C_{10}H_{14}O_3)$ as suggesting the destabilisation and localisation of the ligand $\pi^*$ level on methyl substitution. A parallel result is seen[35] in $Pt(PPh_3)_2\{Cl_2C_2(CN)_2\}$, although it is somewhat surprising to see such large differences in electronic effects between chloro and cyano substituents. An even more impressive delineation of $\sigma$- and $\pi$-bonding effects is shown in the rhodium complexes $Rh(\pi\text{-}C_5H_5)(C_2H_4)(C_2F_4)$ [16] and $Rh(acac)(C_2H_4)(C_2F_4)$ [17], where the mean Rh—$C(C_2H_4)$ bond length is 2.18 Å and that of Rh—$C(C_2F_4)$ is 2.02 Å; both of these values are identical, within experimental error, to the equivalent bond lengths in $Rh(acac)(C_2H_4)_2$ [18] and $RhCl(C_2F_4)(PPh_3)_2$ [19]. There can be little doubt, in view of the decreased Lewis basicity of tetrafluoroethylene *vis à vis* ethylene, that these data illustrate enhanced $\pi$-bonding to the lower lying antibonding levels of tetrafluoroethylene, a feature which we further discuss below in connection with the rotameric structures of olefin–metal complexes.

The pattern of increasing metal–carbon bond order with decreasing carbon–carbon bond order is also well illustrated in the cyclo-octadiene-iridium(I) complexes[24, 25], where the basicity of the (five coordinate) metal ion depends on whether the $\pi$-bonding ligand occupies an axial or equatorial position.

The second major stereochemical feature of coordinated olefins is that they are no longer planar—the substituent groups bend away from the metal so as to give quasi-tetrahedral character to the coordinated carbon atoms. This observation added to the long-standing discussion as to whether the metal–olefin bond is better represented by one or other of the valence structures (1) and (2). The relationship between these two forms had been recognised

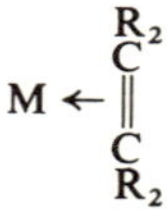

(1)

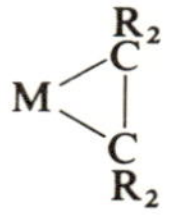

(2)

earlier, particularly for butadiene complexes (*vide infra*)[41]. Structure (1) simply represents the ligand to metal $\sigma$-bond; structure (2) implicitly represents both the $\sigma$- and $\pi$-components of the metal–olefin attachment. It was suggested[42] that there was a relation between the decrease in the carbon–carbon bond length following coordination and the deformation of the bond angle at the coordinated carbon towards $sp^3$ character. Guggenberger and Cramer[16] modified the model somewhat in an attempt to exclude peculiar steric and electronic effects of the olefin substituents but, again, the relationship is not an exact one—it appears that some steric effects are not being allowed for in an adequate way.

Thirdly, we need to comment on the general conformation of the olefin with respect to the coordination plane defined by the metal and other ligands. In Zeise's salt and other platinum(II) complexes, the olefin carbon–carbon bond axis is orthogonal, within a few degrees, to the plane containing, say, the platinum and chlorine atoms. In the nickel(0) and platinum(0) species the ligand is coplanar, to within several degrees, with the metal and phosphorus atoms making up the coordination sphere. No intermediate conformation of an olefin ligand has been observed, although much data are to hand to demonstrate that the rotation of olefins about the metal–olefin bond axis is a process with a very low activation energy. In $Rh(\pi\text{-}C_5H_5)(C_2H_4)_2$ the energy barrier to rotation is 15 kcal $mol^{-1}$, whereas in the perfluoroethylene analogue no rotation, as evidenced by the $^{19}F$ n.m.r. spectra, was observed. In the vinyl fluoride, $Rh(\pi\text{-}C_5H_5)(CH_2{=}CHF)_2$, the barrier to rotation is 16 kcal $mol^{-1}$ [43]. An exhaustive series of investigations of olefin rotation in platinum-(II) complexes has been reported[44] and the energy barriers related to both steric and electronic features of the substituent groups. So far as electronic factors are concerned, Lewis *et al.*[44] take up the earlier suggestion[15] that the out-of-plane olefin conformation in the higher oxidation state complexes is stabilised by $d_{xz}$–$p_z$ hybridisation which is less likely — given a greater d–p energy separation — in the zerovalent species. The essence of this suggestion is that only steric factors can affect the dependence of the σ-bond energy upon rotation; if the remaining ligands in the coordination sphere are regarded as 'innocent', the metal to ligand π-bonding is invariant if it is based on pure metal 'd' orbitals. The latter assumption is unlikely to be justified in detail since stable zerovalent π-complexes of olefins require the other ligands to be tertiary phosphines, phosphites or similar species which are often held to be, at least, weak π-acids. In the rhodium(I) complexes the Rh—C ($C_2F_4$) bond lengths are seen to be 0.15 Å, or less than the values on the ethylene complex; the van der Waals' diameters of hydrogen and fluorine are not very different; it seems reasonable, therefore, to assume that the lack of rotation of the perfluoroethylene ligand now has its origins in electronic factors whose implications are obvious in the stronger metal–carbon bond.

### 5.2.2 Allene complexes

Allene, $H_2C{=}C{=}CH_2$, and several substituted allenes provide complexes with metals in the 0, +1 and +2 oxidation states; Table 5.2 summarises the available structural data.

Only two carbon atoms of the allene are coordinated, the significant distortion of the coordinated allene ligand from linearity ensuring that it is

**Table 5.2 Bond lengths and bond angles in metal–allene complexes**

| *Molecule* | C=C/Å (*coordinated*) | C=C/Å (*uncoordinated*) | M—C/Å | CCC/ degrees |
|---|---|---|---|---|
| $Rh(acac)(H_2C{=}C{=}CMe_2)_2$ [45] | 1.41(1) | 1.30(1) | 2.13 2.07 | 153 |
| $Rh(acac)(Me_2C{=}C{=}CMe_2)_2$ [45] | 1.38(1) | 1.32(1) | 2.18 2.03 | 148 |
| $Rh_2(acac)_2(CO)_2(H_2C{=}C{=}CH_2)$ [47] | 1.39(1) | — | 2.13 2.05 | 145 |
| $RhI(PPh_3)_2(H_2C{=}C{=}CH_2)$ [48] | 1.35(3) | 1.34(3) | 2.17 2.04 | 158 |
| $[PtCl_2(Me_2C{=}C{=}CMe_2)]_2$ [49,50] | 1.36(3) | 1.36(3) | 2.29 2.07 | 151 |
| $Pt(PPh_3)_2(MeHC{=}C{=}CH_2)$ [46] | 1.44 | 1.32 | 2.12 2.05 | 146 |
| $Pt(PPh_3)_2(H_2C{=}C{=}CH_2)$ [51] | 1.48(5) | 1.31(5) | 2.13 2.03 | 142 |
| $Pd(PPh_3)_2(H_2C{=}C{=}CH_2)$ [52] | 1.40(1) | 1.30(1) | 2.12 2.07 | 148 |

formally a two-electron donor only. The metal–carbon bond lengths are obviously non-equivalent, with the central carbon atom lying 0.1 Å or so closer to the metal. The non-linearity of the coordinated allene, like the isoelectronic carbon disulphide ligand, can be related to the general model which envisages that the electron distribution in, and geometries of, ligands coordinated to low-oxidation state metals often approaches that of their low-lying electronically excited states[53,54].

The non-equivalence of the two metal–carbon bonds has been attributed to $\sigma$-orbital radius differences following carbon rehybridisation[55]; this could, however, only contribute a maximum difference of 0.03 Å; another suggestion[47] relates to the possibility of metal d-orbitals overlapping with the $\pi^*$-orbital of the uncomplexed double bond. We note that rehybridisation at the central carbon atom will modify the 'p' electron density in a way which will begin to resemble that in the allyl radical and it is of interest that the dihedral angle between the coordinating plane and the allene plane ranges from 95 to 103° in the non-zerovalent complexes, values which are similar to those observed in $\pi$-allyl complexes (*vide infra*). The conformational difference of the coordinated allene in the zerovalent complex parallels that observed in simple ethylene and substituted alkene complexes (*vide supra*).

### 5.2.3 Alkyne complexes

In contrast to olefin and allene complexes, much of alkyne–transition metal chemistry is dominated by polynuclear complexes which we discuss in Section 5.4.3.3. The structural data for mononuclear species are summarised in Table 5.3. They are seen to be much less accurate than those now to hand for olefin species and, therefore, allow only general remarks on the nature of the metal–ligand bond.

Semi-empirical MO theory suggests that the *cis* bending of substituent groups in coordinated acetylenes is directly proportional to the amount of electron transfer from the ligand $\pi_u$ orbitals to their $\pi_g^*$ counterpart[62]. The bond angles at the coordinated carbon atoms in the platinum(0) and platinum(II) complexes are significantly different and in a way which could have been intuitively expected given the strong $\pi$ basicity of the platinum(0) fragment. As in alkene complexes, the conformation of the alkyne reflects

**Table 5.3 Molecular parameters of metal–alkyne complexes**

| *Molecule* | *Average* M—C/Å | *Average coordinated carbon–carbon bond lengths*/Å | *Mean bond angle,* RCC/degrees |
|---|---|---|---|
| $(p\text{-NC}_6\text{H}_4\text{Me})\text{PtCl}_2(\text{Bu}^t_2\text{C}_2)$ [56] | 2.16(1) | 1.24(2) | 163(1) |
| $(\text{Ph}_3\text{P})_2\text{Pt}(\text{Ph}_2\text{C}_2)$ [57] | 2.04(6) | 1.32(9) | 140 |
| $(\text{CNCMe}_3)_2\text{Ni}(\text{Ph}_2\text{C}_2)$ [58] | 1.90(1) | 1.28(2) | 149(1) |
| $(\pi\text{-C}_5\text{H}_5)\text{Nb}(\text{CO})(\text{Ph}_2\text{C}_2)(\pi\text{-Ph}_4\text{C}_4)$ [59] | 2.25(3) | 1.26(3) | 142(3) |
| $(\pi\text{-C}_5\text{H}_5)\text{Nb}(\text{CO})(\text{Ph}_2\text{C}_2)_2$ [60] | 2.19(2) | 1.35(2) | 138(4) |
| $\text{W}(\text{CO})(\text{Ph}_2\text{C}_2)_3$ [61] | 2.06 | 1.30 | 140 |

the formal oxidation state of the metal, and while calculations[63] place the energy differences between the two conformers as 3–4 eV, there seems little doubt that again the activation energy for rotation is sufficiently small for thermal equilibrium to be established between the two conformers.

Other features of these complexes — particularly their stereochemistries and the rare gas 'rule' for organometallic molecules — have been summarised recently[64].

### 5.2.4 π-Allyl transition metal complexes

An early summary[65] of π-allyl complexes demonstrated the main features of their stereochemistries: in simple symmetrical complexes such as the π-allyl palladium chloride dimer[66], the metal–carbon distances were essentially equivalent and the conformation of the allyl ligand with respect to, say, the palladium–chloride bridge system could be understood in terms of the maximum overlap criterion[67]. In asymmetric complexes — for example, chloro-(2-methylallyl)(triphenylphosphine)palladium(II)[68] — the metal–carbon bond lengths are non-equivalent, reflecting the different *trans* influences of the triphenylphosphine and chloride ligands.

Two features of metal–allyl complexes have emerged more recently and deserve additional emphasis. A number of 2-methylallyl complexes[69] demonstrated the non-planarity of the methyl group with respect to the allyl carbon atoms, the deviation, towards the metal atom, of the methyl carbon atom from the allyl plane being 0.3–0.5 Å. In the π-1,1,3,3-tetramethylallylpalladium chloride dimer[70], each pair of methyl groups is rotated from the plane of the allyl carbon atoms to produce a rather precisely eclipsed conformational arrangement; non-planar substituent groups are also evident in the π-1,3-dimethylallylpalladium chloride dimer[71]. The low-temperature analysis of $[(C_3H_5)PdCl]_2$ shows[66] the hydrogen atoms to be coplanar, within experimental error, with the allylic carbon atoms, whereas in $Ru(\pi\text{-}C_3H_5)_2(PPh_3)_2$ the terminal hydrogen atoms are bent away from the metal[72]. For the ruthenium complex the non-planarity apparently serves to bring non-bonded hydrogen–ruthenium distances into line with normal van der Waals' values, but the ruthenium–carbon and palladium–carbon bond lengths are not greatly different. [Ru—C(central) = 2.13 and Ru—C(terminal) = 2.24 Å; the corresponding Pd—C values are 2.11 and 2.12 Å, with the greater Ru—C(terminal) value presumably reflecting the *trans* influence of the phosphine ligands.] Whether these results can be rationalised by invoking metal-induced σ–π mixing in the allyl ligand and/or steric effects is unanswered at present.

A further problem of considerable interest, which perhaps deserves the traditional Scottish verdict of 'not proven', is highlighted by the very accurate low-temperature structural analysis of bis(cyclopentadienyl)-2,2′-bis(1–3-η-allyl)-dinickel[73] in which the allyl groups are symmetrically bonded to the nickel atoms, but apparently significant differences exist between the five metal–carbon (cyclopentadienyl) bond lengths and between the intracyclopentadienyl carbon–carbon bond distances. Possible localisation of electron density in pentahapto-cyclopentadienyl complexes was discussed some time ago[74]. It was argued that asymmetric bonding is possible whenever there is less than cylindrical symmetry about the metal. There is now much evidence on non-equivalent metal–carbon bonds in cyclopentadienyl complexes which can be rationalised by the suggestion that the degeneracy of the e orbitals of the cyclopentadienyl ring is effectively removed in complexes of low symmetry. But the corresponding asymmetry of carbon–carbon bond lengths is not well defined and further low-temperature analyses are necessary to eliminate misleading data which result from static and/or dynamic disordering of cyclopentadienyl ligands in many crystal structures.

### 5.2.5 Conjugated diene (four electron donor) complexes

(1–4-$\eta$-Butadiene)tricarbonyliron and other dieneiron tricarbonyl complexes have played an important part in developments of theories of bonding between transition metals and unsaturated hydrocarbons. The early structural results were summarised by Churchill and Mason[65] and were provided with an explanation in terms of the 'excited state model' for the geometries of coordinated ligands[53,54]. The essential point of this scheme is that the geometry of butadiene is modified considerably on coordination from the situation described by (3) to ones which are illustrated by (4) and (5). More

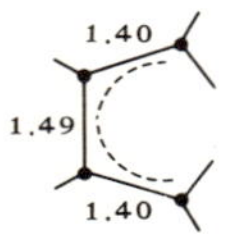


(3)

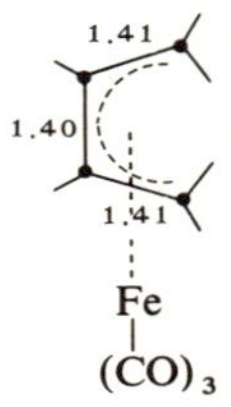


(4)

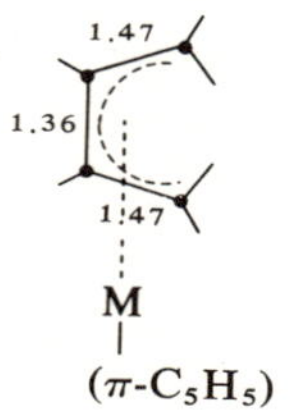


(5)

recent, accurate, structural data confirm these data: Cotton *et al.*[75] show (i) the iron–carbon distances are significantly different, the average value to the 'inner' carbon atoms being 0.08 Å less than that to the 'outer' carbons; (ii) the 1–2 (3–4) carbon–carbon bond length is, on average, 0.02 Å larger than the 2–3 carbon–carbon bond; and (iii) in keeping with earlier remarks, the conformation of the $Fe(CO)_3$ group with respect to the *cis*-diene carbon atoms is always such that the molecule has a fairly precise mirror plane in which the iron atom has effective tetragonal pyramidal symmetry.

An understanding of the diene–metal bonding has had implications on systematic studies of the reactivity of coordinated dienes. The large metal–ligand 'back donation' indicated by the bond lengths of (5) also imply enhanced reactivity of the diene towards electrophiles, whereas more facile nucleophilic attack on the diene is observed in tricarbonyl-metal diene complexes (4) or when the metal has a relatively high oxidation state. *En passant*, we also observe that butadieneiron tricarbonyl is a standard example of the value of the eighteen electron 'rule' in organo-transition metal chemistry, a feature which makes the iron tricarbonyl fragment a valuable reagent for the stabilisation of reactive organic species and for inducing unusual rearrangements in complex organic molecules[76].

### 5.2.6 π-Pentadienyl (five electron donor) complexes

Three accurate structural analyses of cyclohexadienyl complexes[77-79] have been reported and relevant metal–carbon bond lengths are shown in (6)–(8). The isoelectronic iron and manganese tricarbonyl complexes illustrate: (i) the

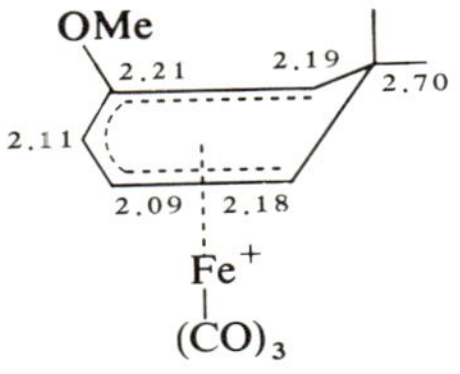


(6)

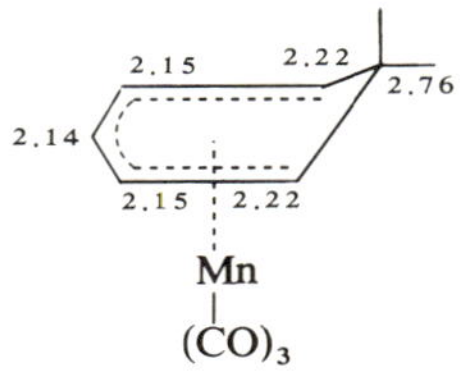


(7)

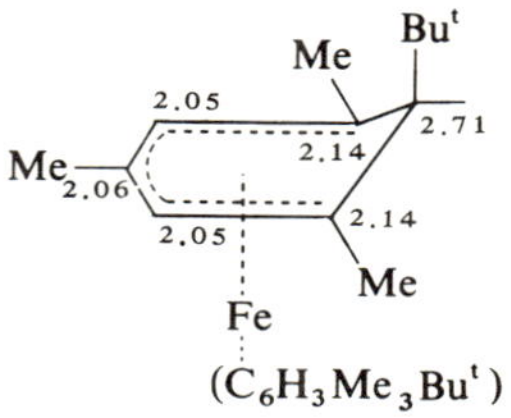


(8)

average Fe–C(O) bond length of 1.82(1) Å is scarcely significantly different from the corresponding Mn—C(O) bond length of 1.79 Å, but the difference should be related to the remaining intramolecular bond lengths discussed below; (ii) apart from the large Fe—C(OMe) bond length — which must be connected with substituent effects — the remaining iron–carbon bond lengths are systematically 0.03–0.04 Å less than their manganese–carbon counterparts. The inference is obvious: $Mn^0$ will have a larger covalent radius than $Fe^I$, which is reflected in the bond lengths to the pentadienyl fragment, but will also be a better π-base with respect to the strongly acidic carbonyl groups; (iii) the mean carbon–carbon bond length in the dienyl fragment of the manganese complex is 1.40 Å, that in the cationic iron tricarbonyl complex is 1.42 Å, while it is 1.41 Å in the bis-pentadienyl species; no single bond length differs significantly from these mean values. The simple pentadienyl cation, radical or anion all have the following Hückel π-bond orders ($P_{ij}$):

$$\underset{1}{C}\overset{0.788}{—}\underset{2}{C}\overset{0.577}{—}\underset{3}{C}\overset{0.577}{—}\underset{4}{C}\overset{0.788}{—}\underset{5}{C}$$

Approximately equal bond lengths in, for example, the radical reflect the Hückel configuration $\psi_1^2\,\psi_2^1\,\psi_3^2$ when $P_{12} = 0.538$ and $P_{23} = 0.5577$, a situation identical to that which obtains for the configuration, $\psi_1^2\,\psi_2^1\,\psi_3^1$, of the cation. Thus, the observed equality of the bond lengths in the coordinated pentadienyl ligand directly illustrates the role of electron donation from the ligand orbital, $\psi_2$, to the metal since a change in the occupancy of the non-bonding ligand orbital, $\psi_3$, has no effect on the lengths of the 'inner' carbon–carbon bonds.

### 5.2.7 Arene (six electron donor) compounds of transition metals

Like ferrocene, the synthesis of bis($\eta$-benzene)chromium played a large part in stimulating research activity in organo-transition metal chemistry. Unlike ferrocene, the question of the nature of the metal–ligand bond was a matter of some controversy, an original suggestion[80] that the molecule had only threefold symmetry being challenged theoretically and seemingly disproved by subsequent structural analyses[81]. But the question of some bond localisation in eighteen electron arene complexes appears to have recurred from the results of x-ray and neutron diffraction analyses, at 78 K, of benzene-chromium tricarbonyl[82]. Both the x-ray and neutron data indicate that there is some bond alternation, three carbon–carbon bonds having a mean value of 1.402 Å, the remainder having an average separation of 1.419 Å. The mean e.s.d. of the bond lengths is claimed as 0.002 Å so that this difference should be judged highly significant. A second interesting feature is that the hydrogen atoms are all displaced, from the plane of the ring, towards the metal by an average of 0.03 Å.

The full details of these analyses will be of considerable interest since the preliminary refinement was based on the centrosymmetric space group $P2_1/m$ and a careful, rigorous analysis will be needed to exclude the non-centrosymmetric space group, $P2_1$, and the possibility that the small molecular distortions are not consequential artefacts. Unequivocal results would certainly add much to our theoretical understanding of these molecules.

By contrast, distortions of arene ligands, in molecules formally containing more than eighteen valence electrons, are well-established and have been recently commented upon[83,84]. Representative of the results are the structural data of hexakis(trifluoromethyl)benzene($\eta$-cyclopentadienyl)rhodium[85] and bis(hexamethylbenzene)ruthenium[86]. With planar symmetrical arene ligands, these complexes would be twenty electron structures but both hexakis-(trifluoromethyl)benzene and one hexamethylbenzene ligand distort to a dihedral form so as to present the metal with an effective four electron, butadiene fragment donor. The driving force for these ligand distortions is identical with those discussed for simple alkenes and alkynes: the total electronic energy of the molecule is reduced by the transfer of localised, strongly antibonding electron density on the metal into lower-lying delocalised orbitals on the ligand with a consequent low-energy-mode deformation of the ligand.

## 5.3 DINUCLEAR ORGANOMETALLIC COMPLEXES

### 5.3.1 General considerations

While enneacarbonyldi-iron(0) was the first metal carbonyl to be characterised by diffraction methods[87], a coherent attack on the development of bonding models in dinuclear organometallics is fairly recent. The emphasis has been placed on complexes with bridging ligands, rather than on molecules such as

$Mn_2(CO)_{10}$[88, 89] and $Mo_2(\pi\text{-}C_5H_5)_2(CO)_6$[90] where the metal atoms are linked simply by a metal–metal bond. The question of bridge–terminal interchange of carbonyl ligands has prompted several structural determinations of carbonyl bridged dinuclear complexes in recent years.

In diborane and more complex boron hydrides, quantitative calculations of molecular parameters have given a reliable picture of the bridge bonds in these molecules[91], but even for the simplest μ-hydrido-transition metal complexes the status of the theoretical description is much less adequate. A simple symmetry based theory[92], which draws analogies between transition metal and main group dinuclear complexes, has, however, proved surprisingly useful for rationalising metal–metal bond lengths and the conformations of the four-membered ring containing the metal and bridging atoms.

Essentially, the model emphasis the point that the polycentric bridge MO can be separated into those which have bonding or antibonding character with respect to the two metals. With the full use of 'd' orbitals, formal metal–metal bond orders, ranging from 0 to 2, can be assigned; an important generalisation is that no complexes with metal–metal bond orders of zero exist for μ-hydrio, μ-alkyl (or aryl) or μ-carbonyl ligands. The metal–metal bond length is also influenced substantially by the electronegativity of the bridging group and by the π-acidity of the terminal ligands so that the metal–metal bond orders cover a range of metal–metal bond lengths — the nature of the bridging ligand in an otherwise isostructural complex may influence the metal–metal bond length by as much as 0.25 Å.

Two other features of the stereochemistries of dinuclear complexes have come to light recently. The dihedral geometries of the $M_2X_2$ ring in bridged complexes may range from ca. 100 to 180° (planar) and may be represented schematically by the four structures shown in (9)–(12). The conformations of

(9)

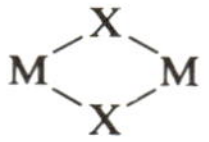

(10)

(11)

(12)

the rings where metal–metal bonding is an important feature, that is (9) and (11), are determined largely by the electronic configuration of the metal and the number of ancillary ligands coordinated to the metals[92]. For example, the complexes $[Mo(\pi\text{-}C_4H_4)Br(CO)_2]_2$[93], $[Cr(\pi\text{-}C_5H_5)(SR)(NO)]_2$[94], $[Co(\pi\text{-}C_5H_5)(PPh_2)]_2$[95] and $[Fe(SR)(NO)_2]_2$[96] all have a formal metal–metal bond order of one, but only $[Co(\pi\text{-}C_5H_5)(PPh_2)]_2$ has the dihedral geometry (9). This can be explained either in terms of a 'bent' metal–metal bond or through a Walsh MO correlation diagram[97]. The latter suggests that the highest occupied MO in $[Co(\pi\text{-}C_5H_5)(PPh_2)]_2$ is stabilised when the dihedral angle is reduced from 180°, because the change in symmetry associated with bending allows it to interact with the lowest empty antibonding MO. The orbitals, derived primarily from the $d_{z^2}$ and $d_{yz}$ metal orbitals, are antibonding with respect to the metals and have the symmetry $b_1$ appropriate to the $C_{2v}$ symmetry of the dihedral $M_2X_2$ moiety. The lowest unoccupied MO in $[Co(\pi\text{-}C_5H_5)(PPh_2)]_2$ is, by inference, destabilised by the angular distortion. That

$[Ni(\pi\text{-}C_5H_5)(PPh_2)]_2$, which has two additional valence electrons, does not have a metal–metal bond and is planar [conformation (10)] supports this view.

The conformations of the non-metal–metal bonded four-membered rings (10) and (12) are by contrast determined by the nature of the bridging groups. Thus $[Pt(SEt)(PPr^n_3)_2]_2$[97] and $[Pt_2Cl_4(PPr^n_3)_2]$[98] have the metals in the same formal oxidation state, but the former has a dihedral angle of 120° and the latter one of 180°. The different π-bonding capabilities of the bridging ligands in these closely related complexes must be held responsible for these different geometries.

### 5.3.2 Asymmetric bridges

Finally, we note that an increasing number of analyses indicate considerable asymmetry in the metal–bridge-ligand bond lengths.

Cotton and his co-workers[99] have shown that the activation energy for the exchange of bridge and terminal carbonyl ligands in complexes such as $[Fe(\pi\text{-}C_5H_5)(CO)_2]_2$ is rather small. Detailed kinetic studies on related complexes have indicated that the carbonyl bridges in the $M_2(CO)_2$ fragment open in the concerted fashion shown in (13) ⇌ (14). x-Ray studies of carbonyl

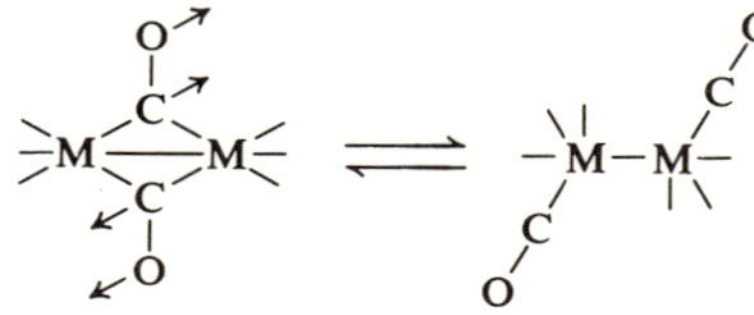

(13) (14)

complexes in the solid state suggest that the potential energy surface connecting the extreme geometries shown is rather featureless since intermediate geometries are 'frozen out' by the presence of asymmetric crystal packing forces. This effect is particularly well demonstrated for the closely related trinuclear iron complexes $Fe_3(CO)_{12}$[100], $Fe_3(CO)_9(PPh_2Me)_3$[101] and $Fe_3(CO)_{11}(PPh_3)$[102], which show different degrees of asymmetry in the solid state.

In complexes which have carbonyl ligands bridging dissimilar metals, both carbonyl ligands may form shorter metal–carbon bonds to one of the metals. This asymmetry is obviously linked to the differences in the electronic properties of the two metals and may also be observed when the terminal ligands, attached to the separate metal atoms, differ in their electronic properties[103].

## 5.4 POLYNUCLEAR (CLUSTER) ORGANOTRANSITION COMPOUNDS

### 5.4.1 Metal carbonyl clusters

Compounds containing metal–metal bonds are common among the halogen, chalcogenide, carbonyl and organometallic derivatives of the transition elements in their lower oxidation states but are almost unknown for higher oxidation states. Neutral metal carbonyl cluster compounds have been definitely established only for the Group VIII metals, Fe, Ru, Os, Co, Rh and Ir. The structures of the polymeric carbonyls $[Re(CO)_4]_n$ [108] and $[Pt(CO)_2]_n$ [109] are not known but presumably contain extended metal–metal linked chains. If anionic and hydrido-carbonyl[110] compounds are considered, the range of metals forming cluster compounds can be extended to include the Group VIA and VIIA metals and nickel.

Following Schäfer and Schnering[111] it can be suggested that metal carbonyl cluster compounds will be formed preferentially when the mean metal–carbonyl bond energy falls below that of the mean metal–metal bond energy. These thermodynamic quantities can be estimated from the heat of formation of the parent mononuclear carbonyl compounds and the enthalpies of vaporisation of the metal (see Table 5.4). From these values, it is clear that cluster formation is favoured for the heavier metals in a group and the middle elements of the transition series.

**Table 5.4**

(*a*) *Heats of atomisation of some transition elements*/kcal $mol^{-1}$ [112]

| | | | | |
|---|---|---|---|---|
| Mn<br>69 | Fe<br>100 | Co<br>102 | Ni<br>102 | Cu<br>81 |
| Tc<br>155 | Ru<br>155 | Rh<br>133 | Pd<br>85 | Ag<br>68 |
| Re<br>186 | Os<br>162 | Ir<br>160 | Pt<br>135 | Au<br>87 |

(*b*) *Metal–carbonyl bond dissociation energies*, $\Delta H°$/kcal $mol^{-1}$ [113]
$1/x\ M(CO)_x(g) \rightarrow 1/x\ M(g) + CO(g)$

| | | |
|---|---|---|
| $Cr(CO)_6$<br>27.1 | $Fe(CO)_5$<br>27.7 | $Ni(CO)_4$<br>35.2 |
| $Mo(CO)_6$<br>35.9 | | |
| $W(CO)_6$<br>42.1 | | |

### 5.4.2 The stereochemistries of polynuclear metal carbonyls

The x-ray structural results for neutral metal carbonyl cluster compounds are summarised in Table 5.5. The simplest cluster compounds $M_3(CO)_{12}$ (M = Fe, Ru or Os) (Figures 5.1 and 5.2) have a triangle of metal atoms.

**Table 5.5 Summary of structural results for metal carbonyl cluster compounds**

| *Compound* | *Metal–metal bond lengths/Å* | *Estimated metal–metal bond length/Å* | *Idealised symmetry* | *Metal polyhedral geometry* | *Figure* | *Ref.* |
|---|---|---|---|---|---|---|
| $Fe_3(CO)_{12}$ | 2.69(1), 2.68(1), 2.55(1) | 2.52 | $C_{2v}$ | Isosceles triangle | 5.1 | 101 |
| $Ru_3(CO)_{12}$ | 2.848(6) | 2.68 | $D_{3h}$ | Equilateral triangle | 5.2 | 114 |
| $Os_3(CO)_{12}$ | 2.877(10) | 2.70 | $D_{3h}$ | Equilateral triangle | 5.2 | 115 |
| $Co_4(CO)_{12}$ | 2.49(2), 2.44–2.53(1) | 2.50 | $C_{3v}$ | Pyramidal | 5.3 | 101 |
| $Rh_4(CO)_{12}$ | 2.73(2), [2.70–2.80(1)] | 2.68 | $C_{3v}$ | Pyramidal | 5.3 | 116 |
| $Ir_4(CO)_{12}$ | 2.68(2) | 2.72 | $T_d$ | Tetrahedral | 5.4 | 117 |
| $Co_2Rh_2(CO)_{12}$ | | | $C_{3v}$ | Isomorphous with $Rh_4(CO)_{12}$ | 5.3 | 118 |
| $Co_2Ir_2(CO)_{12}$ | | | $C_{3v}$ | | | |
| $Rh_6(CO)_{16}$ | 2.776(1) | 2.68 | $T_d$ | Octahedral | 5.5 | 119 |
| $Co_2Rh_4(CO)_{16}$ | | | $T_d$ | Isomorphous with $Rh_6(CO)_{16}$ | | 118 |
| $Co_6(CO)_{16}$ | | | | | | 118 |
| $Os_6(CO)_{18}$ | 2.732–2.836(1) | 2.70 | $C_{2v}$ | Bicapped tetrahedron | 5.6 | 120 |
| $Os_7(CO)_{21}$ | 2.851 (average) | | $C_{3v}$ | Capped octahedron | 5.7 | 121 |

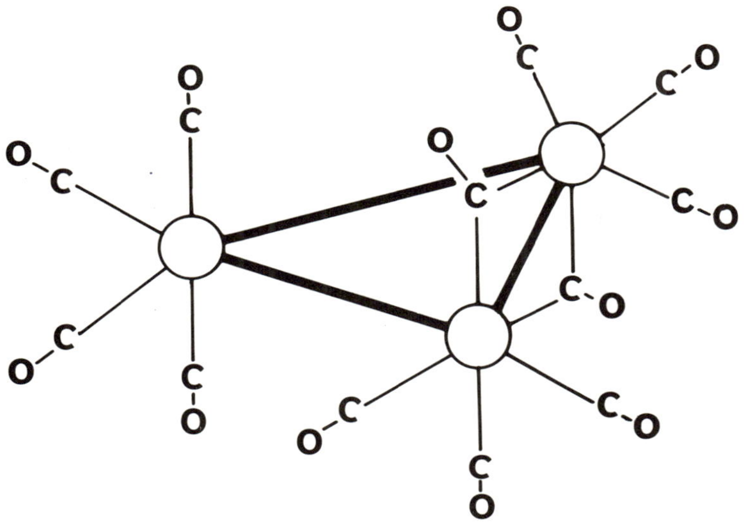


**Figure 5.1** The molecular stereochemistry of $Fe_3(CO)_{12}$ in the disordered crystal

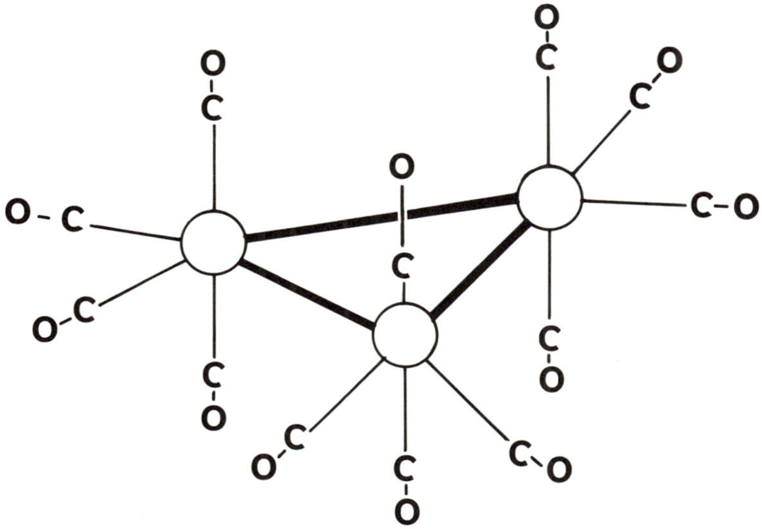


**Figure 5.2** The molecular geometry of the trinuclear carbonyls $M_3(CO)_{12}$ (M = Ru or Os)

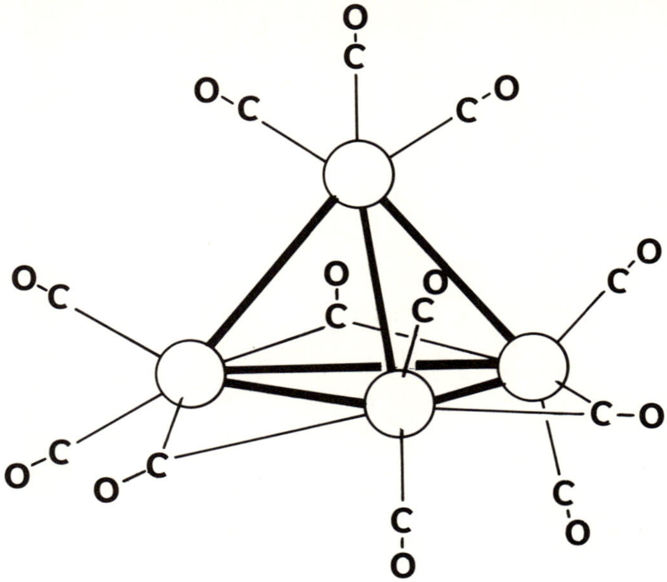

**Figure 5.3** The structure of $M_4(CO)_{12}$ ($M = Co$ or $Rh$) in the crystalline state. The molecules are disordered in the crystal. The iridium complexes $(Ir_4(CO)_9(PPh_3)_3$ and $Ir_4(CO)_{10}(PPh_3)_2$ have a similar arrangement of carbonyl ligands

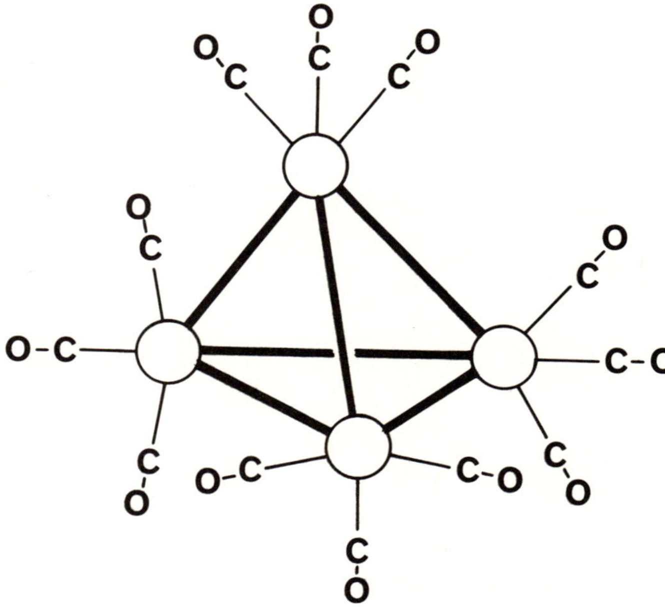

**Figure 5.4** The molecular geometry of $Ir_4(CO)_{12}$, which has all the carbonyl ligands terminally bonded

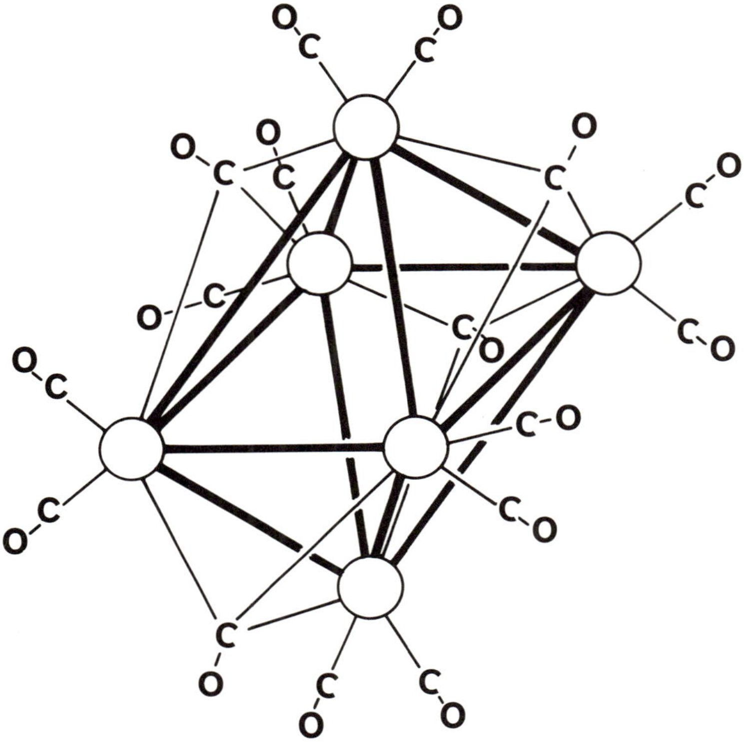

**Figure 5.5** The molecular stereochemistry of $Rh_6(CO)_{16}$. The molecule has $T_d$ symmetry

The other cluster compounds have deltahedral arrangements of metal atoms, i.e. the metal atoms occupy the vertices of polyhedra with almost equilateral triangular faces (see Figures 5.3–5.7). King[122] has suggested that this preference for deltahedra results because a triangle of metal atoms maximises the number of bonding electrons per metal atom and gives the 'minimum maximum metal–metal distance'. It was suggested also that transition metal cluster compounds are unlikely to contain more than six metal atoms because deltahedra with more than six vertices would require unusual angular strain or unusually high coordination numbers. However, recent work by Chini[123], Lewis[124] and their co-workers has shown that these assumptions are not valid: several cluster compounds with more than seven metal atoms and polyhedral clusters with triangular and *square* faces have been reported.

The tetrahedron and octahedron of metal atoms found in $M_4(CO)_{12}$ and $M_6(CO)_{16}$ (M = Co, Rh or Ir) are examples of polyhedra common to many branches of inorganic chemistry. The bicapped tetrahedron and capped octahedral arrangements of metal atoms found in $Os_6(CO)_{18}$ and $Os_7(CO)_{21}$ are less familiar (see Figures 5.6 and 5.7). It is interesting to note that the

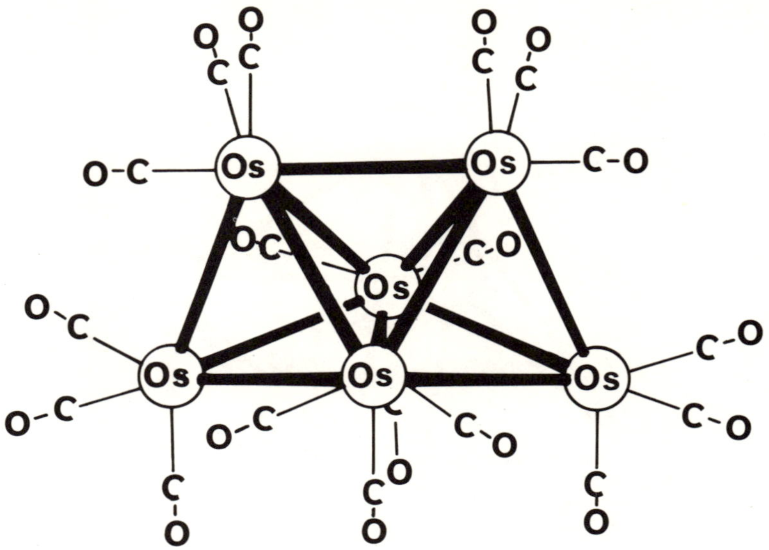


**Figure 5.6** The bicapped tetrahedral metal polyhedral geometry found in $Os_6(CO)_{18}$

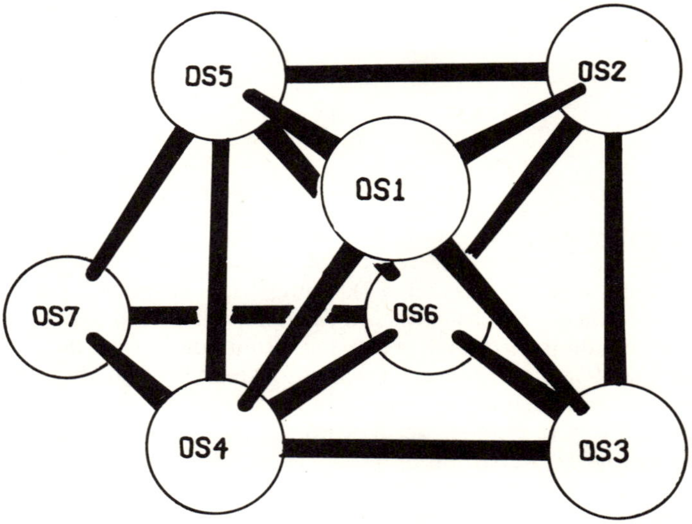


**Figure 5.7** The capped octahedron formed by the osmium atoms in $Os_7(CO)_{21}$

former compound has the same number of edges and faces as a regular octahedron and differs only in the order of the vertices; the capping metal atoms in these compounds lie above the sphere inscribing the metal atoms of the parent regular polyhedron.

The metal–metal distances in these polyhedra are generally somewhat larger than the sum of the metallic radii (for 12 coordination) (see Table 5.5)[123]. The difference is particularly marked for triangular and capped metal cluster compounds. In tetrahedral clusters the metal–metal distances are comparable with the sum of the covalent radii.

In carbonyl complexes of the lighter metals the metal–metal bonds are frequently supplemented by carbonyl bridges. In $Fe_3(CO)_{12}$ two carbonyl ligands bridge one edge of the triangle (Figure 5.1) and in $M_4(CO)_{12}$ the three basal edges of the tetrahedron are bridged (Figure 5.3). In $Rh_6(CO)_{16}$ (Figure 5.5) the carbonyl ligands bridge four of the triangular faces of the octahedron. It has been noted that in the dodecacarbonyl complexes the twelve oxygen atoms of the carbonyl ligands have an icosahedral arrangement which minimises non-bonded repulsions[126]. In $M_3(CO)_{12}$ (M = Ru or Os) and $Ir_4(CO)_{12}$, which have no bridging carbonyl ligands, the oxygen atoms occupy less sterically favourable polyhedra (*viz.* the truncated bipyramid and the truncated tetrahedron, respectively)[9]. The mixed cluster compounds $Co_2Rh_2(CO)_{12}$ and $Co_2Ir_2(CO)_{12}$ are isomorphous with $Co_4(CO)_{12}$ and presumably also have bridging carbonyl groups on the basal edges[118]. The formation of a carbonyl bridge is associated with a shortening of the bridged metal–metal bond (see Table 5.5) in triangular but not in tetrahedral clusters.

### *5.4.2.1 Bonding in metal carbonyl cluster complexes*

The large number of electrons in these molecules precludes rigorous all-electron SCF MO calculations which would establish a reliable picture for the bonding in these compounds. Qualitative and semiqualitative bonding models have been used with varying success to interpret the structural and synthetic results, the simplest bonding description being based on the rare gas 'rule'. This scheme, first formulated by Sidgwick[127], has proved most useful in rationalising the geometries and reactivities of mononuclear transition metal carbonyl and organometallic complexes. It has proved less satisfactory, however, when applied to polymetallic compounds. For tri- and tetra-nuclear carbonyl cluster compounds, canonical forms based on the rare gas 'rule' can easily be suggested as shown in (15) and (16) for $Fe_3(CO)_{12}$ and $Ir_4(CO)_{12}$.

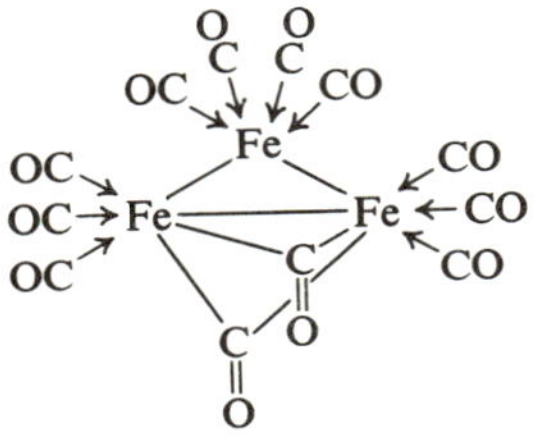

(15)

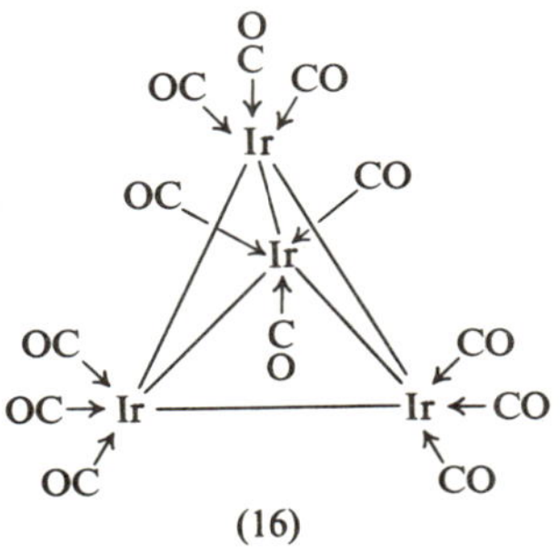

(16)

Metal–metal bonding in these cluster compounds is represented by two-electron two-centre bonds localised along the edges of the triangle and tetrahedron of metal atoms. The canonical forms contain, therefore, a definite topological bias. It is fortunate that for polygons and polyhedra of metal atoms with vertices of order three (the tetrahedron, prisms, etc.) the occupied delocalised skeletal MOs can be transformed into a set of equivalent orbitals[128, 129], each of which is localised on one of the edges. Molecules containing these metal polyhedra have, therefore, MOs which reflect the polyhedral topology in a simple way and the rare gas 'rule' can be applied successfully.

In metal carbonyl cluster compounds which have metal polyhedra with vertices of order four or more (the octahedron, antiprisms, etc.) the occupied MOs can no longer be transformed into a topologically significant set of equivalent orbitals and the inert gas rule or other topological approaches fail to give a good description of the bonding[128]. For example, MO calculations for the octahedral cluster anion $[Co_6(CO)_{14}]^{4-}$ have shown that the 54 valence orbitals of the cobalt atoms give rise to 43 bonding and weakly antibonding MOs which are occupied by the 86 valence electrons and 11 strongly antibonding unoccupied skeletal MOs[130]. As the octahedron has twelve edges, the antibonding MOs have no simple topological significance and cannot be assigned to particular edges according to an equivalent orbital procedure[131]. Canonical forms which assign metal–metal bonds to the twelve edges of the octahedron, therefore, predict a wrong closed-shell electronic configuration (with 84 valence electrons) for this type of molecule.

### *5.4.2.2 Polyhedral skeletal electron pair theory*[132–134]

Polyhedral skeletal electron pair theory retains the condition that each metal atom obeys the rare gas 'rule' but removes the topological bias noted above for the metal–metal bonding. The MO calculations for $[Co_6(CO)_{14}]^{4-}$ suggest that the metal atoms use three hybrid orbitals with predominant s and p character for forming the skeletal molecular orbitals. The d orbitals interact only weakly and, therefore, the cobalt atoms behave as pseudo main group atoms in this ion. Consequently, the results of MO calculations reported for borane anions and for carbon polyhedra[135] can be used to estimate the number of bonding skeletal MOs in metal carbonyl cluster compounds. These calculations suggest the following number of bonding skeletal MOs for metal polyhedra:

| | *Tetrahedron* | *Trigonal bipyramid* | (a) *Octahedron* (b) *Square pyramid* | (c) *Pentagonal bipyramid* (d) *Pentagonal pyramid* | (e) *Dodecahedron* (f) *Trigonal prism* |
|---|---|---|---|---|---|
| Number of bonding skeletal MOs | 6 | 6 | 7 | 8 | 9 |
| Main Group analogue | $P_4$ | $B_5H_5^{2-}$ | (a) $B_6H_6^{2-}$ (b) $B_5H_5^{4-}$ | (c) $B_7H_7^{2-}$ (d) $B_6H_6^{4-}$ | (e) $B_8H_8^{2-}$ (f) $C_6H_6$ (prismane) |

**Table 5.6 The number of electrons contributed for skeletal bonding by metal carbonyl, nitrosyl and cyclopentadienyl fragments**

| | | | | |
|---|---|---|---|---|
| | | Fe(CO) | Co(CO) | Ni(CO) |
| | | −2 | −1 | 0 |
| | | $Fe(CO)_2$ | $Co(CO)_2$ | $Ni(CO)_2$ |
| | | 0 | 1 | 2 |
| $Cr(CO)_3$ | $Mn(CO)_3$ | $Fe(CO)_3$ | $Co(CO)_3$ | $Ni(CO)_3$ |
| 0 | 1 | 2 | 3 | 4 |
| $Cr(CO)_4$ | $Mn(CO)_4$ | $Fe(CO)_4$ | $Co(CO)_4$ | |
| 2 | 3 | 4 | 5 | |
| | | Fe(NO) | Co(NO) | |
| | | −1 | 0 | |
| | $Mn(NO)_2$ | $Fe(NO)_2$ | $Co(NO)_2$ | $Ni(NO)_2$ |
| | 1 | 2 | 3 | 4 |
| | $Mn(\eta\text{-}C_5H_5)$ | $Fe(\eta\text{-}C_5H_5)$ | $Co(\eta\text{-}C_5H_5)$ | $Ni(\eta\text{-}C_5H_5)$ |
| | 0 | 1 | 2 | 3 |

Generally, the number of electrons contributed by $M(CO)_n$ fragments for skeletal bonding in the cluster $[M(CO)_n]_x$ is $m + 2n - 12$ (where $m$ = number of valence electrons of metal M)[134]. For example, the $Fe(CO)_3$ fragment contributes two electrons for skeletal bonding because $n = 3$ and iron has 8 valence electrons. Table 5.6 summarises the number of electrons contributed by a variety of metal carbonyl, cyclopentadienyl and nitrosyl fragments[134].

The skeletal molecular orbitals formed from the $sp^n$ hybrids on the metal atoms also have the correct symmetry to interact with bridging groups lying on the faces and edges of the polyhedron. These bridging groups contribute electrons for skeletal bonding and their contributions are summarised in Table 5.7.

The bridging groups can be main group atoms, e.g. P, S, H, multiple bonded ligands, e.g. CO, $\pi\text{-}C_5H_5$, and even metal carbonyl fragments, e.g. $Os(CO)_3$ in capped polyhedra.

For the polyhedral metal carbonyl cluster compounds, skeletal electron pair counting, according to the rules outlined above, leads to the following conclusions:

| | | |
|---|---|---|
| 6 skeletal electron pairs | $M_4(CO)_{12}$ (M = Co, Rh and Ir) Tetrahedron | $Os_6(CO)_{18}$ Bicapped tetrahedron |
| 7 skeletal electron pairs | $M_6(CO)_{16}$ (M = Co, Rh or Ir) Octahedron | $Os_7(CO)_{21}$ Capped octahedron |

The capped polyhedra have the same number of skeletal electron pairs as the parent platonic polyhedra and the capping $Os(CO)_3$ fragments may be considered as bridging groups, each contributing three orbitals and two electrons for skeletal bonding[120], i.e. $Os(CO)_3$ is pseudo isoelectronic with CO.

**Table 5.7 The number of electrons contributed by bridging groups for skeletal bonding**

| | | | | | |
|---|---|---|---|---|---|
| (*a*) *Edge bridging groups* | | | | | |
| | H, Me, Ph | CO | NO, $PR_2$, SEt | S | |
| | 1 | 2 | 3 | 4 | |
| (*b*) *Bridging groups on triangular faces* | | | | | |
| | H, Me, Ph | CO | NO, P | S, NR | RS, $\eta$-$C_5H_5$ |
| | 1 | 2 | 3 | 4 | 5 |

Polyhedral skeletal electron pair theory suggests that in *triangulo* metal clusters there will be six skeletal electron pairs occupying the skeletal MOs formed from three hybrid orbitals on each metal atom. Symmetry arguments suggest that four of these MOs are bonding and two antibonding (*cf.* cyclo-$C_3H_3^{3-}$)[136] and consequently the highest occupied orbitals are likely to be antibonding between the metals and localised predominantly above and below the triangular plane. This important result has many implications which we will discuss in detail below.

This qualitative electron counting scheme is useful in accounting for the metal polyhedral geometries but a detailed interpretation of metal–metal and metal–ligand bond lengths and spectroscopic properties requires more quantitative calculations; Dahl and his co-workers have reported such schemes for the metal chalcogenide cluster complexes $Co_3S(CO)_9$ [137] and $Fe_4S_4(\pi$-$C_5H_5)_4$ [138].

### 5.4.3 Substituted metal carbonyl cluster compounds

The carbonyl ligands in metal carbonyl cluster compounds can be replaced with tertiary phosphine, alkyne or carbocyclic ligands. The majority of the structural results on these complexes can be accommodated within the qualitative valence models described above.

#### *5.4.3.1 Tertiary phosphine complexes*

Tertiary phosphine complexes, characterised by x-ray crystallography, are given in Table 5.8 and are seen to be structurally closely related to the parent carbonyl cluster compounds. For example, $Fe_3(CO)_{12}$, $Fe_3(CO)_{11}(PPh_3)$ and $Fe_3(CO)_9(PMe_2Ph)_3$ have the same metal and ligand coordination geometries and the corresponding metal–metal bond lengths are very similar. In $Ir_4(CO)_{10}(PPh_3)_2$ and $Ir_4(CO)_9(PPh_3)_3$ the tetrahedral metal geometry of the parent carbonyl is retained, but three of the carbonyl ligands occupy bridging positions in the basal plane and these substituted derivatives accordingly resemble $Co_4(CO)_{12}$ and $Rh_4(CO)_{12}$ more closely. In addition, the metal–metal bond lengths in the substituted derivatives are somewhat longer (0.05 Å) than those reported for the parent carbonyl[141].

**Table 5.8 Summary of x-ray data for tertiary phosphine and phosphite carbonyl cluster complexes**

| | *Metal–metal bond lengths* | *No. of skeletal electron pairs* | *Ref.* |
|---|---|---|---|
| *Triangular metal geometry* | | | |
| $Fe_3(CO)_{11}(PPh_3)$ | 2.565(10)*, 2.702(10)* (average for 2 isomers) | 6 | 103 |
| $Fe_3(CO)_9(PMe_2Ph)_3$ | 2.540(5)*, 2.688(7)* | 6 | 102 |
| $Fe_2Pt(CO)_9(PPh_3)$ | 2.526(3)† 2.605(5)†, 2.780(5)* | 5 | 139 |
| $FePt_2(CO)_5\{P(OPh)_3\}_3$ | 2.550(5)† 2.583(6)†, 2.633(1)‡ | 4 | 140 |
| *Tetrahedral metal geometry* | | | |
| $Ir_4(CO)_9(PPh_3)_3$, $Ir_4(CO)_{10}(PPh_3)_2$ | 2.73(6) | 6 | 141 |
| $Ni_4(CO)_6\{P(CH_2CN)_3\}_4$§ | 2.51(1) | 6 | 142 |
| *Butterfly shape* | | | |
| $Pt_4(CO)_5(PPhMe_2)_4$‖ | 2.750(6)–2.790(7) | 5 | 143 |

* Fe—Fe, † Fe—Pt, ‡ Pt—Pt bond lengths; § see Figure 5.8; ‖ see Figure 5.9

The parent compounds for $Ni_4(CO)_6\{P(CH_2CN)_3\}_4$ [142] and $Pt_4(CO)_5$-$(PMe_2Ph)_4$ [143], which have carbonyl bridges supplementing all the metal–metal bonds, are unknown. The former has a tetrahedron of metal atoms, in agreement with the six skeletal electron pairs assigned to this molecule (see Figure 5.8). The latter compound represents an example of both the polyhedral skeletal electron pair theory and rare gas 'rule' failing to predict the

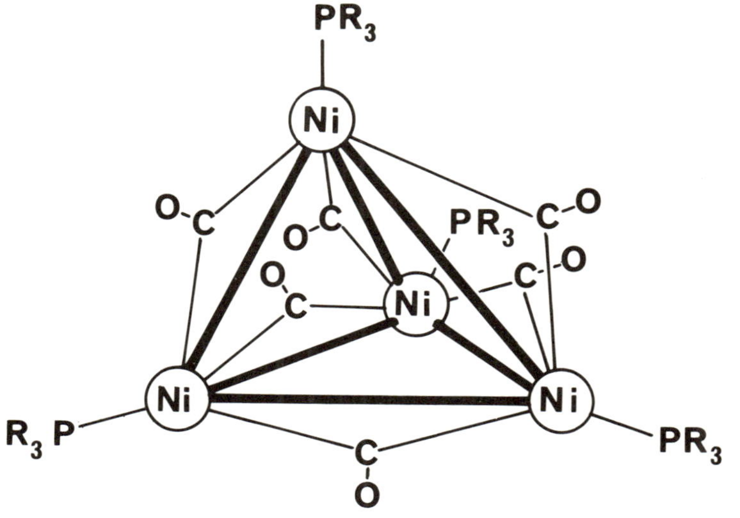


**Figure 5.8** The structure of $Ni_4(CO)_6\{P(CH_2CN)_3\}_4$. The six carbonyl ligands bridge each of the edges of the tetrahedron formed by the nickel atoms

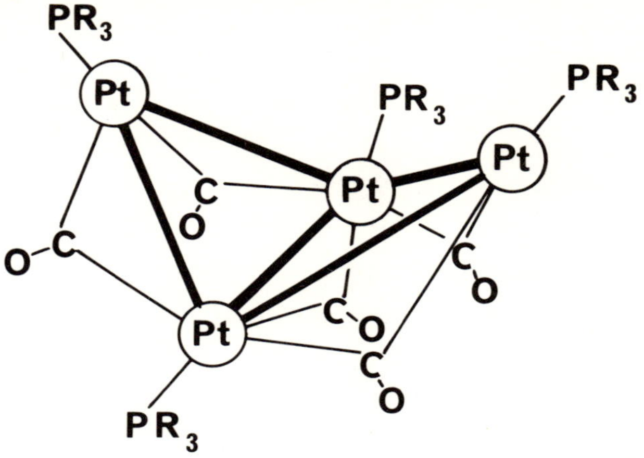


**Figure 5.9** The 'butterfly' geometry exhibited by $Pt_4(CO)_5(PPhMe_2)_4$. Each of the Pt—Pt bonds is supplemented by a carbonyl bridge

'butterfly' geometry found in this molecule (see Figure 5.9). It is interesting to note that in this compound the carbonyl ligands bridge platinum–platinum bonds whose lengths[143] are greater than 2.75 Å.

The mixed metal cluster compounds $PtFe_2(CO)_9(PPh_3)$ [139] and $Pt_2Fe(CO)_5P(OPh_3)_3$ [140] have fewer valence electrons than the $M_3(CO)_{12}$ carbonyls and support the contention that the highest occupied MOs in the triangular cluster compounds are antibonding with respect to metal–metal bonding. The contraction of the metal triangle is best illustrated by comparing the sum of the metal–metal bond lengths with the estimated total bond length from the metallic covalent radii:

| | $Pt_2Fe(CO)_5\{P(OPh)_3\}_3$ | $PtFe_2(CO)_9(PPh_3)$ | $Fe_3(CO)_{12}$ |
|---|---|---|---|
| No. of valence electrons | 44 | 46 | 48 |
| (skeletal electron pairs) | (4) | (5) | (6) |
| Sum of metal–metal bond lengths | 7.76 Å | 7.84 Å | 7.90 Å |
| Estimated sum from metallic radii[125] | 8.08 Å | 7.86 Å | 7.56 Å |
| Difference | −0.32 Å | −0.02 Å | 0.34 Å |

A detailed account of the individual bond lengths in these compounds, however, awaits an analysis of the relative contributions of the platinum and iron valence orbitals to the overall bonding scheme for these triangular clusters. Directional *trans* influences, reminiscent of those in platinum(II) complexes, have been noted for the $PtFe_2(CO)_9(Ph_3P)$ complex[139].

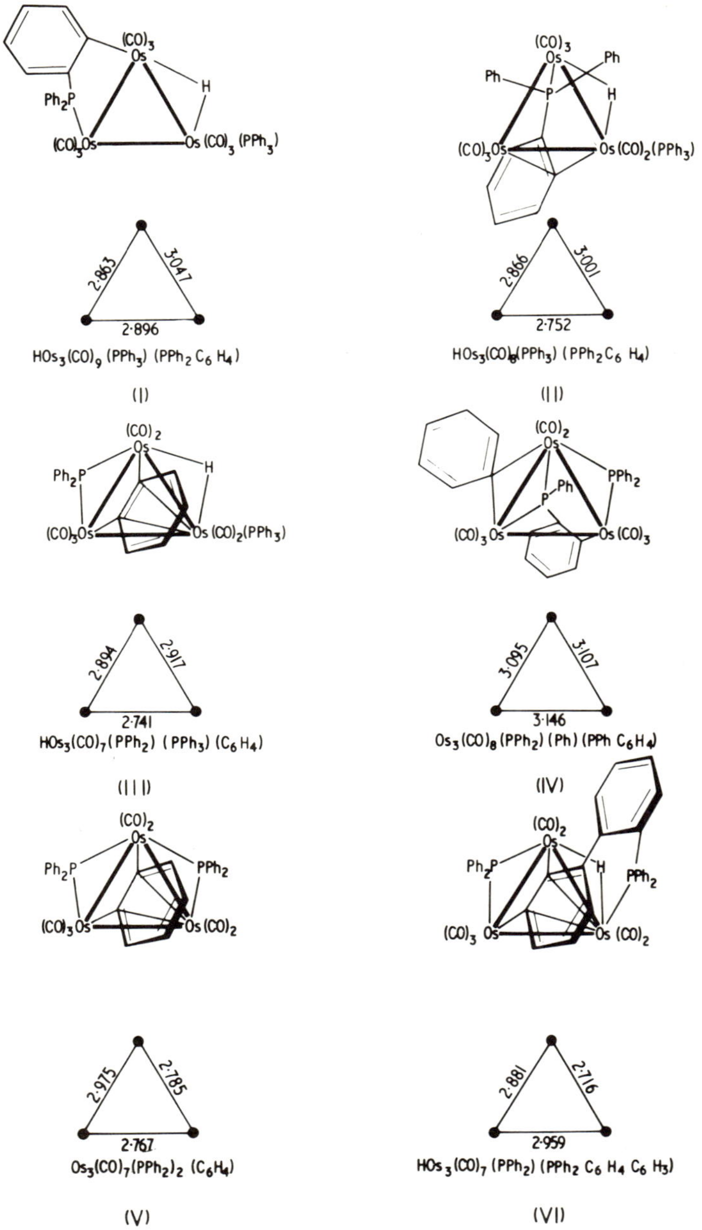


**Figure 5.10** The diagram summarises the x-ray results of the products from the $Os_3(CO)_{12}/PPh_3$ reaction. The triangles below the stereochemical illustrations indicate the Os—Os bond lengths. The standard deviations on these bond lengths are 0.002 Å

### *5.4.3.2 Degradation of triphenylphosphine in osmium carbonyl clusters*

Dodecacarbonyltriosmium(0) reacts with triphenylphosphine in the mole ratio 1:2 to give a mixture of compounds which have been separated by t.l.c. and fully characterised by x-ray crystallography. Three of the ten compounds isolated are normal substituted derivatives $Os_3(CO)_{12-n}(PPh_3)_n$ ($n = 1$, 2 or 3)[144]. Six of the remaining compounds result from successive oxidative-addition reactions and are represented schematically in Figure 5.10[145, 146]. It is likely that (I) is formed by the reaction of a C—H bond of one triphenylphosphine ligand coordinated to one osmium atom with a second osmium atom to form a five-membered chelate ring. Presumably (I) then loses carbon monoxide to form (II) which in turn decomposes to (III), (IV) and (V). Compound (VI) is probably formed from (III) by elimination of hydrogen from the $C_6H_4$ ligand and an *ortho*-carbon–hydrogen bond of the terminal phosphine ligand with the formation of a carbon–carbon bond.

In compounds (I), (II), (III) and (VI) the hydridic hydrogen atoms were not directly located in the x-ray analyses but considerations based on the overall metal–ligand geometries, metal–metal bond lengths and the rare gas 'rule' suggest the positions shown in the figure.

The dimethylphenylphosphine derivative analogous to (V) has been prepared and its proton n.m.r. spectrum indicates that the $C_6H_4$ (benzyne) ligand undergoes rapid rotation relative to the $Os_3$ ring, a process which necessitates a carbonyl transfer between osmium atoms in the cluster compound[147].

### *5.4.3.3 Metal clusters based on alkyne ligands*

Alkynes are very versatile ligands and produce a variety of metal cluster complexes[64]. In many of these complexes the alkyne can be considered as a simple four-electron donor group capping a triangle of metal atoms, but in others it is instructive to consider the complete polyhedron formed between the metal atoms and the acetylenic carbon atoms. This dual behaviour of alkynes depends on whether the ligand caps a triangular face of a metal polyhedron (17) or an edge and face of the polyhedron (18). In the former

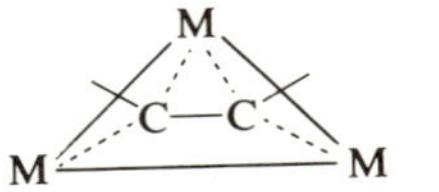


(17)

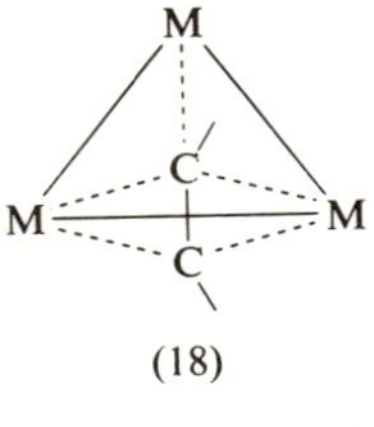


(18)

each carbon atom is bonded to two metal atoms and the four $p_\pi$-electrons have a topological significance because each metal–carbon bond can be assigned an electron pair. Complexes of type (17) can, therefore, be rationalised within the framewo-k of the rare gas 'rule' or polyhedral skeletal electron

**Table 5.9** **Acetylene complexes which have the acetylene ligand capping a triangle of metal atoms**

| *Compound* | *Figure* | *Metal–metal bond lengths/Å* | *No. of skeletal electron pairs* | *Ref.* |
|---|---|---|---|---|
| $Fe_3(CO)_8(PhCCPh)_2$ (violet isomer) | 5.11 | 2.469(3) 2.457(3) 2.592(3) | 6 | 148 |
| $Os_3(CO)_7(PPh_2)_2(C_6H_4)$ | 5.10(V) | 2.767(2) 2.785(2) 2.975(2) | 6 | 144 |
| $Os_3H(CO)_7(PPh_2)(PPh_3)(C_6H_4)$ | 5.10(III) | 2.741(2) 2.894(2) 2.917(2) | 6 | 145 |
| $Os_3H(CO)_7(PPh_2)(PPh_2C_6H_4C_6H_3)$ | 5.10(VI) | 2.716(2) 2.881(2) 2.959(2) | 6 | 145 |
| $Ru_3(CO)_9(C_8H_{12})H_2$ | 5.12 | 2.995(2) 2.725(2) 2.845(2) | 6 | 149, 150 |

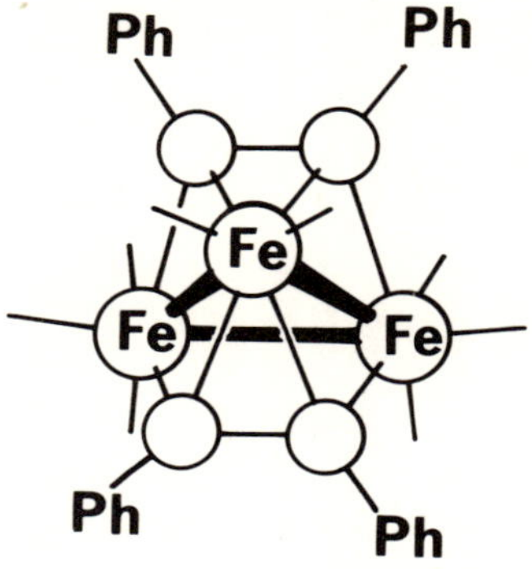


**Figure 5.11** The geometry of $Fe_3(CO)_8(Ph_2C_2)_2$ (violet isomer). The faces of the triangle of metal atoms are capped by the two diphenylacetylene ligands

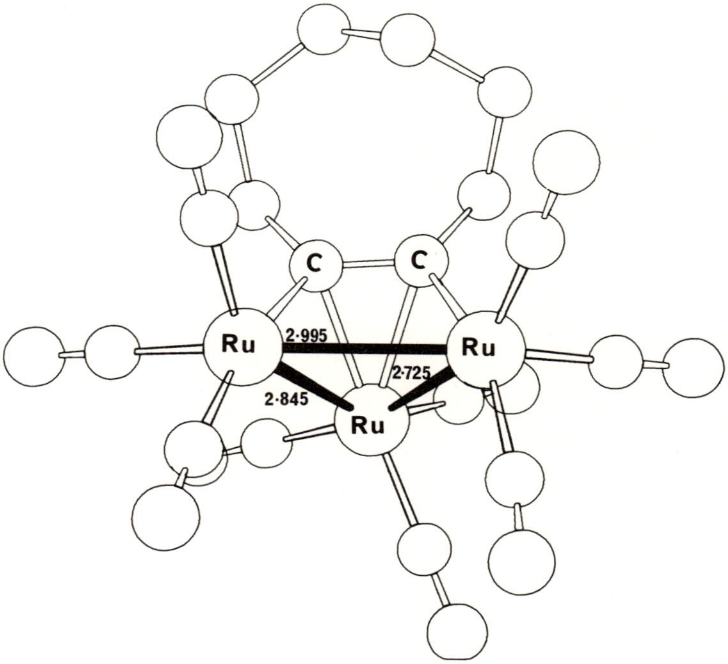


**Figure 5.12** The molecular stereochemistry of $Ru_3H_2(CO)_9(C_8H_{12})$. The ruthenium–ruthenium bond lengths are indicated (e.s.d. 0.002 Å)

**Table 5.10 Acetylene complexes which have the acetylene ligand capping a face and an edge of the metal polyhedron**

| *Compound* | *Ref.* | *Metal–carbon polyhedron*[134, 154] | *No. of skeletal electron pairs* | *Figure* |
|---|---|---|---|---|
| $Co_2(CO)_6(Ph_2C_2)$ | 151 | Tetrahedron ($Co_2C_2$) | 6 | |
| $Fe_3(CO)_9(Ph_2C_2)$ | 152 | Trigonal bipyramid ($Fe_3C_2$) | 6 | 5.13 |
| $Co_4(CO)_{10}(Et_2C_2)$ | 153 | Octahedron ($Co_4C_2$) | 7 | 5.14 |
| $Fe_3(CO)_8(Ph_2C_2)_2$ (violet isomer) | 148 | Pentagonal bipyramid ($Fe_3C_4$) | 8 | 5.15 |

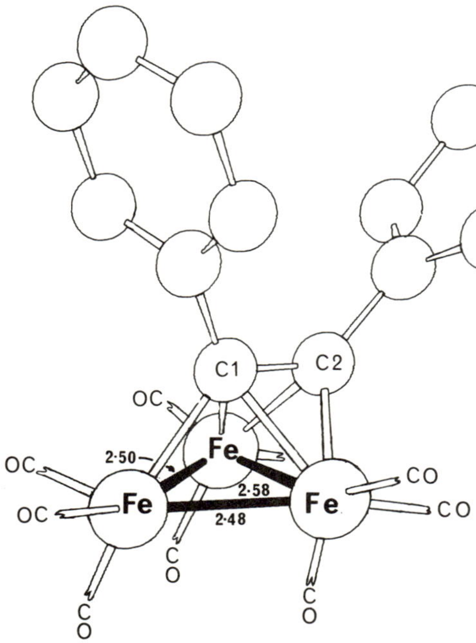


**Figure 5.13** The molecular geometry of $Fe_3(CO)_9(Ph_2C_2)$. The iron and acetylenic carbon atoms define a distorted trigonal bipyramid

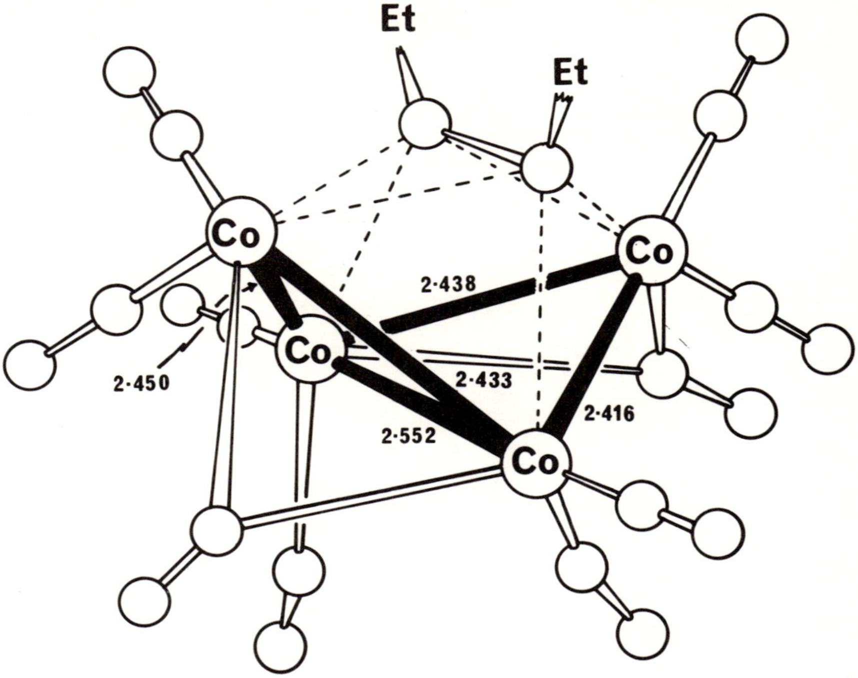


**Figure 5.14** Stereochemistry and metal–metal bond lengths in $Co_4(CO)_{10}(Et_2C_2)_2$. The cobalt and acetylenic carbon atoms define a distorted octahedron. The four cobalt atoms adopt a 'butterfly' geometry

pair theory (see Table 5.9). It should be added that the benzyneosmium complexes encountered in Section 5.4.3.2 can be classified as substituted alkyne complexes. In (18), one carbon is attached to three metal atoms and the other to two metal atoms; consequently, there is no longer a topological link between the number of carbon orbitals and electrons and the number of metal–carbon bonds. It is therefore necessary to consider a more delocalised bonding scheme based on the total polyhedron formed between the metal and acetylenic carbon atoms. Thus, $Fe_3(CO)_9(Ph_2C_2)$, $Co_4(CO)_{10}(Et_2C_2)$ and $Fe_3(CO)_8(Ph_2C_2)_2$ (black isomer) can best be described as trigonal bipyramidal, octahedral and pentagonal bipyramidal metal–carbon polyhedra, respectively (see Figures 5.13–5.15)[148, 151-153]. Each of these complexes has the predicted number of skeletal electron pairs as shown in Table 5.10. $Ru_4(CO)_{11}(C_8H_{10})$ (Figure 5.16) has a structure which is closely related to that of $Co_4(CO)_{10}(Et_2C_2)$.

In $Os_3(CO)_7(Ph_2C_2)(Ph_4C_4)$ [155] one diphenylacetylene ligand caps the osmium triangle according to coordination mode (17) and the other two $Ph_2C_2$ fragments form a pentagonal pyramidal metal–carbon polyhedron with two of the osmium atoms (see Figure 5.17). $Os_3(CO)_8(Ph_4C_4)$ [156] has a very closely related molecular geometry. In addition, one of the ligand phenyl

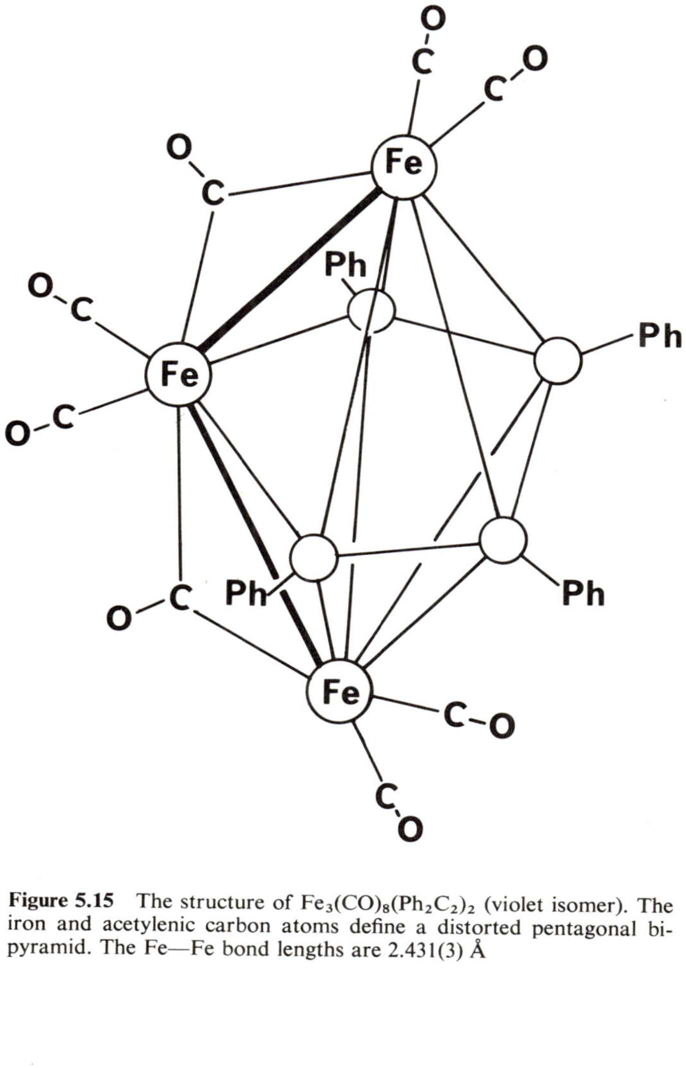


**Figure 5.15** The structure of $Fe_3(CO)_8(Ph_2C_2)_2$ (violet isomer). The iron and acetylenic carbon atoms define a distorted pentagonal bipyramid. The Fe—Fe bond lengths are 2.431(3) Å

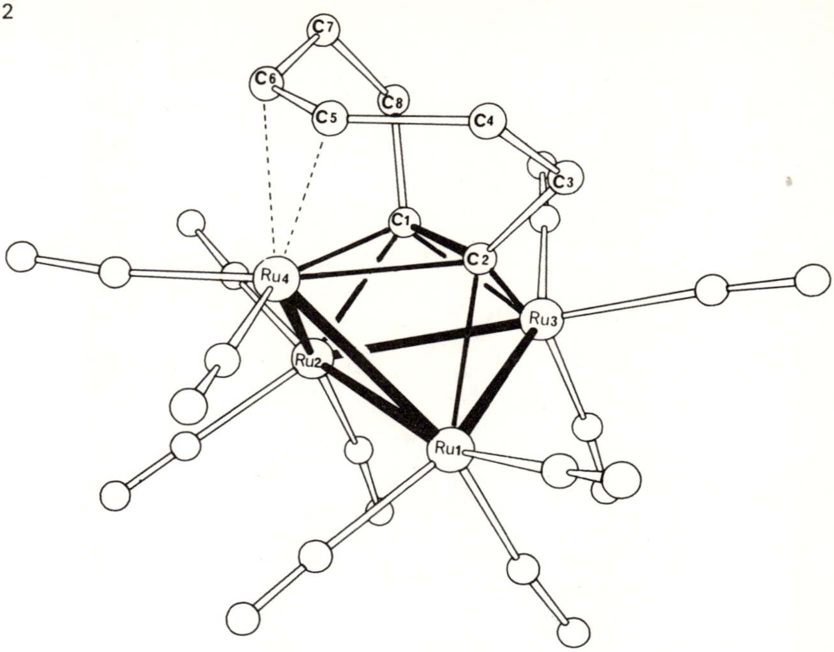


**Figure 5.16** The coordination geometry of $Ru_4(CO)_{11}(C_8H_{10})$ which closely resembles that of $Co_4(CO)_{10}(Et_2C_2)$ (Figure 5.14)

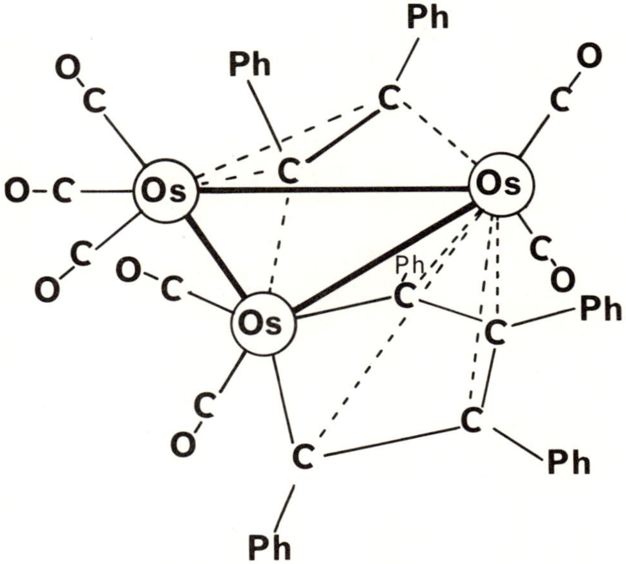


**Figure 5.17** The molecular stereochemistry of $Os_3(CO)_7(Ph_2C_2)$-$(Ph_4C_4)$

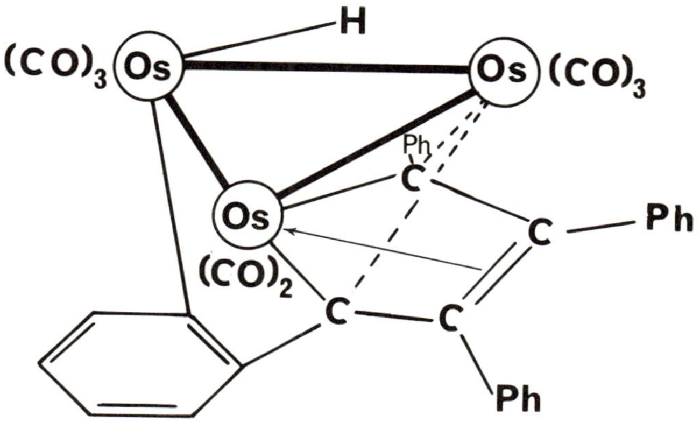


**Figure 5.18** The molecular geometry of $Os_3(CO)_8(Ph_4C_4)$

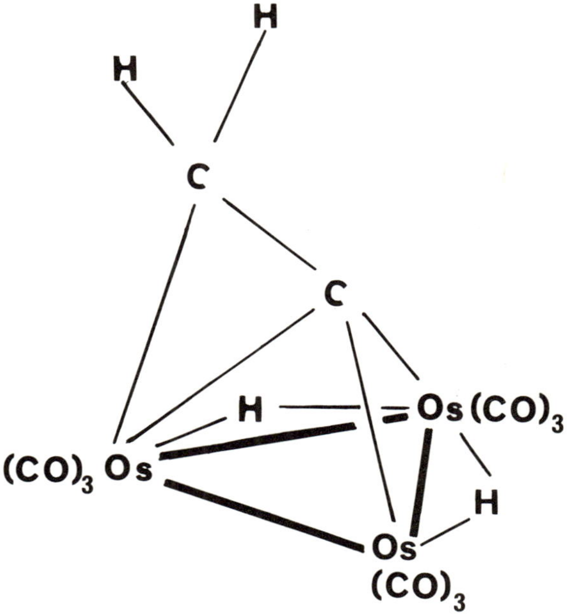


**Figure 5.19** The suggested structure for the vinylidene complex $Os_3H_2(CCH_2)(CO)_9$

rings has undergone an oxidative-addition reaction to give an *ortho*-metallated bridging group and hydride bridging two osmium atoms (Figure 5.18).

Even ethylene undergoes a rearrangement reaction with $Os_3(CO)_{12}$ to give $Os_3H_2(CCH_2)$, as shown in Figure 5.19[157]. The fluxional behaviour of this molecule is best accounted for by the arrangement of atoms shown in Figure 5.19.

### 5.4.3.4 *Cyclopentadienyl, arene and allylic carbonyl cluster compounds*

Examples of cyclopentadienyl metal carbonyl compounds are given in Table 5.11. Two isomers of the *triangulo* cluster $Rh_3(\eta\text{-}C_5H_5)_3(CO)_3$ have been identified[158, 159]: the more symmetrical isomer has the shorter average metal–metal bond lengths. In $Ni_3(\eta\text{-}C_5H_5)_3(CO)_2$ the carbonyl ligands lie above and below the triangular plane of metal atoms; polyhedral skeletal electron pair theory suggests that the unpaired electron in this molecule occupies an anti-bonding metal–metal MO, a conclusion which is supported by the shorter average bond length in $Ni_2Co(\eta\text{-}C_5H_5)_3(CO)_2$, which has one less valence electron[160, 161].

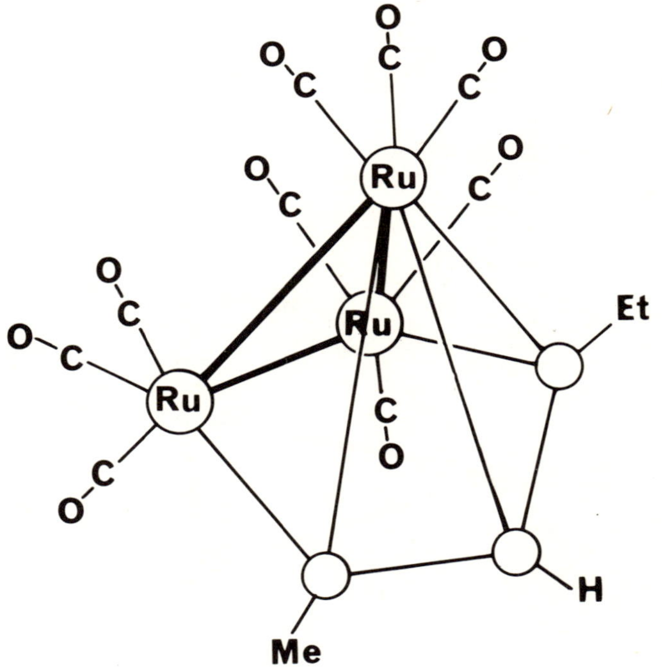


**Figure 5.20** The molecular stereochemistry of $Ru_3H(CO)_9(C_6H_9)$. The ruthenium and acetylenic carbon atoms define a pentagonal pyramid. The hydrogen atom was not located in the x-ray structural determination

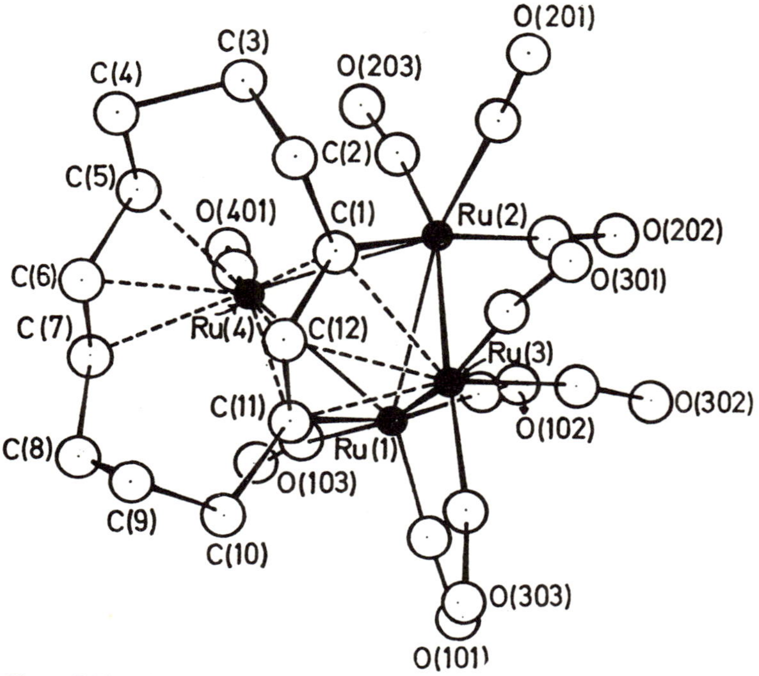


**Figure 5.21** The structure of $Ru_4(CO)_{10}(C_{12}H_{16})$. (Reproduced from Belford *et al.*[174], by courtesy of the Chemical Society)

$Fe_4(\eta\text{-}C_5H_5)_4(CO)_4$ has a tetrahedron of bonded iron atoms with each face bridged by a carbonyl ligand. The cation $[Fe_4(CO)_4(\eta\text{-}C_5H_5)_4]^+$ formed on oxidation of the parent compound shows a distortion of the iron tetrahedron towards $D_{2d}$ symmetry, the average iron–iron distance of 2.48 Å being slightly smaller for the cation. Dahl and his co-workers have rationalised this change in terms of a qualitative MO scheme which suggests that the electron lost on oxidation has been removed from a MO which is slightly antibonding between the metal atoms and the carbon and oxygen atoms[138, 163].

$Rh_2(\eta\text{-}C_5H_5)_2Fe_2(CO)_8$ and $RhFe_3(\eta\text{-}C_5H_5)(CO)_{11}$ are isovalence-electronic with $Fe_4(CO)_4(\eta\text{-}C_5H_5)_4$ and have distorted tetrahedra of metal atoms. The carbonyl bridges in these compounds are markedly asymmetric, perhaps for reasons similar to those mentioned earlier for dinuclear species[164-166].

Arene, azulene, cycloheptatriene and cyclo-octadiene derivatives of the metal carbonyl cluster compounds have been reported and the structural results are summarised in Table 5.12. For the substituted $Ru_3(CO)_{12}$ derivatives it is interesting to note that the average metal–metal bond lengths are considerably longer than that found for the parent carbonyl. The organo ligands mentioned above are relatively poor Lewis acids and hence less able to remove anti-bonding electron density from the metal triangle. In this context it is pertinent to note that the carbonyl bridged metal–metal bond in $Ru_3(CO)_7(C_{10}H_8)$[169,170] is 0.2 Å shorter than the non-bridged metal–metal bonds.

**Table 5.11 Summary of x-ray data for carbonyl, nitrosyl and cyclopentadienyl metal cluster compounds**

| | *Metal–metal bond lengths/Å* | *No. of skeletal electron pairs* | *Ref.* |
|---|---|---|---|
| *Triangular metal clusters* | | | |
| *sym*-$Rh_3(CO)_3(\eta\text{-}C_5H_5)_3$ | 2.62(3) | 6 | 158 |
| *asym*-$Rh_3(CO)_3(\eta\text{-}C_5H_5)_3$ | 2.62, 2.71, 2.66(1) | 6 | 159 |
| $Ni_3(CO)_2(\eta\text{-}C_5H_5)_3$ | 2.389(2) | $6\frac{1}{2}$ | 160 |
| $CoNi_2(CO)_2(\eta\text{-}C_5H_5)_3$ | 2.358(2) | 6 | 161 |
| $Mn_3(NO)_4(\eta\text{-}C_5H_5)_3$ | 2.48 | 6 | 162 |
| *Tetrahedral metal clusters* | | | |
| $Fe_4(CO)_4(\eta\text{-}C_5H_5)_4$ | 2.520(6) | 6 | 163 |
| $Fe_4(CO)_4(\eta\text{-}C_5H_5)_4^+$ | 2.467(2)–2.506(2) | $5\frac{1}{2}$ | 138, 163 |
| $Rh_2Fe_2(CO)_8(\eta\text{-}C_5H_5)_2$ | 2.648(3) (Rh—Rh) 2.539(7) (Fe—Fe)<br>2.570(5) 2.598(5) (Fe—Rh) | 6 | 165 |
| $RhFe_3(CO)_{11}(\eta\text{-}C_5H_5)$ | 2.568(3), 2.615(3), 2.607(3) (Rh—Fe)<br>2.586(3), 2.553(3), 2.594(3) (Fe—Fe) | 6 | 166 |

**Table 5.12 Summary of x-ray results for organo-substituted metal carbonyl cluster complexes**

| | *Ru—Ru* | *Description of metal cluster* | *Figure* | *Ref.* |
|---|---|---|---|---|
| *Arene* | | | | |
| $Ru_6C(CO)_4$(mesitylene) | 2.854(7)–2.956(7) | Octahedral metal cluster with mesitylene ligand replacing 3 CO | | 167 |
| *Cyclo-octadiene* | | | | |
| $Ru_3(CO)_4(C_8H_8)_2$ | 2.947(2)<br>2.928(2)<br>2.782(2) | Two cyclo-octadiene ligands replacing 8 CO from $Ru_3(CO)_{12}$ | | 168 |
| *Azulene* | | | | |
| $Ru_3(CO)_7(C_{10}H_8)$ | 2.937(4)<br>2.949(4)<br>2.740(4) | Ruthenium triangle with 1 bridge CO Azulene ligand over metal triangle to provide $10p_\pi$ electrons for cluster bonding | | 169 |
| $Ru_4(CO)_9\{4,6,8\text{-}(Me_3C_{10}H_5)\}$ | 2.702(5)–2.902(5)*<br>2.698(3)–2.908(3)† | Ruthenium tetrahedron; substituted azulene arched across largest face of an irregular tetrahedron | | 170 |
| *Cycloheptatriene* | | | | |
| $Ru_3(CO)_6(C_7H_7)(C_7H_9)$ | 2.840, 2.836<br>2.864 | Ruthenium triangle; 2 carbonyl and $C_7H_9$ ring bridging edges | 5.22 | 171 |
| *Cyclododecatrienyl* | | | | |
| $HRu_3(CO)_9(C_{12}H_{15})$ | 2.78(4)<br>2.78(4)<br>2.93(4) | The metal and allylic carbon atoms define a distorted pentagonal pyramid, consistent with the presence of 8 skeletal electron pairs | | 172 |
| *Hexadienyl* | | | | |
| $HRu_3(CO)_9(C_6H_9)$ | 2.783(5)<br>2.786(5)<br>2.964(6) | | 5.20 | 173 |
| $Ru_4(CO)_{10}(C_{12}H_{16})$ | 2.775(5)–2.850(6) | Ruthenium and allylic carbon atoms define a distorted pentagonal bipyramid (8 skeletal electron pairs) | 5.21 | 174 |

* Monoclinic form; † triclinic form

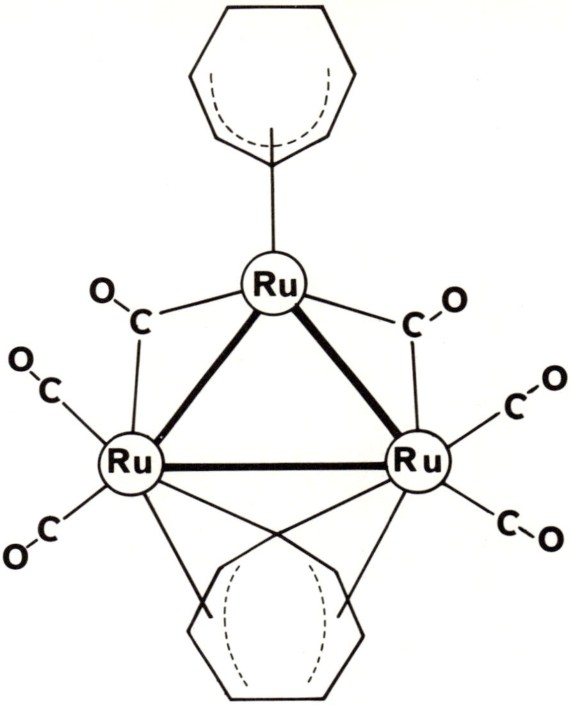


**Figure 5.22** The molecular structure of the cycloheptatriene ruthenium complex $Ru_3(CO)_6(C_7H_7)(C_7H_9)$

$Ru_3H(CO)_9(C_{12}H_{15})$ and $Ru_3H(CO)_9(C_6H_9)$ have very similar geometries with a π-allylic fragment acting as a five-electron donor. The dimensions of the Ru(allylic $C_3$) fragment (see Figure 5.20) are consistent with the $C_3$ system π-bonding to one ruthenium atom and σ-bonding to the other two metal atoms. In $Ru_4(CO)_{10}(C_{12}H_{16})$ the open $C_{12}$ ring is attached to the $Ru_4$ 'butterfly' ring by means of an allyl group which is σ-bonded to the cluster hinge at both ends and π-bonded to the cluster tips on both sides. There is a further attachment from the ring to one of the butterfly rings by a second π-allyl group (Figure 5.21). Allylic moieties are also found in the cycloheptatriene complex $Ru_3(CO)_6(C_7H_7)(C_7H_9)$ (Figure 5.22). Azulene and methyl-substituted azulene complexes with *triangulo* and *tetrahedro* ruthenium clusters are known; the azulene is arched across the triangle of metal atoms and effectively donates ten π-electrons for skeletal bonding[169, 170].

### *5.4.3.5 Metal carbonyl anions*

The metal polyhedra observed in metal carbonyl cluster anions are essentially the same as those found in the isoelectronic neutral complexes (see Table 5.13). For example, the complexes $Co_6(CO)_{16}$, $Co_6(CO)_{15}^{2-}$ and $Co_6(CO)_{14}^{4-}$

**Table 5.13 Summary of x-ray results for metal carbonyl anion cluster compounds**

| *Compound* | *Metal geometry* | *No. of skeletal electron pairs* | *Metal–metal bond lengths/Å* | *Ref.* |
|---|---|---|---|---|
| $[Fe(py)_6][Fe_4(CO)_{13}]$ | Trigonal pyramidal | 6 | 2.50(3) (basal)<br>2.59(3) (axial) | 176 |
| $[NBu^t_4][Re_4(CO)_{16}]$ | Two fused equilateral triangles | 7 | 2.956(7)–3.024(7) | 177 |
| $[(Ph_3P)_2N][Mo_2Ni_3(CO)_{16}]$ | Trigonal bipyramidal | 8 | 2.327(10)–2.350(8) (Ni—Ni)<br>3.045(5)–3.172(5) (Mo—Ni) | 178 |
| $[(Ph_3P)_2N][W_2Ni_3(CO)_{16}]$ | | | 2.324(12)–2.360(12) (Ni—Ni)<br>3.032(5)–3.182(5) (W—Ni) | 178 |
| $Cs_2Co_6(CO)_{15}(H_2O)_3$ | Octahedral | 7 | 2.454(4)–2.532(4) | 179 |
| $K_4Co_6(CO)_{14}(H_2O)_6$ | Octahedral | 7 | 2.466(2)–2.534(2) | 180 |
| $[NBu^n_4][Rh_6(CO)_{15}I]$ | Octahedral | 7 | 2.717(5)–2.772(4) | 181 |
| $[NMe_4]_2[Rh_{12}(CO)_{30}]$* | Two linked octahedra | 7 | 2.68(1)–2.84(1) | 182 |
| $[NMe_4]_3[Rh_7(CO)_{16}]$ | Capped octahedron | 7 | 2.72(1)–2.81(1) | 183 |

* Molecular structure shown in Figure 5.23

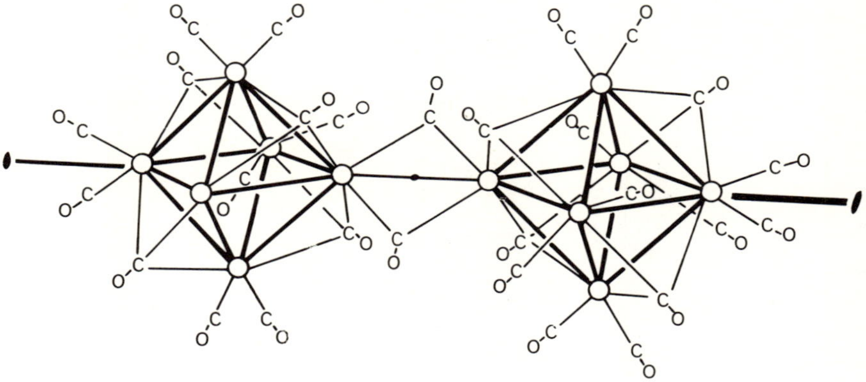

**Figure 5.23** The molecular structure of the $[Rh_{12}(CO)_{30}]^{2-}$ anion which contains two $Rh_6(CO)_{15}^-$ octahedral fragments joined by a metal–metal bond and carbonyl bridges. (Modified from Albano and Bellon[182], by courtesy of North Holland Publishing Co.)

have similar octahedral arrangements of metal atoms. The distribution of carbonyl ligands is, however, very different for these compounds. $Co_6(CO)_{16}$ has four carbonyl ligands bridging faces of the octahedron [by analogy with $Rh_6(CO)_{16}$, see Figure 5.5], $Co_6(CO)_{15}^{2-}$ has three edge and three face bridging carbonyls[179], and $Co_6(CO)_{14}^{4-}$ has eight face bridging carbonyls[180]. These results suggest that the additional negative charge is distributed primarily on the faces and edges of the polyhedron and supports Chini's contention that face bridging carbonyls are more effective at withdrawing electron density than edge bridging carbonyls, which in turn are more effective than terminal carbonyls in polynuclear species[9].

The anions $Re_4(CO)_{16}^{2-}$ and $W_2Ni_3(CO)_{16}^{2-}$ have one and two electron pairs respectively more than that expected for a tetrahedron and trigonal bipyramid polyhedra of metal atoms. The metal geometry found in $Re_4(CO)_{16}^{2-}$ is anticipated for such an electron-rich polyhedron and may be derived from a tetrahedron by breaking one edge bond and flattening the dihedron thus formed[134]. In $W_2Ni_3(CO)_{16}^{2-}$ the additional electron pairs cause an axial expansion of the trigonal bipyramid; the two tungsten atoms are at different distances from the triangle of nickel atoms to give the molecule $C_{3v}$ rather than $D_{3h}$ symmetry. Dahl and his co-workers have put forward a more detailed qualitative MO scheme to account for this distortion[178].

### 5.4.4 Metal carbonyl hydride cluster complexes

The complete stereochemical characterisation of metal carbonyl hydrides, including hydrogen atom locations, has been achieved in only a limited number of cluster complexes. The hydrogen atoms have been located by x-ray methods in $Mn_3H(CO)_{10}(B_2H_6)$ [185] and $Mn_3H_3(CO)_{12}$ [186] (see Figures 5.24 and 5.25). In $Mn_3H_3(CO)_{12}$ the hydrogen atoms are coplanar with the $Mn_3$ plane within $\pm0.2$ Å and are displaced from the centres of the Mn—Mn bonds by *ca.* 0.7 Å. The metal–hydrogen bond lengths average to 1.72(3) Å, which is directly comparable with that reported for the metal bridging hydrogen distance in $Mn_3H(CO)_{10}(B_2H_6)$ (1.65 Å). The fact that these distances are slightly longer than the terminal Mn—H distance in $HMn(CO)_3$ [1.60(1) Å] may reflect the lower bond order (formally one half) for the M—H bridge bonds. The average Mn—Mn bond length of 3.111(2) Å in $Mn_3H_3(CO)_{12}$

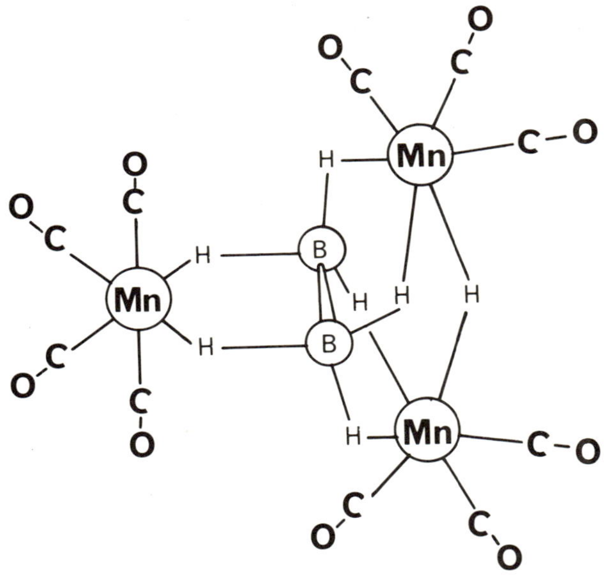


**Figure 5.24** The structure of $Mn_3H(B_2H_6)(CO)_{10}$

is substantially longer than the Mn—Mn bond length in $Mn_2(CO)_{10}$ (2.923 Å). Bridging hydrogen atoms have been located also in several bimetallic complexes[187-190].

Structural information on metal hydride cluster complexes can be obtained from $^1$H n.m.r. studies of solutions of the complexes in the nematic phase of a liquid crystal. Buckingham *et al.* have calculated Ru—H distances of 1.905 Å for the hydride $Ru_3H_3(CCH_3)(CO)_9$ on the basis of the geometry shown in Figure 5.26. Because the experimental dipolar couplings yield only ratios of internuclear distances (neglecting vibrational corrections), estimates for the C—C, Ru—C, Ru—Ru and H—H distances were made[191].

In the absence of definitive x-ray, neutron or n.m.r. measurements, the positions of the hydrogen atoms may be inferred from the asymmetric disposition of the remaining ligands or from asymmetric metal–metal bond

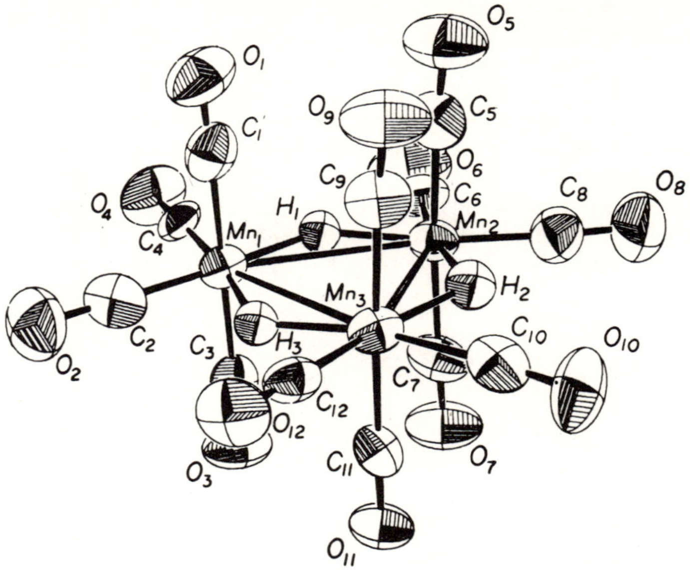


**Figure 5.25** The molecular geometry of $Mn_3H_3(CO)_{12}$. The hydrogen atoms were located in the x-ray structural determination. (Reproduced from Kirtley *et al.*[186], by courtesy of the American Chemical Society.)

lengths. In $Re_3H_2(CO)_{16}^-$ [192] and $Re_3H(CO)_{10}^{2-}$ [193] the metal atoms lie at the vertices of isosceles triangles and the hydrogen atoms are assigned to the longer edges. In $Ru_4H_2(CO)_{13}$ [194] and $FeRu_3H_2(CO)_{13}$ [195] the hydrogen atoms are thought to bridge two edges of the tetrahedra of metal atoms, a suggestion which is supported by the presence of asymmetric carbonyl bridges in these complexes. A symmetrical distribution of hydrogen atoms is predicted for $Re_4H_6(CO)_{12}^{2-}$ [196]. In $Fe_3H(CO)_{11}^-$ [197] one of the edges of the metal triangle is thought to be bridged by hydrogen and a carbonyl ligand. Similarly, $OsH(SEt)(CO)_{10}$ [198] is thought to have an osmium–osmium bond doubly bridged by H and SEt ligands. The osmium hydride cluster compounds which are isolated from the reaction of triphenylphosphine and $Os_3(CO)_{12}$ have been discussed earlier (Section 5.4.3.2). $Os_3(CO)_{10}H_2$ can be formulated with a double metal–metal bond according to the rare gas 'rule' and this bond is found to be 0.14 Å shorter than the other osmium–osmium bonds.

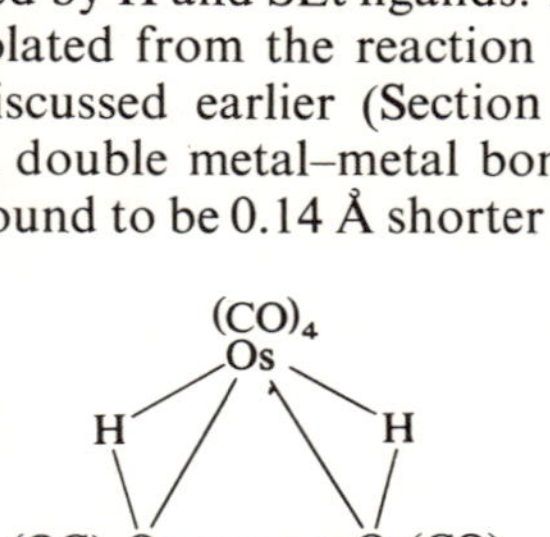


(19)

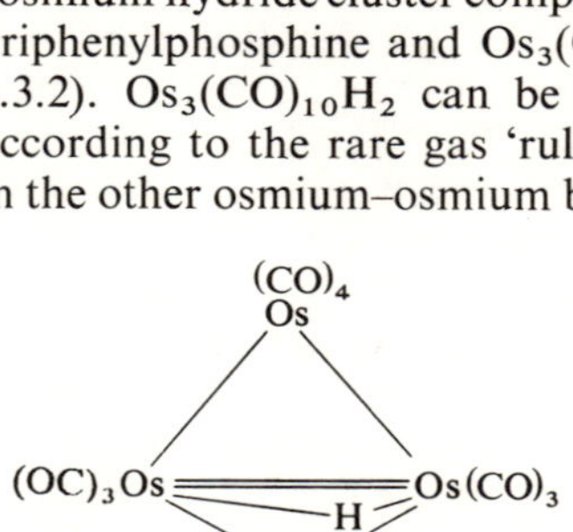


(20)

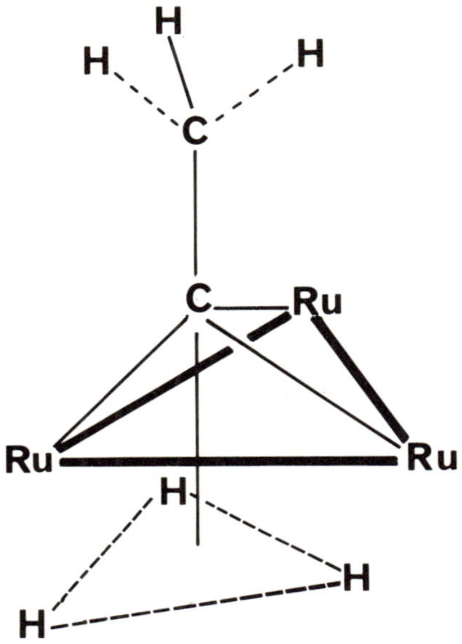


**Figure 5.26** The assumed molecular geometry used to calculate the metal–hydrogen distances from dipolar coupling n.m.r. data in the complex $Ru_3H_3CCH_3(CO)_9$

In this compound the hydrogen atoms can be associated with the two single osmium–osmium bonds (19) or the double bond (20). The latter is preferred, however, by analogy with $Os_3H(SEt)(CO)_{10}$[199].

In $Ru_6H_2(CO)_{18}$[200, 201] and $Rh_3H(\pi\text{-}C_5H_5)_3$[202, 203] the hydridic hydrogen atoms are thought to lie above or in the triangular faces. In $Ru_6H_2(CO)_{18}$ the distribution of metal–metal bond lengths suggests that the hydrogens bridge two *trans* triangular faces[204].

Neutron inelastic scattering studies[204] are also helpful for establishing the positions of hydrogen atoms relative to the metal skeleton. From such measurements it has been possible to suggest that the hydrogen atom in $Fe_3CoH(CO)_{12}$ lies at the centre of the metal tetrahedron. A similar suggestion has been made for $Fe_4H(CO)_{13}^-$[205] on the basis of Mössbauer and i.r. studies. It is pertinent to note in this context that neutron diffraction studies on $Nb_6HI_{11}$ have established unequivocally that the hydrogen atom lies at the centroid of the metal octahedron[206].

### 5.4.5 Metal chalcogenide and pnicogenide cluster complexes

The metal carbonyl cluster complexes and the substituted derivatives discussed above generally have closed polyhedra of metal atoms. The introduction of electron-rich sulphur or phosphorus atoms into the metal cluster has many interesting consequences; for example, more open polyhedral

**Table 5.14 Metal–metal bond lengths for some triangular metal carbonyl chalcogenides $MM'_2(CO)_9X$ (M = Co or Fe; M′ = Co; X = S, Se or Te)**

| *Compound* | *Metal–metal bond length*/Å | M—X/Å | *Ref.* |
|---|---|---|---|
| $Co_3(CO)_9S$ | 2.637(3) | 2.139(4) | 209 |
| $FeCo_2(CO)_9S$ | 2.554(3) | 2.159(4) | 208 |
| $Co_3(CO)_9Se$ | 2.616(1) | 2.282(1) | 207 |
| $FeCo_2(CO)_9Se$ | 2.577(1) | 2.285(1) | 207 |
| $FeCo_2(CO)_9Te$ | 2.598(2) | 2.466(1) | 207 |

networks of metal atoms are observed and the complexes show interesting redox properties.

Dahl and his co-workers have studied the complexes $Co_2MX(CO)_9$ (M = Co or Fe; X = S, Se or Te) in which a triangle of metal atoms is capped by a chalcogenide atom[207-209]. All the complexes have approximately $C_{3v}$ symmetry and in the mixed metal complexes this means that the metal atoms are randomly disordered. The mixed metal complexes have the 6 skeletal electron pairs expected for a bonded triangle of metal atoms but the tricobalt complexes have an additional electron which occupies an antibonding skeletal MO. This antibonding character is reflected in longer metal–metal bond lengths for the tricobalt complex and confirmed by more detailed MO calculations and e.s.r. measurements. The average metal–metal distance is also very dependent on the chalcogenide and increases regularly in the series S < Se < Te (see Table 5.14). Several bicapped triangular metal clusters

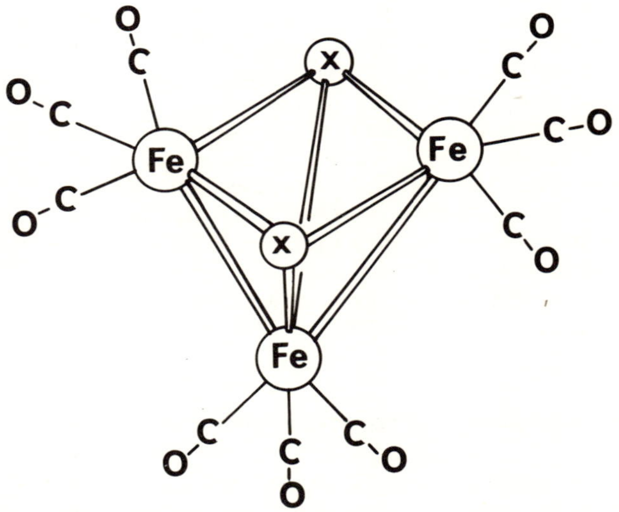


**Figure 5.27** The molecular structure of $Fe_3(CO)_9X_2$ (X = S, Se, NMe or NNPh)

**Table 5.15 Metal–metal bond lengths for the cluster complexes $M_3L_{3n}X_2$**

| | *Metal–metal bond length(s)/Å* | *Skeletal symmetry* | *No of skeletal electron pairs* | *No. of unpaired electrons* | *Ref.* |
|---|---|---|---|---|---|
| *Carbonyl complexes* | | | | | |
| $Fe_3(CO)_9As_2$ | 2.62(1) [3]* | $D_{3h}$ | 6 | 0 | 210 |
| $Fe_3(CO)_{10}(NSiMe_3)$ | 2.535(2) [3] | $D_{3h}$ | 6 | 0 | 211 |
| $Fe_3(CO)_9S_2$ | 2.60(1) [2] 3.37(1) [1] | $C_{2v}$ | 7 | 0 | 212 |
| $Fe_3(CO)_9Se_2$ | 2.65(2) [2] 3.51(2) [1] | $C_{2v}$ | 7 | 0 | 213 |
| $Fe_3(CO)_9(NMe)_2$ | 2.46(1) [2] 3.04(1) [1] | $C_{2v}$ | 7 | 0 | 214 |
| $Fe_3(CO)_9(NNCPh_2)_2$ | 2.38 [2] 3.06 [1] | $C_{2v}$ | 7 | 0 | 215 |
| *Cyclopentadienyl complexes* | | | | | |
| $CoNi_2(CO)_2(\eta\text{-}C_5H_5)_3$ | 2.358(2) [3] | $D_{3h}$ | 6 | 0 | 161 |
| $Co_3(CO)(O)(\eta\text{-}C_5H_5)_3$ | 2.365(4) [3] | $C_{3v}$ | 6 | 0 | 216 |
| $Co_3(CO)(S)(\eta\text{-}C_5H_5)_3$ | 2.452(2) [3] | $C_{3v}$ | 6 | 0 | 216 |
| $Ni_3(CO)_2(\eta\text{-}C_5H_5)_3$ | 2.389(2) [3] | $D_{3h}$ | $6\frac{1}{2}$ | 1 | 160 |
| $Co_3S_2(\eta\text{-}C_5H_5)_3^+$ | 2.591(average) | | | | |
| | 2.474(2) [1] 2.649(1) [2] | $C_{2v}$ | $6\frac{1}{2}$ | 1 | 216 |
| $Co_3S_2(\eta\text{-}C_5H_5)_3$ | 2.687(3) [3] | $D_{3h}$ | 7 | 2 at R.T.<br>0 below 173 K | 216 |
| $Ni_3S_2(\eta\text{-}C_5H_5)_3$ | 2.801(5) [3] | $D_{3h}$ | $8\frac{1}{2}$ | 1 | 217 |

* The number in square brackets represents the number of metal–metal bond lengths which have been averaged

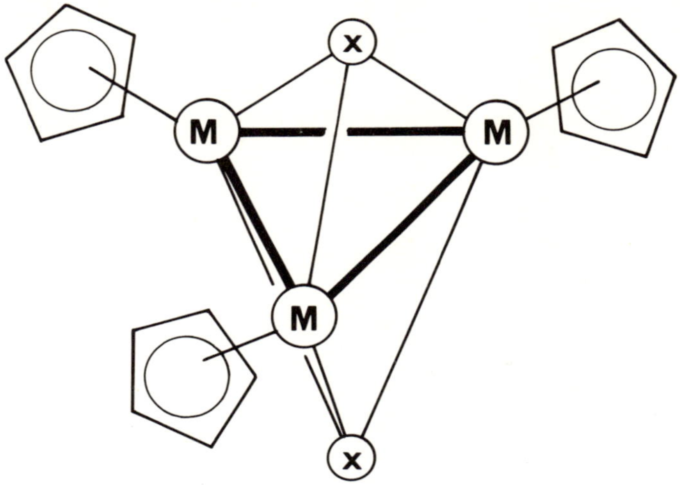


**Figure 5.28** The molecular structures of $M_3X_2(\eta\text{-}C_5H_5)_3$ (X = S, O or CO; M = Co or Ni)

with the general formula $M_3L_{3n}X_2$ ($n = 1$, L = $\eta$-$C_5H_5$; $n = 3$, L = CO; X = S, O, NR or CO) have been studied (see Table 5.15) and demonstrate interesting structural variations. The complexes $Fe_3(CO)_9X_2$ (X = S, Se, NMe or NNPh) have 7 skeletal electron pairs: the antibonding character of the additional electron pair is large and apparently localised on one edge of the triangle, since the $Fe_3$ triangle opens to give an angular Fe—Fe—Fe moiety with two normal Fe—Fe bond lengths and one long (non-bonded) distance (see Figure 5.27)[214, 215].

In contrast, the capped cyclopentadienyl complexes $M_3X_2(\eta\text{-}C_5H_5)_3$ (Figure 5.28) generally react to the additional antibonding electron density by a symmetrical expansion of the metal triangle. For related complexes the average metal–metal bond length increases regularly from 2.452 Å for $Co_3CO(S)(\eta\text{-}C_5H_5)_3$ (6 skeletal electron pairs) to 2.801 Å for $Ni_3S_2(\eta\text{-}C_5H_5)_3$ [$8\frac{1}{2}$ skeletal pairs (see Table 5.15)]. Only for $Co_3S_2(\eta\text{-}C_5H_5)_3^+$, which has two metal–metal bonds 0.18 Å longer than the third metal–metal bond, is a significant distortion from threefold symmetry observed[160, 161, 216, 217].

Dahl and Frisch have suggested, on the basis of qualitative MO arguments, that the strongly antibonding skeletal MOs have $a_2'$ and $e'$ symmetries. If the latter be the more stable, the asymmetric metal triangle in $Co_3S_2(\eta\text{-}C_5H_5)_3^+$ can be accounted for on the basis of a Jahn–Teller distortion of the $^2E'$ ground state. The ground state spin multiplicity of $Co_3S_2(\eta\text{-}C_5H_5)_3$ changes at 173 K, which suggests that the $a_2'$ and $e'$ levels are very close in energy[216].

While emphasising the metal geometries in these complexes, it is also possible to gain some insight into the overall metal chalcogenide or pnicogenide geometry from the polyhedral skeletal electron approach. For example, As and $Co(CO)_3$ are isoskeletal electronic, each contributing three electrons for skeletal bonding. With this in mind, the series of tetrahedral metal arsenide

**Table 5.16 Summary of x-ray data for tetranuclear metal chalcogenides, pnicogenides and oxides**

| *Compound* | *Metal–metal bond lengths*/Å* | *Symmetry*[227] | *No. of skeletal electron pairs* | *Description of metal geometry* | *Description of $M_4X_4$ geometry* | *Ref.* |
|---|---|---|---|---|---|---|
| $Fe_4S_3(NO)_7$ | 2.71(1) [3] | $C_{3v}$ | 9 | Compressed | | 223 |
| | 3.57(2) [3] | | | trigonal pyramid | | |
| $Fe_4S_4(\eta\text{-}C_5H_5)_4$ | 2.650(6) [2] | $D_{2d}$ | 10 | Elongated | | 224 |
| | 3.363(6) [4] | | | tetragonal | Tetracapped | |
| | | | | disphenoid† | disphenoid | 225 |
| $Co_4P_4(\eta\text{-}C_5H_5)_4$ | 2.504(2) [2] | $D_{2d}$ | 10 | | | |
| | 3.630(2) [4] | | | | | |
| $Co_4S_4(\eta\text{-}C_5H_5)_4^+$ | 3.330(5) [2] | $D_{2d}$ | $11\frac{1}{2}$ | Compressed | | 226 |
| | 3.172(5) [4] | | | tetragonal | Tetracapped | |
| | | | | disphenoid | disphenoid | |
| $Co_4S_4(\eta\text{-}C_5H_5)_4$ | 3.295(2) [6] | $< D_{2d}$ | 12 | Distorted | Distorted | 227 |
| $Re_4(SCH_3)_4(CO)_{12}$ | 3.854–3.957(3) | $< D_{2d}$ | | null tetrahedron‡ | cube | 228 |
| $Co_4Sb_4(CO)_{12}$ | 4.115(4) [6] | $T_d$ | 12 | Null tetrahedron | Cube | 229 |
| $Os_4O_4(CO)_{12}$ | 3.222(6) [6] | $T_d$ | | | | 230 |

* The figure in square brackets represents the number of metal–metal bonds with this length

† For a definition of disphenoid, see Ref. 231

‡ A null tetrahedron is a tetrahedron of metal atoms with metal–metal bond lengths greater than those expected for normal covalent bonds

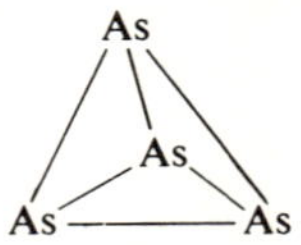

(21)

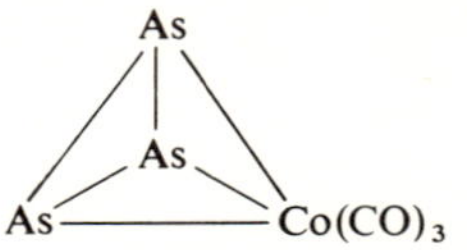

(22)

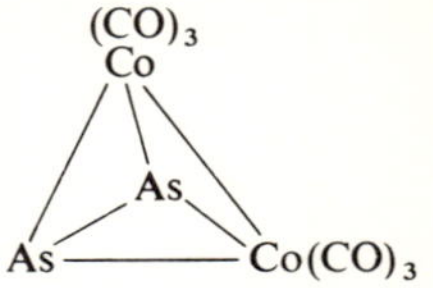

(23)

structures (21)–(23) can be readily rationalised and the compound As-$\{Co(CO)_3\}_3$ can be predicted with some degree of certainty[218, 219].

Dahl and his co-workers have studied in some detail the electronic and structural properties of tetrahedral metal cluster complexes with chalcogenide and pnicogenide bridging ligands, research in this area being stimulated greatly when it was discovered that the $F_4S_4$ cluster unit is a major feature of the bacterial redox proteins known as the ferredoxins[220-222].

All the cluster compounds listed in Table 5.16 have more than the six skeletal electron pairs associated with a closed tetrahedron of metal atoms. For this series of compounds it appears that the additional electron pairs cause selective breaking of metal–metal bonds[134], a feature of triangular metal chalcogenide cluster compounds. Thus $Fe_4S_3(NO)_7$ has nine skeletal electron pairs and can be described as a compressed trigonal prism with three iron–iron bonds[223] and $Fe_4S_4(\pi\text{-}C_5H_5)_4$ which has ten skeletal electron pairs has a disphenoid of metal atoms with only two iron–iron bonds[224]. The wide range of molecules with twelve skeletal electron pairs have no metal–metal bonds. For molecules with unpaired electrons, the perfect pairing approximation is no longer valid and more symmetrically distorted metal geometries are observed.

More complex chalcogenide complexes are listed and briefly described in Table 5.17.

### 5.4.6 Metal carbonyl carbide cluster complexes

Five transition metal carbide complexes have been characterised by x-ray crystallographic techniques. These complexes have two important features. Firstly they have a 'bare' carbon atom covalently bonded to all the metal atoms of the cluster and, secondly, the carbon atom frequently lies at the centre of a closed polyhedron of bonded metal atoms. These compounds can be described as carbon-cluster coordination compounds.

In $Ru_6C(CO)_{17}$[241], $Ru_6C(CO)_{14}(Me_3C_6H_3)$[167] and $Fe_6C(CO)_{16}^{2-}$[242] the carbon atom lies at the centre of an octahedron of metal atoms. The metal–metal and metal–carbon bond lengths in these compounds are not unlike those found in other metal cluster and organometallic compounds and indicate that the metal polyhedron is being neither distorted nor expanded to accommodate the carbon atom. In $Rh_6C(CO)_{15}^{2-}$[123] the carbon atom lies at the origin of the trigonal prism of metal atoms—a unique metal cluster polyhedral geometry (see Figure 5.29). The carbon atom in $Fe_5C(CO)_{15}$[243] forms bonds of similar length to all the iron atoms of the cluster and lies

**Table 5.17 Summary of x-ray data for metal carbonyl chalcogenide compounds**

| *Compound* | *Brief description of molecular structure* | *Metal–metal bond length*/Å | *Ref.* |
|---|---|---|---|
| $Co_3(CO)_4(SEt)_5$ | Triangle of cobalt atoms; 5 SEt and 1 CO bridges | 2.55(1), 2.48(1) | 232 |
| $Co_4(CO)_4(SEt)_8$ | Rectangle of cobalt atoms; 8 SEt bridges | 2.498(5), 3.132(9) | 233 |
| $Co_5(CO)_{10}(SEt)_5$ | Triangle of $Co_3(SEt)_4(CO)_5$ attached to $Co_2(SEt)_2(CO)_5$ by 2 SEt bridges | 2.43(1), 2.53(1) | 234 |
| $Co_6(CO)_{11}S(SEt)_4$ | Triangle of $Co_3(SEt)_4(CO)_6$ attached to $Co_3S(CO)_5$ triangle by 3 SEt bridges | | 235 |
| $\{SCo_3(CO)_7\}_2S_2$ | Two $Co_3S(CO)_7$ triangles linked by a tetradentate $S_2$ bridge | 2.47(1)–2.53(1) | 236 |
| $[S_2Co_4(CO)_{10}]$ | Square of cobalt atoms with two S atoms bridging the square face | | 233 |
| $Co_4S_6(\pi\text{-}C_5H_5)_4$ | Distorted null tetrahedron of $Co(\pi\text{-}C_5H_5)$ fragments interconnected by two triply bridging disulphide groups and two triply bridging sulphur atoms located above the four triangular faces | | 237 |
| $\{Re_2Mo(\pi\text{-}C_5H_5)(CO)_8\}S\{SMo(\pi\text{-}C_5H_5)(CO)_3\}$ | Irregular $Re_2Mo$ triangle capped by sulphide and $—SMo(CO)_3(\pi\text{-}C_5H_5)$ ligands | 2.985(1) (Re—Re)<br>2.909(1) (Re—Mo) | 238 |
| $[Mo_3(\pi\text{-}C_5H_5)_3S_4]^+$ | Equilateral molybdenum triangle capped by sulphide ligand. Each edge is also bridged by sulphide ligands | 2.812(1) | 239 |
| $Ni_5(\pi\text{-}C_5H_5)_4S_4$ | The nickel and sulphur atoms form two distorted $Ni_3S_2$ trigonal bipyramids which are connected to each other by a corner sharing of one common equatorial atom, giving this atom square-planar coordination | 2.495(3)<br>2.979(2) | 240 |

**Table 5.18 Summary of x-ray results for carbide cluster complexes**

| *Compound* | *Metal–metal bond lengths/Å* | *Metal–carbon bond lengths/Å* | *Metal geometry* | | *Ref.* |
|---|---|---|---|---|---|
| $Fe_5C(CO)_{15}$ | 2.59–2.67 | 1.89(2)–1.96(2) | Square pyramidal | 7 skeletal electron pairs | 243 |
| $Ru_6C(CO)_{17}$ | 2.827(5)–3.034(5) | 2.05(7) | Octahedral | 7 skeletal electron pairs | 241 |
| $Ru_6C(CO)_{14}(Me_3C_6H_3)$ | 2.854(7)–2.956(7) | 1.88(4)–2.12(4) | Octahedral | 7 skeletal electron pairs | 167 |
| $[Fe_6C(CO)_{16}]^{2-}$ | 2.553(10)–2.743(10) | 1.82(4)–1.97(4) | Octahedral | 7 skeletal electron pairs | 242 |
| $[Rh_6C(CO)_{15}]^{2-}$* | 2.776(3) (basal)<br>2.817(2) (axial) | 2.13(1) | Trigonal prismatic | 9 skeletal electron pairs | 123 |

* see Figure 5.29

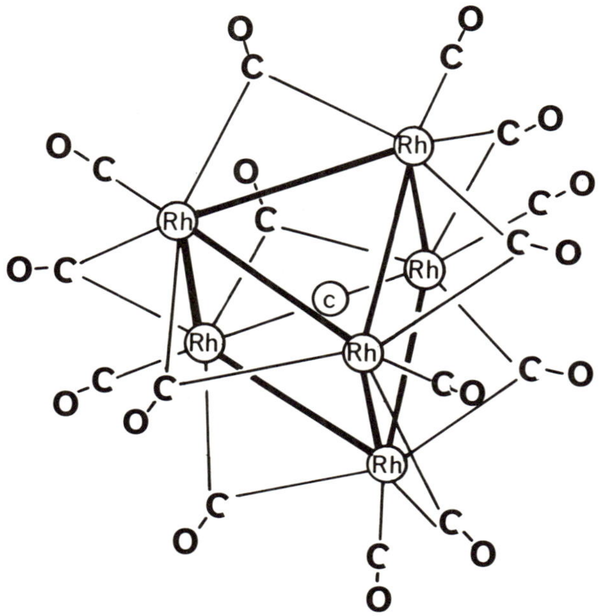


**Figure 5.29** The molecular structure of the carbide anion cluster $Rh_6C(CO)_{15}^{2-}$. (Modified from Albano *et al.*[123], by courtesy of the Chemical Society.)

slightly below the centre of the basal plane of the square pyramid of iron atoms.

It is assumed that the unique carbon atom in these cluster compounds contributes all its valence electrons for skeletal bonding; these compounds can be readily rationalised by the polyhedral skeletal electron pair approach (see Table 5.18).

'Bare' tin atoms coordinated to metal atoms are relatively common. However, in these complexes the tin atom is presumably too large to be accommodated inside a closed metal polyhedron and more open structures result. For example, $Fe_4Sn(CO)_{16}$[244] has a disphenoid of iron atoms with the tin atom lying at the centre. The presence of two iron–iron bonds in this molecule is consistent with the presence of ten skeletal electron pairs [*cf.* $Fe_4S_4(\eta\text{-}C_5H_4)_4$].

## References


1. Cotton, F. A. (1966). *Quart. Rev. Chem. Soc.*, **20,** 389
2. Penfold, B. (1968). *Perspectives in Structural Chemistry*, Vol. 2, 71 (J. A. Ibers and J. Dunitz, editors) (New York: Wiley)
3. King, R. B. (1972). *Progress In Inorganic Chemistry*, Vol. 15, 287 (S. T. Lippard, editor) (New York: Interscience)
4. Johnston, R. D. (1970). *Advances Inorganic Radiochemistry*, Vol. 13, 471 (H. J. Emeleus and A. G. Sharpe, editors) (London: Academic Press)

5. Chini, P. (1968). *Inorg. Chim. Acta Rev.*, **2,** 31
6. Abel, E. W. and Stone, F. G. A. (1969). *Quart. Rev. Chem. Soc.*, **23,** 325
7. Cotton, J. D. (1972). *Specialist Periodical Report — Organometallic Chemistry*, Vol. 1, 194 (E. W. Abel and F. G. A. Stone, editors) (London: Chemical Society)
8. Connelly, N. G. (1973). *Specialist Periodical Report — Organometallic Chemistry*, Vol. 2, 234 (E. W. Abel and F. G. A. Stone, editors) (London: Chemical Society)
9. Chini, P. (1970). *Fourth International IUPAC Conference on Organometallic Chemistry*, 489 (M. I. Bruce and F. G. A. Stone, editors) (London: Butterworths)
10. Johnston, R. D. (1972). *MTP International Review of Science, Inorganic Chemistry Series One*, Vol. 6, 1 (M. J. Mays, editor) (London: Butterworths)
11. Jarvis, J. A., Kilbourn, B. T. and Owston, P. G. (1971). *Acta Crystallogr.*, **B27,** 366; Hamilton, W. C., Klanderman, K. A. and Spratley, R. (1969). *Acta Crystallogr.*, **A25,** 5172
12. Benedetti, E., Corradini, P. and Pedone, C. (1969). *J. Organometal. Chem.*, **18,** 203
13. Whitla, W. A., Powell, H. M. and Venanzi, L. (1966). *Chem. Commun.*, **310,**
14. Spagna, R. and Zambonelli, L. (1971). *J. Chem. Soc. A*, 2544
15. Mason, R. and Robertson, G. B. (1969). *J. Chem. Soc. A*, 492
16. Guggenberger, L. J. and Cramer, R. (1972). *J. Amer. Chem. Soc.*, **94,** 3779
17. Evans, J. A. and Russell, D. R. (1971). *Chem. Commun.*, 197
18. Hitchcock, P. B., McPartlin, M. and Mason, R. (1969). *Chem. Commun.*, 1367
19. Hitchcock, P. B., McPartlin, M. and Mason, R. (1969). *Chem. Commun.*, 1367
20. Nave, C. and Truter, M. R. (1973). *J. Chem. Soc. Dalton Trans.*, 2202
21. Muir, L. M., Muir, K. W. and Ibers, J. A. (1969). *Discuss. Faraday Soc.*, **47,** 84
22. Muir, L. M., Muir, K. W. and Ibers, J. A. (1969). *Discuss. Faraday Soc.*, **47,** 84
23. Ricci, J. S. and Ibers, J. A. (1971). *J. Amer. Chem. Soc.*, **93,** 2391
24. Churchill, M. R. and Bezman, S. A. (1973). *Inorg. Chem.*, **12,** 260
25. Churchill, M. R. and Bezman, S. A. (1973). *Inorg. Chem.*, **12,** 531
26. Luxmoore, A. R. and Truter, M. R. (1962). *Acta Crystallogr.*, **15,** 1117
27. Pedone, C. and Sirigu, A. (1968). *Inorg. Chem.*, **7,** 2164
28. Robertson, G. B. and Whimp, P. O. (1973). *J. Chem. Soc. Dalton Trans.*, 2454
29. Kruger, C. and Tsay, Y. H. (1972). *J. Organometal. Chem.*, **34,** 387
30. Dreissing, W. and Dietrich, H. (1968). *Acta Crystallogr.*, **B24,** 108
31. Cheng, P. T., Cook, C. D., Nyburg, S. C. and Wan, K. Y. (1971). *Inorg. Chem.*, **10,** 2210
32. Guggenberger, L. J. (1973). *Inorg. Chem.*, **12,** 499
33. Stalick, J. K. and Ibers, J. A. (1970). *J. Amer. Chem. Soc.*, **92,** 5333
34. Browning, J. and Penfold, B. R. (1973). *J. Chem. Soc. Chem. Commun.*, 198
35. McAdams, A., Francis, J. N. and Ibers, J. A. (1971). *J. Organometal. Chem.*, **29,** 149
36. Cheng, P. T., Cook, C. D., Koo, C. H., Nyburg, S. C. and Shiomi, M. T. (1971). *Acta Crystallogr.*, **B27,** 1904
37. Bombieri, G., Forsellini, E., Panattoni, C., Grazioni, R. and Bandoli, G. (1970). *J. Chem. Soc. A*, 1313
38. Francis, J. N., McAdam, A. and Ibers, J. A. (1971). *J. Organometal. Chem.*, **29,** 131
39. Dewar, M. J. S. (1951). *Bull. Chem. Soc. Fr.*, **18,** C71, Chatt, J. and Duncanson, L. A. (1953). *J. Chem. Soc.*, 2939
40. Mason, R., Mingos, D. M. P., Rucci, G. and Connor, J. A. (1972). *J. Chem. Soc. Dalton Trans.*, 1729
41. Kettle, S. F. A. and Mason, R. (1966). *J. Organometal. Chem.*, **5,** 97
42. Stalick, J. K. and Ibers, J. A. (1970). *J. Amer. Chem. Soc.*, **92,** 5444
43. Cramer, R. and Reddy, G. S. (1973). *Inorg. Chem.*, **12,** 346
44. Ashley-Smith, J., Douek, Z., Johnson, B. F. G. and Lewis, J. (1974). *J. Chem. Soc. Dalton Trans.*, 128
45. Racànelli, P., Pantini, G., Immirzi, A., Allegra, G. and Pari, L. (1969). *Chem. Commun.* 361
46. Okamoto, K., unpublished work, cited in Ref. 52
47. Hewitt, T. G. and De Boer, J. J. (1971). *J. Chem. Soc. A*, 817
48. Kashiwagi, T., Yasuoka, N., Kasai, N. and Kakudo, M. (1969). *Chem. Commun.*, 317
49. Hewitt, T. G., Anzenhofer, K. and De Boer, J. J. (1969). *J. Organometal. Chem.*, **18,** 91
50. Hewitt, T. G., Anzenhofer, K. and De Boer, J. J. (1969). *Chem. Commun.*, 312
51. Kadonago, M., Yasuoka, N. and Kasai, N. (1971). *Chem. Commun.*, 1597

52. Okamoto, K., Kai, Y., Yasuoka, N. and Kasai, N. (1974). *J. Organometal. Chem.*, **65,** 427
53. Mason, R. (1968). *Nature*, **217**, 543
54. McWeeny, R., Mason, R. and Towl, A. D. C. (1969). *Discuss. Faraday Soc.*, **47,** 21
55. Shaw, B. L. and Stringer, A. J. (1973). *Inorg. Chim. Acta Rev.*, **7,** 1
56. Davies, G. R., Hewertson, W., Mais, R. H. B., Ouston, P. G. and Patel, C. G. (1970). *J. Chem. Soc. A*, 1873
57. Glanville, J. O., Stewart, J. M. and Grim, S. O. (1967). *J. Organometal. Chem.*, **7,** C9
58. Dickson, R. S. and Ibers, J. A. (1972). *J. Organometal. Chem.*, **36,** 191
59. Gusev, A. I. and Struchov, Y. T. (1969). *Zh. Strukt. Khim.*, **10,** 515
60. Gusev, A. I. and Struchov, Y. T. (1969). *Zh. Strukt. Khim.*, **10,** 294
61. Laine, R. M., Moriarty, R. E. and Bau, R. (1972). *J. Amer. Chem. Soc.*, **94,** 1402
62. Blizzard, A. C. and Santry, D. P. (1968). *J. Amer. Chem. Soc.*, **90,** 5749
63. Nelson, H. H., Wheelock, K. S., Cusacks, L. S. and Jonassen, H. B. (1969). *J. Amer. Chem. Soc.*, **91,** 7005
64. Mason, R. and Thomas, K. M. (1974). *Proc. N.Y. Acad. Sci.*, **239,** 225
65. Churchill, M. R. and Mason, R. (1967). *Adv. Organometal. Chem.*, **5,** 93
66. Smith, A. E. (1965). *Acta Crystallogr.*, **18,** 331
67. Kettle, S. F. A. and Mason, R. (1966). *J. Organometal. Chem.*, **5,** 573
68. Mason, R. and Russell, D. R. (1966). *Chem. Commun.*, 26
69. Mason, R. and Wheeler, A. G. (1968). *J. Chem. Soc. A*, 2549
70. Mason, R. and Wheeler, A. G. (1968). *J. Chem. Soc. A*, 2543
71. Davies, G. R., Mais, R. H. B., O'Brien, S. and Owston, P. G. (1967). *Chem. Commun.*, 1151
72. Smith, A. E. (1972). *Inorg. Chem.*, **11,** 2306
73. Smith, A. E. (1972). *Inorg. Chem.*, **11,** 165
74. Bennett, M. J., Churchill, M. R. Gerloch, M. and Mason, R. (1964). *Nature*, **201,** 1318
75. Cotton, F. A., Day, V. W., Frenz, B. A., Hardcastle, K. I. and Troup, J. M. (1973). *J. Amer. Chem. Soc.*, **95,** 4522
76. Mason, R. (1971). *Special Lectures XXIII IUPAC Congress*, **6,** 31
77. Bailey, P. M. and Mason, R., unpublished observations
78. Churchill, M. R. and Scholer, F. R. (1969). *Inorg. Chem.*, **8,** 1950
79. Mathew, M. and Palenik, G. J. (1972). *Inorg. Chem.*, **11,** 2809
80. Jellinek, F. (1960). *Nature*, **185,** 871
81. Cotton, F. A., Dollase, W. H. and Wood, J. S. (1960). *J. Amer. Chem. Soc.*, **85,** 153; Ibers, J. A. (1964). *J. Chem. Phys.*, **40,** 3129
82. Rees, B. and Coppens, P. (1972). *J. Organometal. Chem.*, **42,** C102
83. Mingos, D. M. P. (1971). *Nature Phys. Sci.*, **229,** 193
84. Mason, R. (1972). *Chem. Soc. Rev.*, **1,** 431
85. Churchill, M. R. and Mason, R. (1966). *Proc. Roy. Soc.*, **A292,** 61
86. Hattner, G., Lange, S. and Fischer, E. O. (1972). *Angew. Chem. Int. Edn. Engl.*, **10,** 556
87. Powell, H. M. and Ewens, R. V. G. (1939). *J. Chem. Soc.*, 286
88. Dahl, L. F. and Rundle, R. E. (1963). *Acta Crystallogr.*, **16,** 419
89. Bennett, M. J. and Mason, R. (1965). *Nature*, **205,** 761
90. Wilson, F. C. and Schoemaker, D. P. (1958). *J. Chem. Phys.*, **27,** 809
91. Switkes, E., Stevens, R. M., Lipscomb, W. N. and Newton, M. D. (1969). *J. Chem. Phys.*, **51,** 2085
92. Mason, R. and Mingos, D. M. P. (1973). *J. Organometal. Chem.*, **50,** 53
93. Mathew, M. and Palenik, G. J. (1969). *Can. J. Chem.*, **47,** 205
94. McPhail, A. T. and Sim, G. A. (1968). *J. Chem. Soc. A*, 1858
95. Coleman, J. M. and Dahl, L. F. (1967). *J. Amer. Chem. Soc.*, **89,** 542
96. Thomas, J. T., Robertson, J. H. and Cox, E. G. (1958). *Acta Crystallogr.*, **11,** 599
97. Mason, R. and Mingos, D. M. P. unpublished results
98. Hall, M. C., Jarvis, J. A. J., Kilbourn, B. T. and Owston, P. G. (1972). *J. Chem. Soc. Dalton Trans.*, 1544
99. Black, M., Mais, R. H. B. and Owston, P. G. (1969). *Acta Crystallogr.*, **B25,** 1760
100. Adam, R. D. and Cotton, F. A. (1973). *J. Amer. Chem. Soc.*, **95,** 6589
101. Wei, C. H. and Dahl, L. F. (1969). *J. Amer. Chem. Soc.*, **91,** 1351
102. McDonald, W. S., Muss, J. R., Raper, G., Shaw, B. L., Greatrex, R. and Greenwood, N. N. (1969). *Chem. Commun.*, 1295

103. Dahm, D. J. and Jacobsen, R. A. (1966). *Chem. Commun.*, 496
104. Cotton, F. A., Frenz, B. A. and Kruczynski, L. (1973). *J. Amer. Chem. Soc.*, **95,** 951
105. Einstein, F. W. B. and Jones, R. D. G. (1971). *J. Chem. Soc. A*, 3359
106. Epstein, E. F. and Dahl, L. F. (1970). *J. Amer. Chem. Soc.*, **92,** 493
107. Churchill, M. R., Wormald, J., Knight, K. and Mays, M. J. (1971). *J. Amer. Chem. Soc.*, **93,** 3073
108. Osbourne, A. G. and Stiddard, M. H. B. (1965). *J. Organometal. Chem.*, **3,** 340
109. Booth, G., Chatt, J. and Chini, P. (1965). *Chem. Commun.*, 639
110. Kaesz, H. D. and Saitlant, R. B. (1972). *Chem. Rev.*, **72,** 231
111. Schäfer, H. and Schnering, H. G. (1964). *Angew. Chem.*, **76,** 883
112. (1965–66). *Handbook of Chemistry and Physics* (Cleveland, Ohio: Chemical Rubber Publ. Co.)
113. Abel, E. W. (1963). *Quart. Rev. Chem. Soc.*, **17,** 139
114. Mason, R. and Rae, A. I. B. (1968). *J. Chem. Soc. A*, 778
115. Corey, E. R. and Dahl, L. F. (1962). *Inorg. Chem.*, **1,** 521
116. Wei, C. H. (1969). *Inorg. Chem.*, **8,** 2384
117. Wei, C. H., Wilkes, G. R. and Dahl, L. F. (1967). *J. Amer. Chem. Soc.*, **89,** 4792
118. Martinengo, S., Chini, P., Albano, V. G., Cariati, F. and Salvatori, T. (1973). *J. Organometal. Chem.*, **59,** 379
119. Corey, E. R., Dahl, L. F. and Beck, W. (1963). *J. Amer. Chem. Soc.*, **85,** 1202
120. Mason, R., Mingos, D.M. P. and Thomas, K. M. (1973). *J. Amer. Chem. Soc.*, **95,** 3802
121. Mason, R. and Thomas, K. M. (1973). Unpublished results
122. King, R. B. (1972). *J. Amer. Chem. Soc.*, **94,** 95
123. Albano, V. G., Sansoni, M., Chini, P. and Martinengo, S. (1973). *J. Chem. Soc. Dalton Trans.*, 651
124. Eady, C. R., Johnson, B. F. G. and Lewis, J. (1972). *J. Organometal. Chem.*, **37,** C39
125. Wells, A. F. (1962). *Structural Inorganic Chemistry, 3rd Edn.*, 979 (London: Oxford University Press)
126. Corradini, P. and Paiaro, G. (1966). *Ric. Sci.*, **36,** 365
127. Sidgwick, N. V. and Bailey, R. W. (1934). *Proc. Roy. Soc.*, **A144,** 521
128. Kettle, S. F. A. (1966). *J. Chem. Soc. A*, 1013; (1967). *J. Chem. Soc. A*, 314
129. Kettle, S. F. A. (1965). *Theor. Chim. Acta*, **3,** 282
130. Mingos, D. M. P. (1974). *J. Chem. Soc. Dalton Trans.*, 123
131. Lipscomb, W. N. (1973). *Accounts Chem. Res.*, 257
132. Wade, K. (1971). *Chem. Commun.*, 792
133. Wade, K. (1972). *J. Inorg. Nucl. Chem. Lett.*, **8,** 559
134. Mingos, D. M. P. (1972). *Nature Phys. Sci.*, **236,** 99
135. Lipscomb, W. N. and Hoffman, R. (1962). *J. Chem. Phys.*, **36,** 2179
136. Cotton, F. A. (1970). *Chemical Applications of Group Theory, 2nd Edn.*, (New York: John Wiley)
137. Strouse, C. E. and Dahl, L. F. (1969). *Discuss. Faraday Soc.*, **47,** 93
138. Trinh-Toan, Fehlhammer, W. P. and Dahl, L. F. (1972). *J. Amer. Chem. Soc.*, **94,** 3389
139. Mason, R. and Zubieta, J. A. (1974). *J. Organometal. Chem.*, **66,** 289
140. Albano, V. G., Ciani, G., Bruce, M. I., Shaw, G. and Stone, F. G. A. (1972). *J. Organometal. Chem.*, **42,** C99
141. Albano, V. G., Bellon, P. L. and Scatturin, V. (1967). *Chem. Commun.*, 730
142. Bennett, M. J., Cotton, F. A. and Winquist, B. H. C. (1967). *J. Amer. Chem. Soc.*, **89,** 5366
143. Vranka, R. G., Dahl, L. F., Chini, P. and Chatt, J. (1969). *J. Amer. Chem. Soc.*, **91,** 1574
144. Bradford, C. W. and Nyholm, R. S. (1973). *J. Chem. Soc. Dalton Trans.*, 529
145. Bradford, C. W., Nyholm, R. S., Gainsford, G. J., Guss, J. M., Ireland, P. R. and Mason, R. (1972). *J. Chem. Soc. Chem. Commun.*, 87
146. Gainsford, G. J., Guss, J. M., Ireland, P. R., Mason, R., Bradford, C. W. and Nyholm, R. S. (1972). *J. Organometal. Chem.*, **40,** C70
147. Deeming, A. J., Nyholm, R. S. and Underhill, M. (1972). *J. Chem. Soc. Chem. Commun.*, 224
148. Dodge, R. P. and Schomaker, V. (1965). *J. Organometal. Chem.*, **3,** 274
149. Gainsford, G. J., Mason, R. and Thomas, K. M., to be published; quoted in Ref. 64
150. Canty, A. J., Johnson, B. F. G. and Lewis, J. (1972). *J. Organometal. Chem.*, **43,** C35

151. Sly, W. G. (1959). *J. Amer. Chem. Soc.*, **81,** 18
152. Blount, J. F., Dahl, L. F., Hoogzand, C. and Hübel, W. (1966). *J. Amer. Chem. Soc.*, **88,** 292
153. Dahl, L. F. and Smith, D. L. (1962). *J. Amer. Chem. Soc.*, **84,** 2450
154. Wade, K. (1972). *J. Inorg. Nucl. Chem. Lett.*, **8,** 563
155. Ferraris, G. and Gervasio, G. (1973). *J. Chem. Soc. Dalton Trans.*, 1933
156. Ferraris, G. and Gervasio, G. (1972). *J. Chem. Soc. Dalton Trans.*, 1057
157. Deeming, A. J. and Underhill, M. (1973). *J. Chem. Soc. Chem. Commun.*, 277
158. Mills, O. S. and Paulus, E. F. (1967). *J. Organometal. Chem.*, **10,** 331
159. Paulus, E. F., Fischer, E. O., Fritz, H. P. and Schuster-Woldon. (1967). *J. Organometal. Chem.*, **10,** P3
160. Hock, A. and Mills, O. S. (1961). *Advances in the Chemistry of Coordination Compounds* (E. Kirschner, editor) (New York: Macmillan)
161. Uchtman, V. A. and Dahl, L. F. (1969). *J. Amer. Chem. Soc.*, **91,** 3763
162. Elder, R. C., Cotton, F. A. and Schunn, R. A. (1967). *J. Amer. Chem. Soc.*, **89,** 3645
163. Dahl, L. F., Trinh-Toan, Neuman, M. A., Toan, T. and Fehlhammer, W. P. (1972). *J. Amer. Chem. Soc.*, **94,** 3383
164. Churchill, M. R. and Veidis, M. V. (1970). *Chem. Commun.*, 1470
165. Churchill, M. R. and Veidis, M. V. (1971). *J. Chem. Soc. A*, 2170
166. Churchill, M. R. and Veidis, M. V. (1971). *J. Chem. Soc. A*, 2995
167. Mason, R. and Robinson, W. R. (1968). *Chem. Commun.*, 468
168. Bennett, M. J., Cotton, F. A. and Legzdins, P. (1967). *J. Amer. Chem. Soc.*, **89,** 6797
169. Churchill, M. R., Scholer, F. R. and Wormald, J. (1973). *Inorg. Chem.*, **12,** 191
170. Churchill, M. R., Gold, K. and Bird, P. H. (1969). *Inorg. Chem.*, **8,** 1956
171. Bau, R., Burt, J. C., Knox, S. A. R., Laine, R. M., Phillips, R. P. and Stone, F. G. A. (1973). *J. Chem. Soc. Chem. Commun.*, 726
172. Woodward, P. and Cox, A. (1971). *J. Chem. Soc. A*, 3599
173. Evans, M., Hursthouse, M. B., Randall, E. W., Rosenberg, E., Milone, L. and Valle, M. (1972). *J. Chem. Soc. Chem. Commun.*, 545
174. Belford, R., Woodward, P. and Taylor, H. P. (1972). *J. Chem. Soc. Dalton Trans.*, 2425
175. Parkins, A. W., Fischer, E. O., Hatner, G. and Regler, D. (1970). *Angew. Chem. Int. Ed. Engl.*, **9,** 633
176. Doedens, R. J. and Dahl, L. F. (1966). *J. Amer. Chem. Soc.*, **88,** 4847
177. Churchill, M. R. and Bau, R. (1968). *Inorg. Chem.*, **7,** 2606
178. Ruff, J. K., White, R. P. Jr. and Dahl, L. F. (1971). *J. Amer. Chem. Soc.*, **93,** 2159
179. Albano, V. G., Chini, P. and Scatturin, V. (1968). *J. Organometal. Chem.*, **15,** 423
180. Albano, V. G., Bellon, P. L., Chini, P. and Scatturin, V. (1969). *J. Organometal. Chem.*, **16,** 461
181. Albano, V. G., Bellon, P. L. and Sansoni, M. (1971). *J. Chem. Soc. A*, 678
182. Albano, V. G. and Bellon, P. L. (1968). *J. Organometal. Chem.*, **19,** 405
183. Albano, V. G., Bellon, P. L. and Ciani, G. F. (1969). *Chem. Commun.*, 1024
184. Kaesz, H. D. (1973). *Chem. Brit.*, 344
185. Kaesz, H. D., Fellman, W., Wilkes, G. and Dahl, L. F. (1965). *J. Amer. Chem. Soc.*, **87,** 2753
186. Kirtley, J. W., Olsen, J. P. and Bau, R. (1973). *J. Amer. Chem. Soc.*, **95,** 4532
187. Doedens, R. J., Robinson, W. T. and Ibers, J. A. (1967). *J. Amer. Chem. Soc.*, **89,** 4323
188. Churchill, M. R. and Wei-Yang Ni, S. (1973). *J. Amer. Chem. Soc.*, **95,** 2150
189. Churchill, M. R. and Wei-Yang Ni, S. (1973). *J. Chem. Soc. Chem. Commun.*, 691
190. Fallani, G., Midollini, S. and Sacconi, L. (1973). *J. Amer. Chem. Soc.*, **95,** 2021
191. Buckingham, A. D., Yesinowski, J. P., Carty, A. J. and Rest, A. J. (1973). *J. Amer. Chem. Soc.*, **95,** 2734
192. Churchill, M. R., Bird, P. H., Kaesz, H. O., Bau, R. and Fontal, B. (1968). *J. Amer. Chem. Soc.*, **90,** 7135
193. Kirtley, S. W., Kaesz, H. D., Churchill, M. R. and Knobler, C., in the press; quoted in Ref. 184
194. Yawney, P. B. W. and Doedens, R. J. (1972). *Inorg. Chem.*, **11,** 838
195. Gilmore, C. J. and Woodward, P. (1971). *J. Chem. Soc. A*, 3453
196. Kaesz, H. D., Fontal, B., Bau, R., Kirtley, S. W. and Churchill, M. R. (1969). *J. Amer. Chem. Soc.*, **91,** 1021
197. Dahl, L. F. and Blount, J. F. (1965). *Inorg. Chem.*, **4,** 1373

198. Mason, R. (1971). *IUPAC XXIIIrd Int. Congr. Pure Appl. Chem.*, **6**, 31
199. Duckworth, V. F. and Mason, R., unpublished results
200. Churchill, M. R., Wormald, J., Knight, J. and Mays, M. J. (1970). *Chem. Commun.*, 458
201. Churchill, M. R., Wormald, J., Knight, J. and Mays, M. J. (1971). *J. Amer. Chem. Soc.*, **93**, 5670
202. Fischer, E. O., Mills, O. S., Paulus, E. F. and Waverik, H. (1967). *Chem. Commun.*, 643
203. Fischer, E. O. and Paulus, E. F. (1968). *J. Organometal. Chem.*, **11**, 587
204. White, J. W. and Wright, C. J. (1971). *J. Chem. Soc. A*, 2843
205. Farmery, K., Kilner, M., Greatrex, R. and Greenwood, N. N. (1969). *J. Chem. Soc. A*, 2339
206. Simon, A. (1967). *Z. Anorg. Allg. Chem.*, **355**, 311
207. Strouse, C. E. and Dahl, L. F. (1971). *J. Amer. Chem. Soc.*, **93**, 6032
208. Stevenson, D. L., Wei, C. H. and Dahl, L. F. (1971). *J. Amer. Chem. Soc.*, **93**, 6027
209. Wei, C. H. and Dahl, L. F. (1967). *Inorg. Chem.*, **6**, 1229
210. Delbaere, L. T. J., Kruczynski, L. J. and McBridge, D. W. (1973). *J. Chem. Soc. Dalton Trans.*, 307
211. Barnett, B. L. and Kruger, C. (1971). *Angew. Chem.*, **83**, 969
212. Wei, C. H. and Dahl, L. F. (1965). *Inorg. Chem.*, **4**, 493
213. Dahl, L. F. and Sutton, P. W. (1963). *Inorg. Chem.*, **2**, 1067
214. Doedens, R. J. (1969). *Inorg. Chem.*, **8**, 570
215. Baikie, P. E. and Mills, O. S. (1967). *Chem. Commun.*, 1228
216. Frisch, P. D. and Dahl, L. F. (1972). *J. Amer. Chem. Soc.*, **94**, 5082
217. Vahrenkamp, H., Uchtman, V. A. and Dahl, L. F. (1967). *J. Amer. Chem. Soc.*, **40**, 3272
218. Foust, A. S., Foster, M. S. and Dahl, L. F. (1969). *J. Amer. Chem. Soc.*, **91**, 5633
219. Foster, M. S. and Dahl, L. F. (1969). *J. Amer. Chem. Soc.*, **91**, 5631
220. Sieber, L. C., Adman, E. and Jensen, L. H. (1972). *Nature*, **235**, 40
221. Carter, C. W., Freer, S. T., Xuong, Ng. H., Aldon, R. A. and Kraut, J. (1972). *Cold Spring Harbor Symp. Quant. Biol.*, **36**, 381
222. Mason, R. and Zubieta, J. A. (1973). *Angew. Chem. Int. Ed. Engl.*, **12**, 310
223. Johanison, G. and Lipscomb, W. N. (1958). *Acta Crystallogr.*, **11**, 594
224. Wei, C. H., Wilkes, R. G., Treichel, P. M. and Dahl, L. F. (1966). *Inorg. Chem.*, **5**, 900
225. Simon, G. L. and Dahl, L. F. (1973). *J. Amer. Chem. Soc.*, **95**, 2175
226. Simon, G. L. and Dahl, L. F. (1973). *J. Amer. Chem. Soc.*, **95**, 2164
227. Averill, B. A., Herskovitz, T., Holm, R. H. and Ibers, J. A. (1973). *J. Amer. Chem. Soc.*, **95**, 3523
228. Harrison, W., Marsh, W. C. and Trotter, J. (1972). *J. Chem. Soc. Dalton Trans.*, 1009
229. Foust, A. S. and Dahl, L. F. (1970). *J. Amer. Chem. Soc.*, **92**, 7337
230. Bright, D. (1970). *Chem. Commun.*, 1169
231. Coxeter, H. S. M. (1948). *Regular Polytopes* (London: Methuen)
232. Wei, C. H. and Dahl, L. F. (1968). *J. Amer. Chem. Soc.*, **90**, 3960
233. Wei, C. H. and Dahl, L. F. (1967). *Amer. Crystallogr. Assoc. Prog., Winter Meeting*, **15**, 63
234. Wei, C. H. and Dahl, L. F. (1968). *J. Amer. Chem. Soc.*, **90**, 3969
235. Wei, C. H. and Dahl, L. F. (1968). *J. Amer. Chem. Soc.*, **90**, 3977
236. Stevenson, D. L., Magnuson, V. R. and Dahl, L. F. (1967). *J. Amer. Chem. Soc.*, **89**, 3727
237. Uchtman, V. A. and Dahl, L. F. (1969). *J. Amer. Chem. Soc.*, **91**, 3756
238. Vergamani, P. J., Vahrenkamp, H. and Dahl, L. F. (1971). *J. Amer. Chem. Soc.*, **93**, 6326
239. Vergamani, P. J., Vahrenkamp, H. and Dahl, L. F. (1971). *J. Amer. Chem. Soc.*, **93**, 6327
240. Vahrenkamp, H. and Dahl, L. F. (1969). *Angew. Chem. Int. Ed. Engl.*, **8**, 144
241. Sirigu, A., Bianchi, M. and Benedetti, E. (1969). *Chem. Commun.*, 546
242. Churchill, M. R., Wormald, J., Knight, J. and Mays, M. J. (1971). *J. Amer. Chem. Soc.*, **93**, 3073
243. Braye, E. H., Dahl, L. F., Hübel, W. and Wampler, D. L. (1962). *J. Amer. Chem. Soc.*, **84**, 4633
244. Lindley, P. F. and Woodward, P. (1967). *J. Chem. Soc. A*, 382

# 6
# Structures of Natural Products — Alkaloids

**A. McL. MATHIESON**
C.S.I.R.O., Clayton, Victoria



## 6.1 INTRODUCTION

In Series One, natural product structures classified as terpenes and steroids were reviewed[1]. A further group, that of the alkaloids, is treated here. As in the previous article, it has been found convenient, for reasons of space, for reference to be made to figures in an earlier work by the author[2]. In such cases the figures are distinguished by the letter P.

As before, the literature search has been greatly simplified by the generous assistance of the Crystallographic Data Centre at Cambridge University, U.K., which provided up-to-date lists of the relevant papers from their files. Lists up to 1969 are available in print[3].

## 6.2 INDOLE ALKALOIDS

The basic chemical grouping is the indole group with the side chain —C—C—N—C (1). Examples closely related to this group are eserine (physostigmine) (2)[4] which involves *cis* fusion of the B/C rings, ring C being described as in a half-chair form (in appearance closer to envelope). The alkaloid from

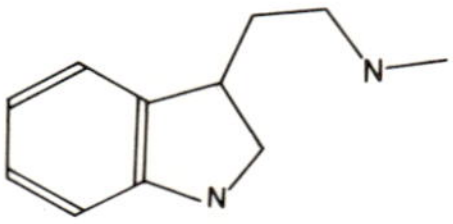


(1)

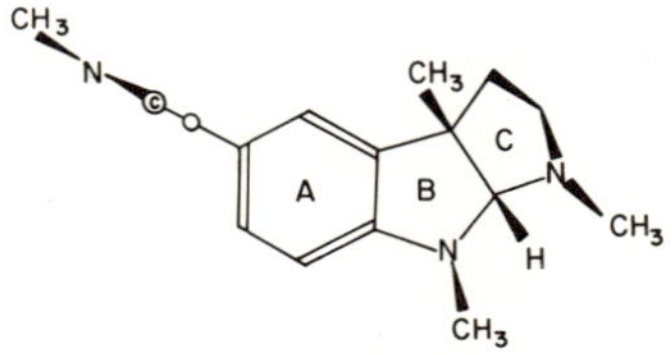


(2)

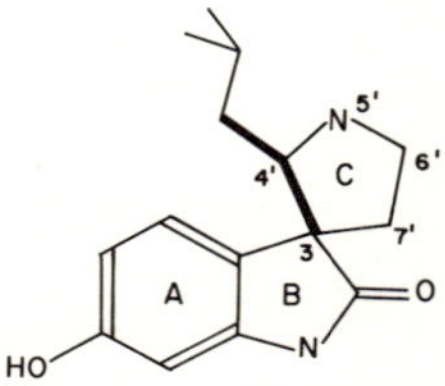


(3)

*Eleagnus commutata* (3)[5] has a spiro B/C ring junction. Despite the conjunction of the benzene ring and the lactam group, ring B is not quite planar, C(3) deviating by 0.11 Å from the plane of the other four, themselves tilted 2.8° from the plane of ring A. Ring C is in a twist-form with C(3) −0.20 Å and C(4′) +0.38 Å from the C(6′)C(7′)N(5′) plane. This alkaloid was isolated as the racemate, presumably indicative of the alternative modes of the ring closure available to the precursor. On the basis of increasing complexity, mention may be made first of the α- and β-methiodides of hunterburnine (P13)[6], which are epimeric at N(12), while mitragynine hydroiodide (P14)[7] has similar structural features. Both *syn-* and *anti-*de-ethyleburnamonine (4)

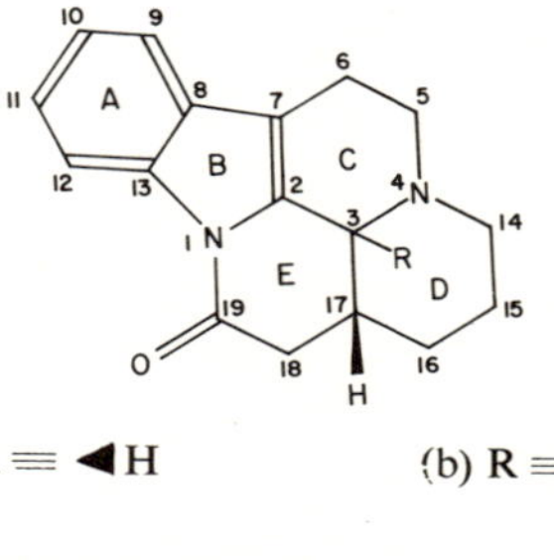


(4) (a) R ≡ ◀H (b) R ≡ ▶H

have been studied[8] to determine the junctions of the D/E rings. In both cases the presence of the double bond between rings B and C and the lactam system shared between rings B and E constrains near-planarity up to C(6), C(3) and

C(18). Rings C and E in both cases have the sofa form; ring D is in the chair form in both but the overall shapes of the molecules differ. In the *anti-trans* structure the molecule is generally flattish, while in the *syn-cis* case ring D is essentially at right angles to the mean plane of the remainder of the molecule. (Note that both structures were studied in the racemic form.)

As a possible reference structure for aspidospermine, iboga and vinca alkaloids, cleavamine methiodide (P5)[9] has been analysed, as has ibogaine hydrobromide (5)[10]. Structurally related to the last compound are akuammidine methiodide (P12)[11] and macusine-A iodide (P11)[12] which differ in the inversion of the substituents —$CH_2OH$ and —$CO_2CH_3$ at C(11a). A similar but more elaborate ring structure occurs in macrosalhine bromide (6)[13], the

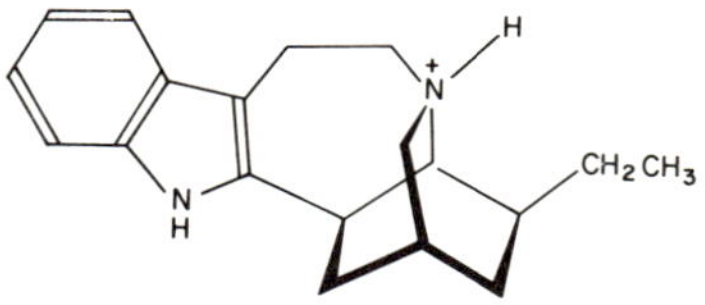


(5)

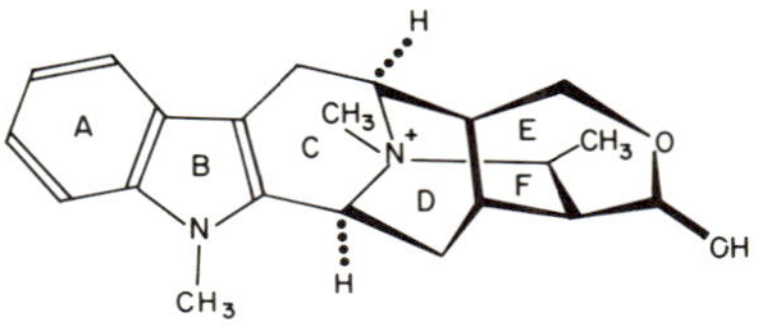


(6)

ring conformations being mainly defined by the influence of the cage unit. A representative quaternary base from *Hunteria eburnea* Pichon, hunteracine, analysed as the bromide (7)[14], also has a cage structure associated with N(12); voaphylline hydroxyindolenine hydroiodide (8)[15] has a similar spiro junction

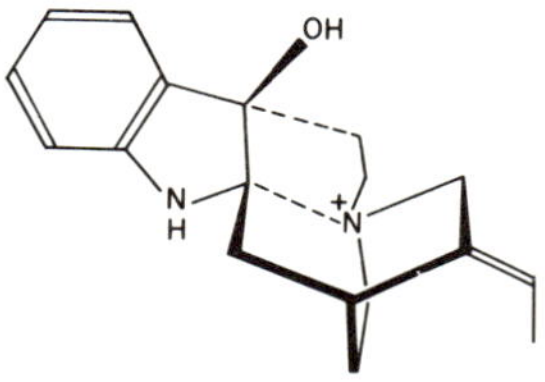


(7)

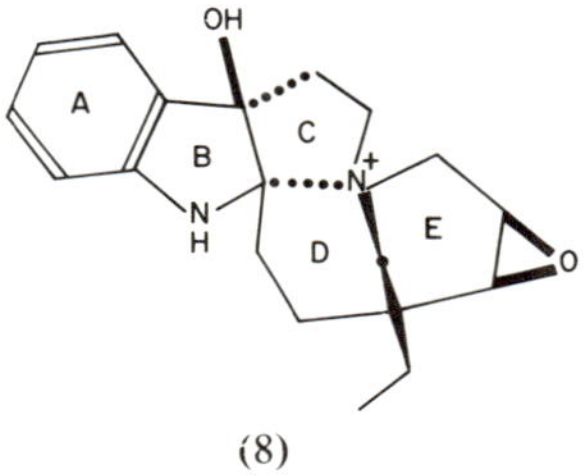


(8)

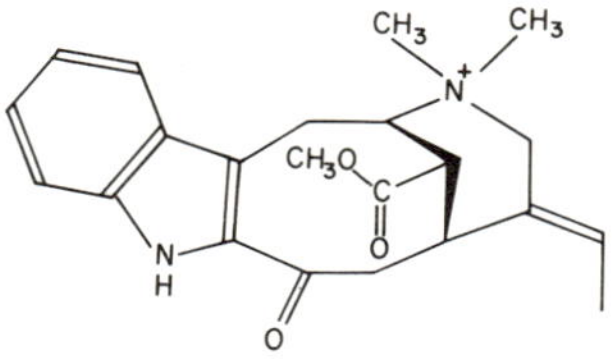


(9)

at C(2). In this compound ring D is in the chair form, owing to the disposition of the epoxy group. Further examples involving ring-closure of N(12) and C(2) are echitamine bromide (P8)[16] and corymine (P9)[17], which are closely similar in structure except for an additional ring closure in the latter compound. In this area may be grouped also vobasine (9)[18].

Five aspidospermine compounds have been studied. The early analysis of (−)-aspidospermine $N^{B}$-methiodide (P4)[19a] was placed on an absolute basis as (10)[19b]. However, the structure of $N^{A}$-acetyl-7-ethyl-5-de-ethyl-aspidospermidine $N^{B}$-methiodide (P7)[20] has also been established absolutely and is of the opposite chirality. This latter compound was prepared from dihydrocleavamine on the assumption that configuration at C(5) in (10) was

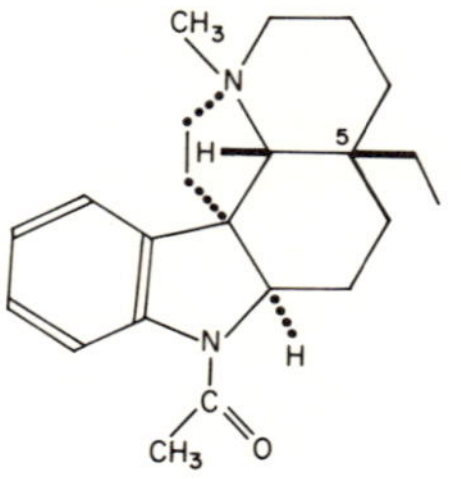


(10)

retained and this appears to accord with the absolute structure of cleavamine (P5)[9]. No explanation for this discrepancy has been presented. In these compounds, rings C, D and E are *cis* fused, ring D being in the boat form. In (−)-17-*O*-methylaspidospermine $N^{B}$-hydrobromide[21], ring D is in a chair form, as it is in (±) deacetylaspidospermine $N^{A}$-hydroiodide[22], studied specifically to establish the form of this ring when $N^{B}$ was not methylated. In this compound ring C is in a distorted half-chair form. (±)-1-Acetyl-3-methylaspidospermidine $N^{B}$-methiodide[23] has also been analysed and, since there are two non-equivalent molecules, their close similarity in conformational details is evidence that, for molecules of this type, intermolecular forces play only a minor role on the distribution of strain throughout the molecule. In this work[23] a comparison of conformational features, including torsional angles, for all the aspidospermine structures then available is detailed. A closely related structure, kopsanone $N^{B}$-methiodide (11)[24] reveals rings A and B distorted from strict planarity as a result of steric requirement of fusion with

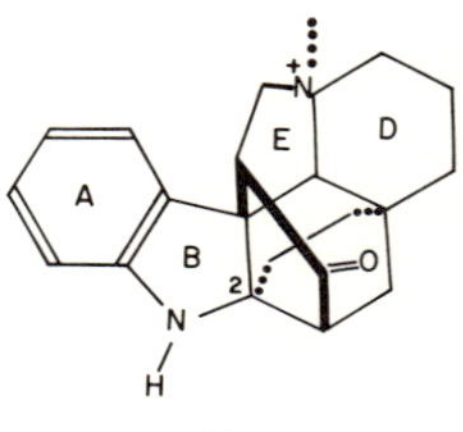


(11)

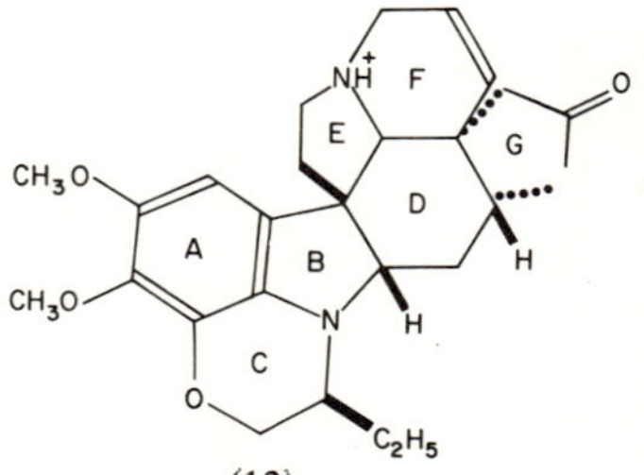


(12)

the cage-like component of the molecule. Rings B and C are both 2-envelopes. Ring D is in the chair form. Obscurinevine hydrobromide (12)[25] is an L-shaped molecule with the ABC ring system in one plane and the DEFG ring system roughly at right angles.

Vallesamidine $N^{B}$-methiodide (13)[26] differs from the usual aspidospermine skeleton only in that C(19) is attached to the indole nucleus at C(2) rather than at C(12). Meloscin $N^{B}$-methobromide (14)[27] is regarded as derived from

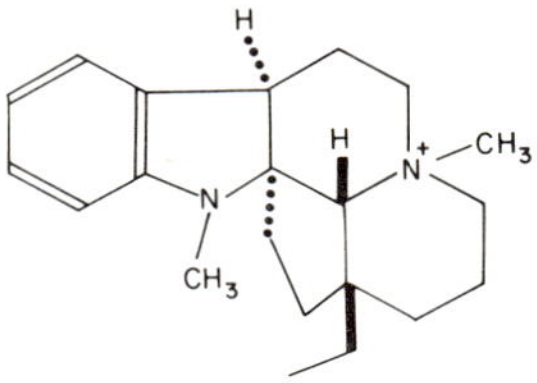


(13)

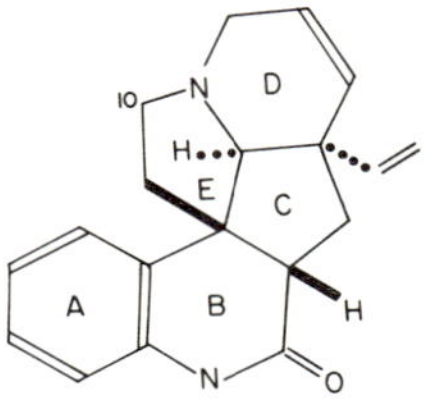


(14)

a precursor with an aspidospermine skeleton. Ring B appears to be in a skew chair form with ring A and the amide group near-planar, while ring D is a distorted boat. The oxindole alkaloid rauvoxinine methiodide (15)[28] has ring

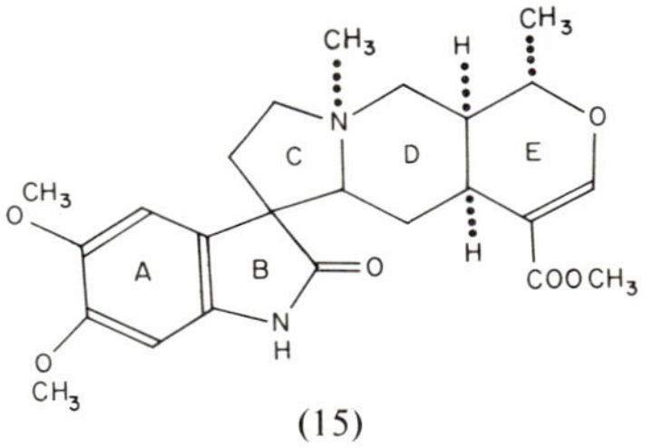


(15)

E a probable half-chair and ring D a chair. Other compounds with a spiro B/C ring junction are gelsemine [P17(a)][29], $N$-methylgelsemicine hydroiodide [P17(b)][30] and alkaloid C [P17(c)][31].

The structure of strychnine (P16)[32] with its complex ring fusion was a major step in the application of x-rays to natural products, as also was the

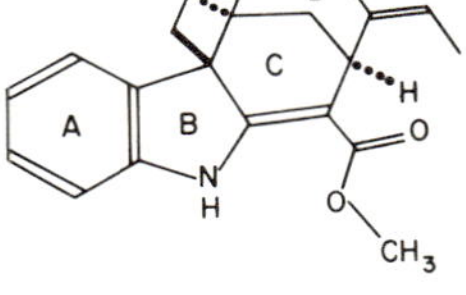


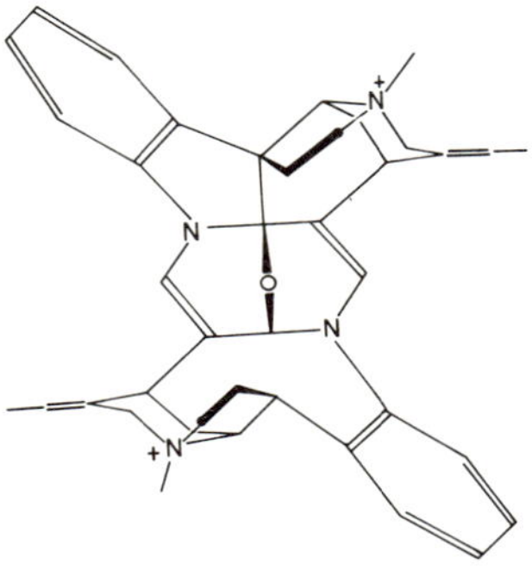


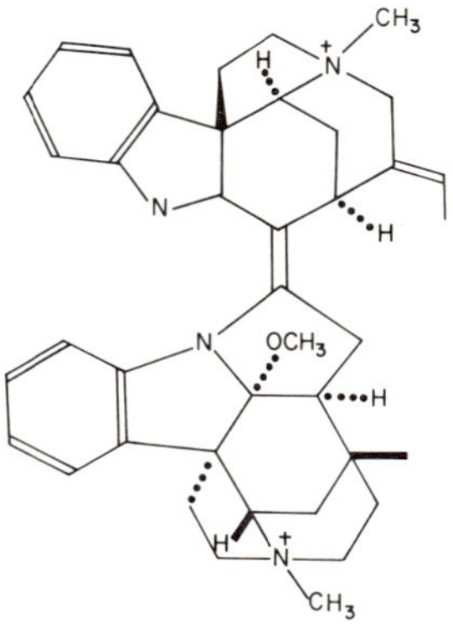


(16) (17) (18)

establishment of its absolute chirality[33]. Sewarine *N*⁴-methiodide (16)[34] is a structure related to that of strychnine in which ring C has a twisted boat form, ring D is also in a boat form while ring E is in an envelope form with C(5) below the plane of the other four atoms. A bis-indole structure with two-fold internal symmetry involving units which have similarities to that of sewarine is C-curarine di-iodide (17)[35]. In this structure, the quaternary N· · ·N distance is 8.5 Å and it is presumed that this feature is related to its curare action by competing with acetylcholine for receptor sites, *cf.* Pauling[36]. The structure of caracurine-II dimethiodide (P21)[37] involves fusion of two strychnine-like units with two-fold symmetry. Another compound with similar internal symmetry and also with certain ring features akin to those in strychnine is the calebash alkaloid calebassin-C di-iodide (P22)[38]. Treatment of this alkaloid with hot mineral acid yields anhydro-isocalebassine, whose methyl ether di-iodide (18)[39] has been analysed. As is evident, conversion involves a radical change in the central part of the molecule.

The examples above derive from the linkage of two indole units of the same structure. Examples involving two different units are haplophytine dihydrobromide (19)[40], which involves an aspidospermine moiety and one which resembles eburnamine, leurocristine (vincristine) methiodide (P18)[41], with a cleavamine unit and one of an aspidospermine type, and villalstonine (P19)[42], of which one component matches the structure of pleiocarpamine and the other has certain similarities to the structure of macrosalhine (6)[13]. Isocinchophyllamine (20)[43] also involves two dissimilar moieties. The two rigid half-molecules are oriented at an angle of approximately 75°. Rings C and C′

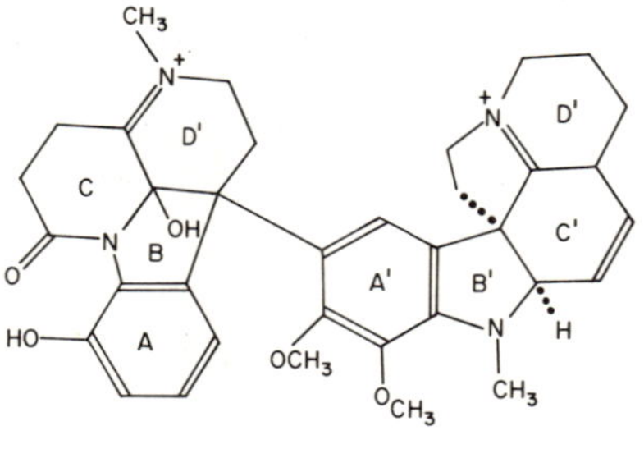


(19)

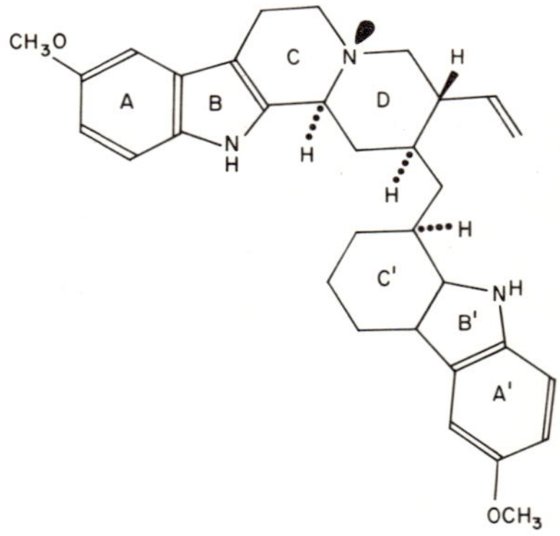


(20)

are in the half-chair form, while ring D is in the chair form. The two moieties in dibromovobtusine (21)[44] are very similar in stereochemistry. The main difference relates to the C rings, ring C′ being in a chair form while ring C appears to have a slightly distorted boat form.

Calycanthine dihydrobromide [P20(a)][45] involves a redistribution of bonds between the indole units, whereas in chimonanthine dihydrobromide [P20(b)][46] there is a single bond between the two units. In hodgkinsine trimethiodide (22)[47], different features occur. Two units are linked as in chimonanthine

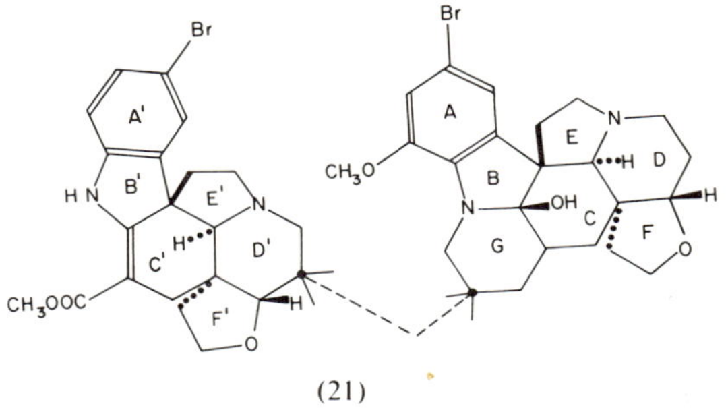


(21)

but here they are opposite in chirality and hence internally compensated, while the third unit is attached at C(8) of the benzene ring of the second unit.

An indole alkaloid with a high degree of oxygenation is dichotine hydrobromide (23)[48]. In the free base the $N^B$ nitrogen is bonded to the carbonyl carbon to give a zwitterion. Protonation occurs on the carbonyl oxygen and a C—N full bond is formed.

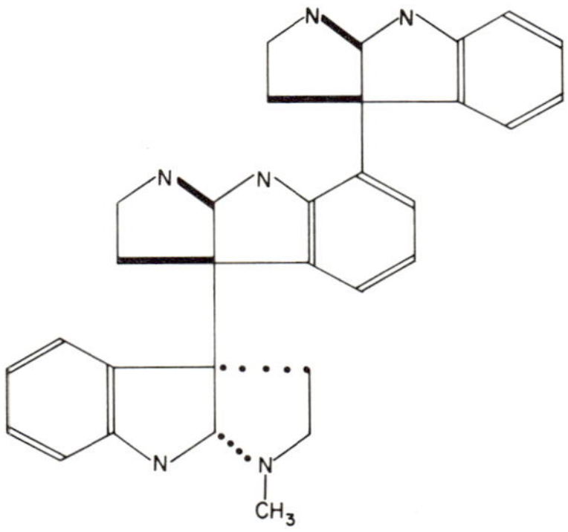


(22)

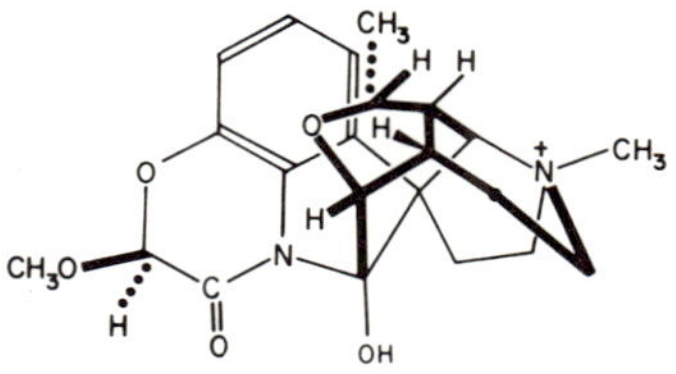


(23)

The unsubstituted parent compound of the yohimbin family, yohimbane, has been studied[49] as the hydrobromide (24). Ring system AB is essentially planar, ring C is in the half-chair form and rings D and E are in the chair form. Yohimbine hydrochloride (25)[50] has a similar ring system. 7α-Acetoxy-7*H*-yohimbine (P15)[51] and reserpine (26)[52] are similar in that rings C, D and E are

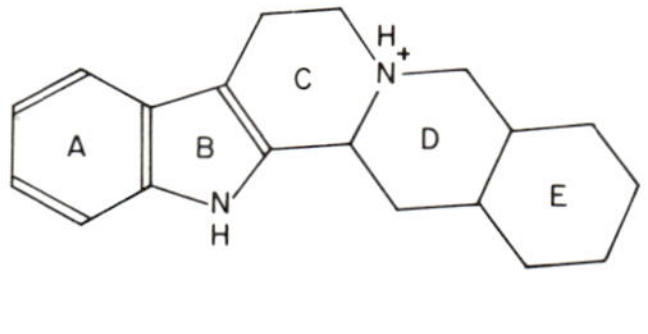


(24)

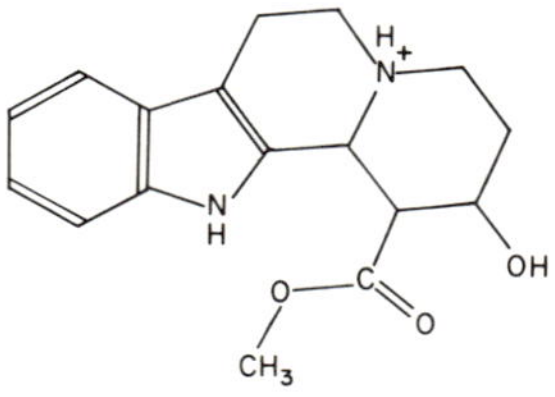


(25)

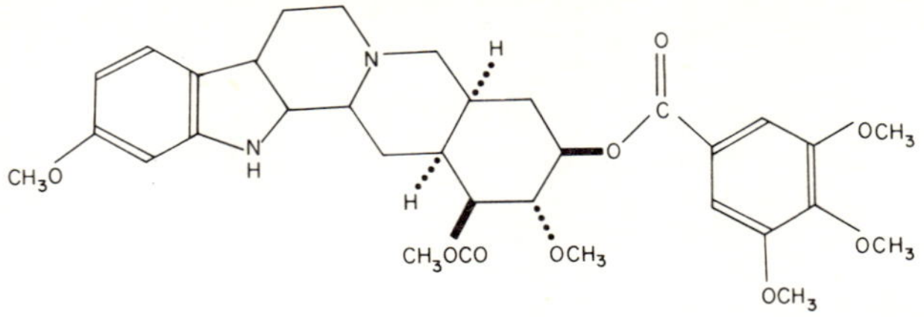


(26)

in the chair form. It is noted, however, that the conformation of rings C and D about the common C—N bond is different in the two compounds. Serpentinine dihydrobromide (27)[53] is a bis-indole alkaloid involving two modified yohimbine units. In this derivative form, rings A′, B′ and C′ are mainly coplanar as far as C(14′) and C(21′). Rings D′ and E′ are in the half-chair form. For the other moiety, rings A, B and C are coplanar as far as C(3) and C(6) while ring C is in the half-chair and D in the boat form. In lysergic acid

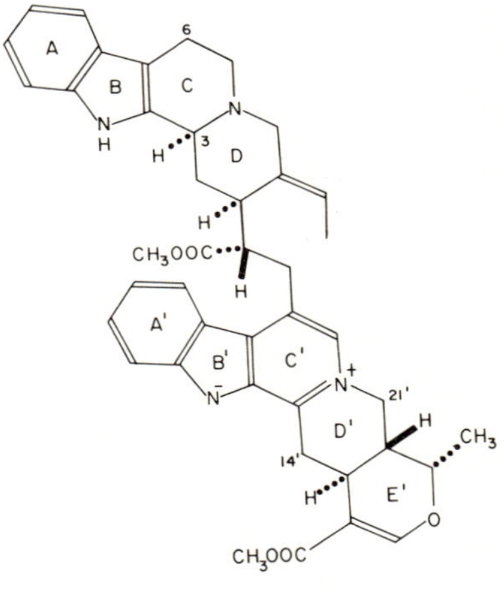


(27)

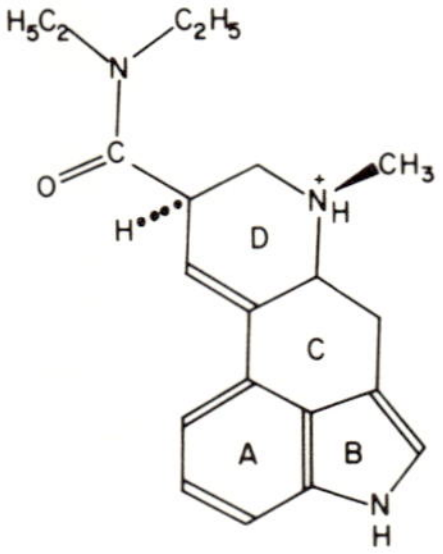


(28)

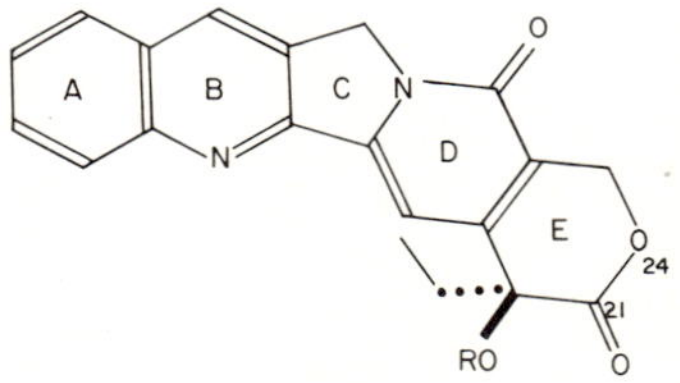


(29)

diethylamide *o*-iodobenzoate (28)[54], ring C takes approximately a sofa conformation with C(5) out-of-plane: the side chain, essentially planar except for the methyl groups, lies approximately at right angles to the mean D ring plane. In camptothecin iodoacetate (29)[55], rings A to D are closely parallel while C(21) and O(24) deviate from this plane by 0.69 and 0.73 Å so that ring E has a boat form.

## 6.3 DITERPENE ALKALOIDS

The structures in this group generally involve fused ring systems and can be arranged in three sub-groups. In the first, the nitrogen is involved only in one ring and ring B is six-membered, lucidusculine hydroiodide (P26)[56] belonging to this group. Ring A is in the boat form owing to hydrogen bonding between the hydroxy group and the nitrogen. In the case of denudatine (30) both the base[57a] and its methiodide[57b] have been analysed. Here ring A has the chair form and the CD ring system forms a cage. Ring B is in a boat form bridged

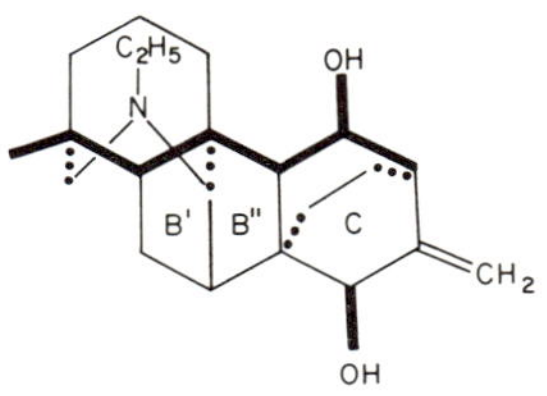


(30)

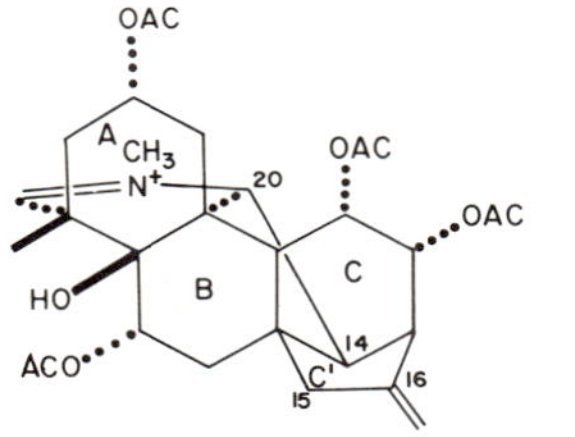


(31)

by C(20). A derivative of anopterine, tetra-acetylazomethineanopteryl alcohol methiodide (31)[58] involves a C(20)—C(14) link. Ring A′ is in a half-chair form while ring C is in a chair form with C(15)—C(16) bridging. An unusual feature is the hydroxy group at C(5).

In the second group, ring B is seven-membered. One such structure is that of an intermediary in the synthesis of delphinine, determined to establish the configuration of the methoxy group at C(1). This compound, delphinine aromatic product acid oxalate (32)[59], shows that ring A is in the chair form with the C(1) methoxy group *cis* to the N bridge. Ring B′ is in a distorted envelope form while ring B″ is in a distorted half-chair form.

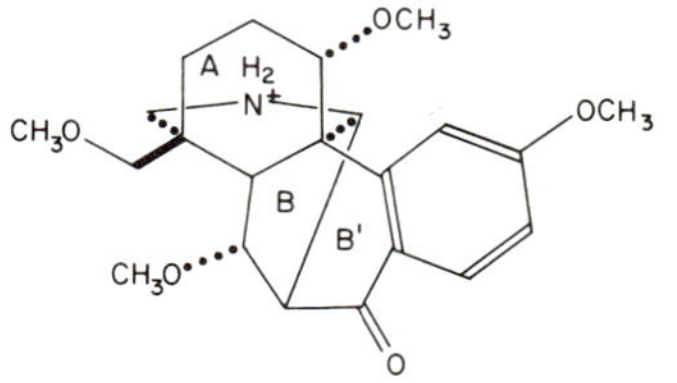


(32)

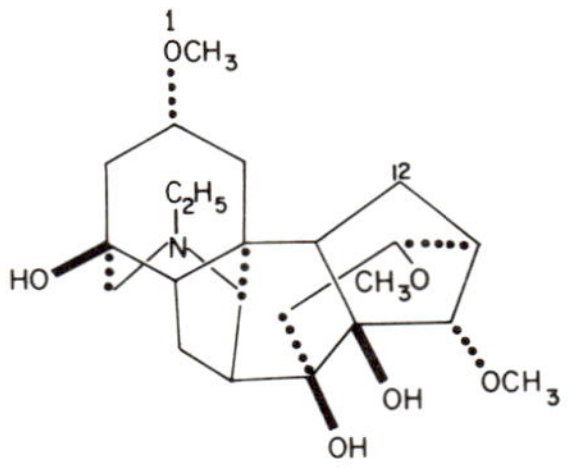


(33)

Lappaconine hydrobromide (33)[60] is also in the second group. In this structure, ring A is said to be a strained and flattened boat, owing to the intramolecular bond between O(1) and N and the avoidance of contact O(1)· · · C(12) if ring A had a chair form. A similar feature was evident in heteratisine hydrobromide (P25)[61] and in delcosine hydrobromide[62]. Others in this group are deoxymethylenelycoctonine hydroiodide (P23)[63] and demethanolaconinone hydroidide (P24)[64]. In these latter compounds, however, this condition does not hold and ring A is in the chair form.

A number of compounds in the third group, where N is common to three rings, has been studied. Hetisine hydrobromide (P27)[65] was the first member studied, illustrating the close approximation to plane symmetry not only of the bicyclo[2,2,2]octane system but also of the BB′CC′ ring system. This is modified in 2-deoxyhetisine cyclopropyl derivative methiodide (34)[66] as a result of rearrangement following lithium aluminium hydride reduction, but holds for some other members such as kobusine methiodide (35)[67], anhydroignavinol methiodide (36)[68] and hypognavinol methiodide (37)[68b]. The remaining examples involve a keto group: spiradine A methiodide (38)[69],

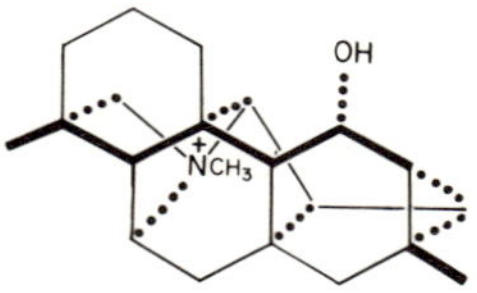


(34)

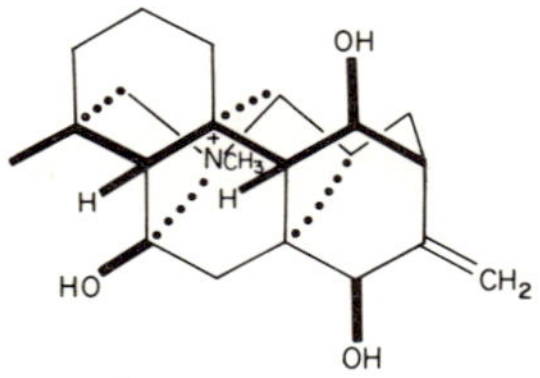


(35)

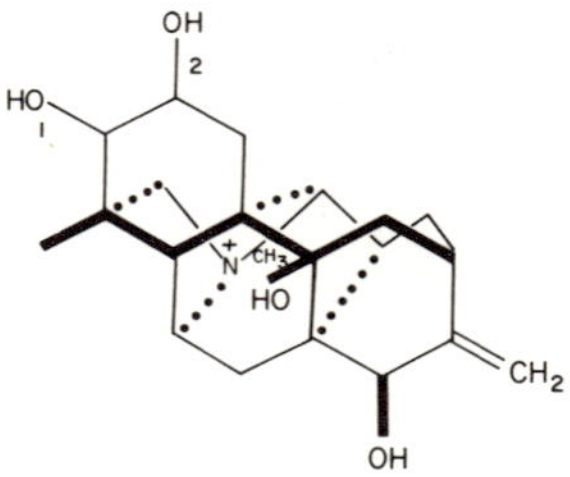


(36) 1 ≡ ▬▬ 2 ≡ ....    (37) 1 ≡ ....  2 ≡ ▬▬

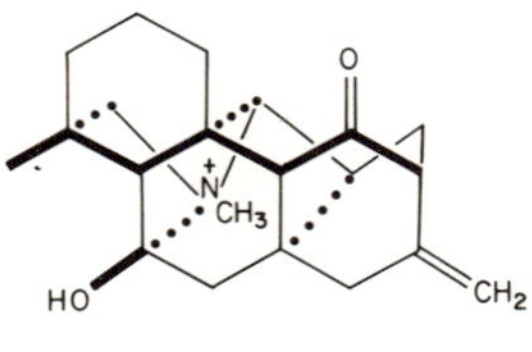


(38)

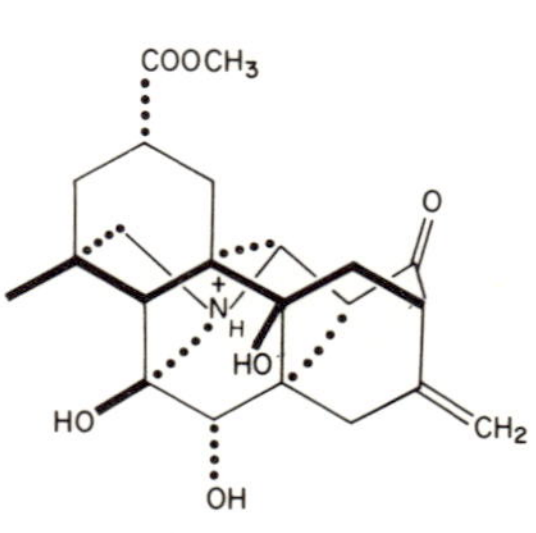


(39)

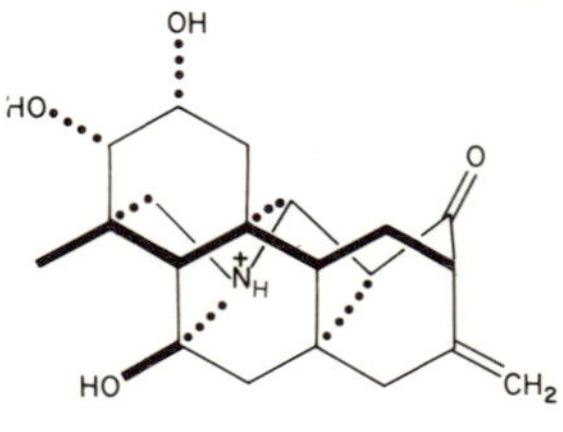


(40)

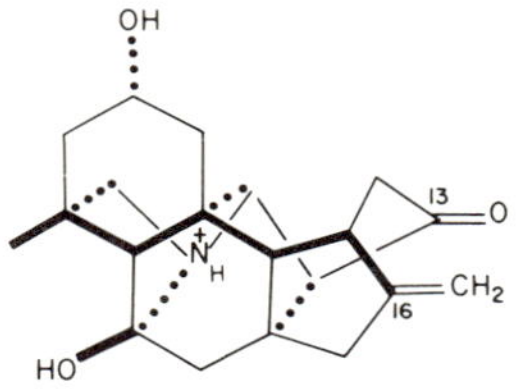


(41)

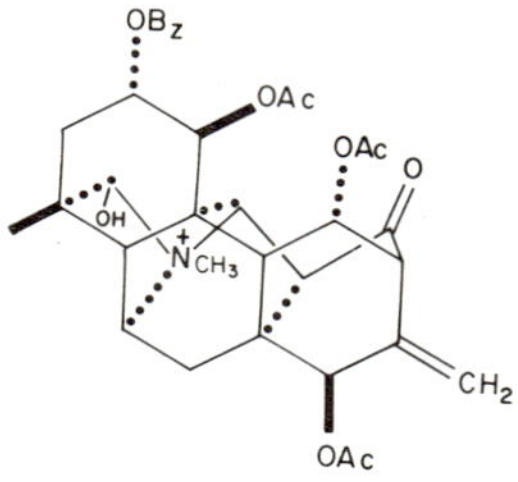


(42)

miyaconitine hydrobromide (39)[70], hetidine hydroiodide (40)[71], delnudine hydrochloride (41)[72] and vakognavine hydriodide (42)[73]. In delnudine the C═O and C═$CH_2$ groups lie rather close together, C(13)—C(16) = 2.90 Å, and oriented in similar directions and the u.v. absorption indicates their interaction. A similar effect has been noted in the case of the miyaconitine derivative. It is suggested in the case of delnudine that this interaction causes ring D to have a chair form whereas it is boat-shaped in hetisine. In the cases where a fourth bond to $N^+$ is formed on creation of an ionic derivative, such as hetidine, miyaconitine and vakognavine, the structure of the original base has to be deduced by independent evidence.

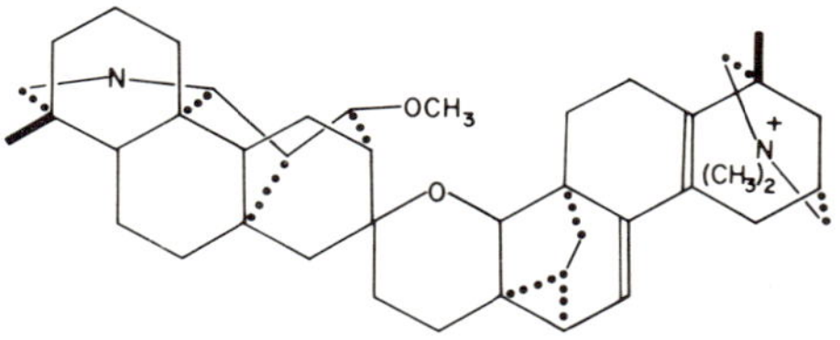


(43)

Staphisine monomethiodide (43)[74] is a dimer structure whose units are of the atisine type. A rearrangement in the B-unit has apparently given rise to the conjugated diene system.

## 6.4 PROTOPINE AND STRUCTURALLY SIMILAR ALKALOIDS

Both cryptopine (44a)[75] and protopine (44b)[76] have been analysed. In each case the ten-membered ring is buckled with C═O, N—$CH_3$ *cisoid* bringing N(7) and C(14) to within ca. 2.57 Å. Also, the grouping around the nitrogen atom and the N(7)· · ·C(14) direction is close to tetrahedral and bond formation readily occurs with acid salts, as in the case of coulterpine hydrobromide (45)[77] where rings B and C appear to be half-chair forms. Of similar structure are the tetrahydroprotoberberine alkaloids capaurine hydrobromide (46)[78], capaurimine *p*-bromobenzoate (47)[79] and the hydrobromide of a position isomer of capaurimine (48)[80]. In these cases the B/C rings are *cis*-fused because of repulsion between the oxygen substituent at C(1) and the C(13) hydrogen.

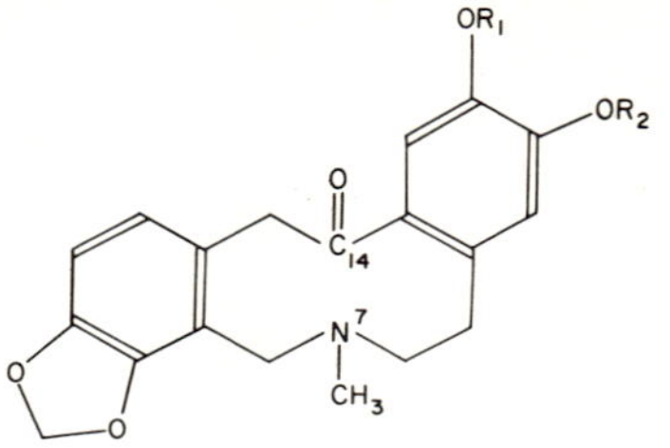


(44) (a) $R_1 = R_2 = CH_3$
(b) $R_1 = CH_2 = R_2$

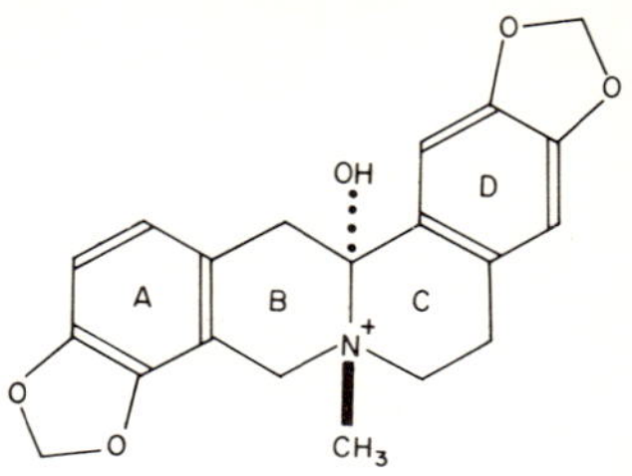


(45)

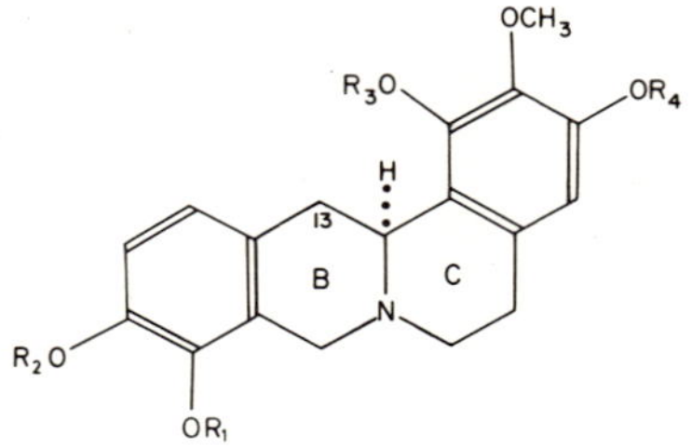


(46) $R_1 = R_2 = CH_3$, $R_3 = H$, $R_4 = CH_3$

(47) $R_1 = CH_3$, $R_2 = -COOC_6H_4Br$, $R_3 = H$, $R_4 = CH_3$

(48) $R_1 = H$, $R_2 = R_3 = CH_3$, $R_4 = OH$

Two tetrahydroisoquinoline structures may be mentioned here: anhalonine hydrobromide (49)[81] and (+)-*O*-methylanhalonidine hydrobromide (50)[81]. A simple benzyltetrahydroisoquinoline, typical of precursors of many of the following compounds, has been analysed: coclaurine hydrobromide (51)[82]. Three aporphine structures have been studied, bulbocapnine methiodide[83], isoboldine hydrobromide (52a)[84] and leucoxine hydrobromide (52b)[85], but virtually no structural details have been recorded. Analysis of (−)-*O*-methyllaurepukin (53)[86] shows the oxygen in the equatorial and the methyl in the axial positions. The dihedral angle between the two benzene rings is ~30°. In the case of thalphenine iodide (54)[87] the further constraint of a methyleneoxy bridge reduces this angle to ~18°. In cularine methiodide (55)[88] the

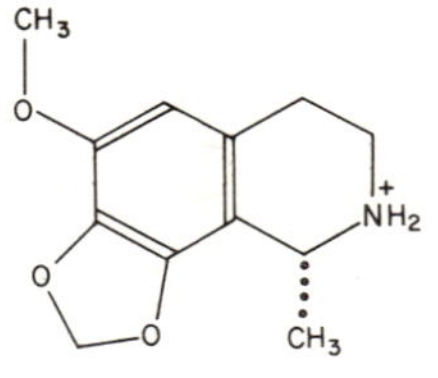


(49)

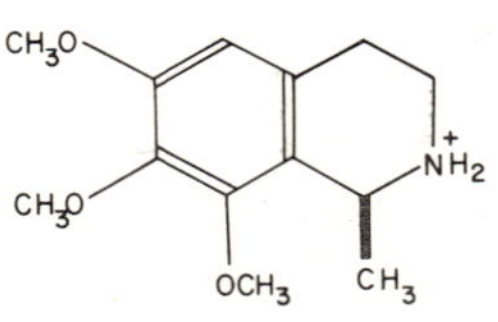


(50)

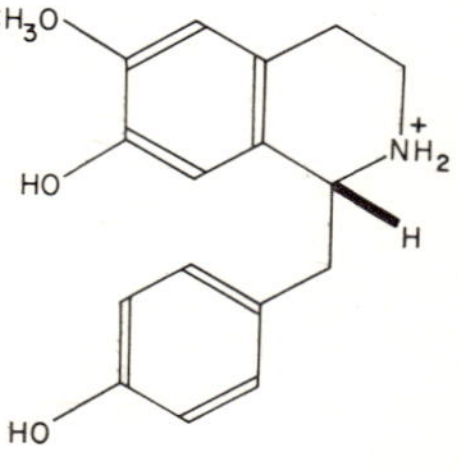


(51)

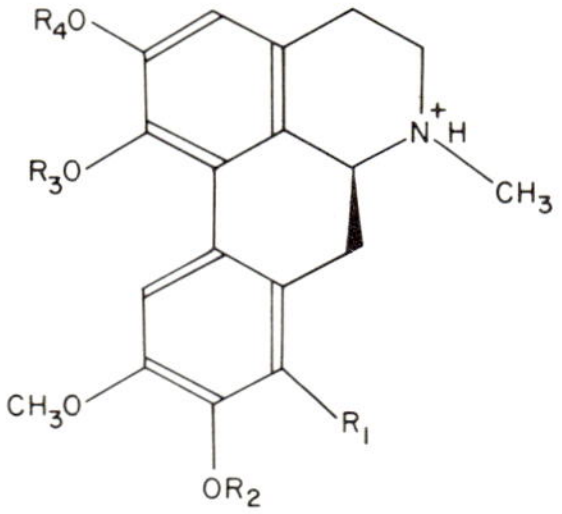


(52) (a) $R_1 = R_2 = R_3 = H$, $R_4 = CH_3$
(b) $R_1 = OH$, $R_2 = CH_3$, $R_3 =$ CH $R_4$

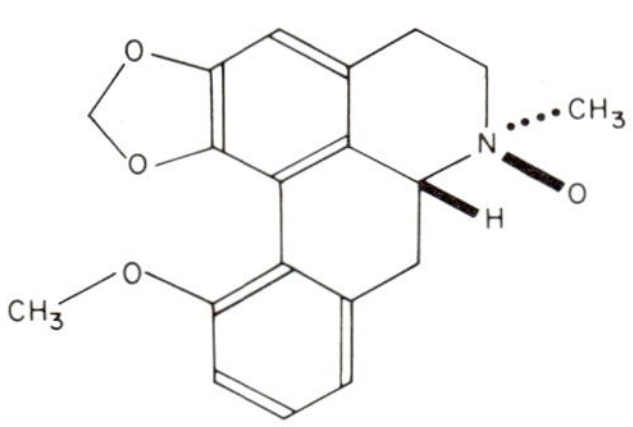


(53)

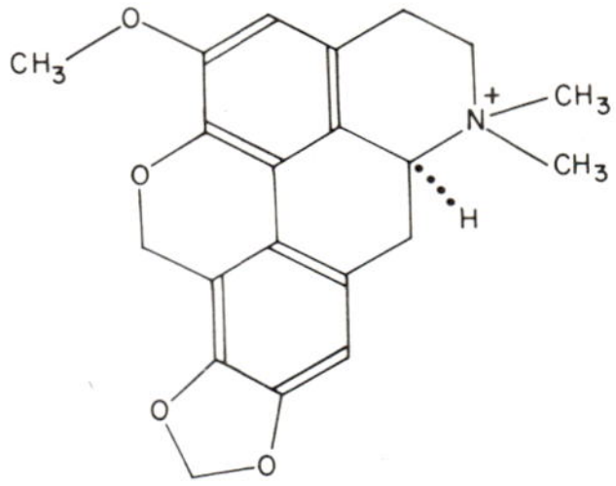


(54)

benzene rings are separated by an ether oxygen and the dihedral angle between the A and D rings is of the order of 57°. The dihydro-oxepin ring is of a twist-boat shape, the oxygen atom being the bow.

Dimeric structures in this group are (+)-tubocurarine chloride (56a)[89] and its trimethylated derivative (56b)[90]. Although these structures are compact, there are minor conformational differences, the main difference relating to the non-phenolic tetrahydroisoquinoline ring. In the tubocurarine molecule the distance between the nitrogen atoms is ~9 Å whereas in the trimethylated species it is 10.7 Å. In both structures the two positive centres are on opposite sides of the molecule so that attachment of both to a receptor may be regarded as unlikely. Interpretation in detail of the physiological action of these two

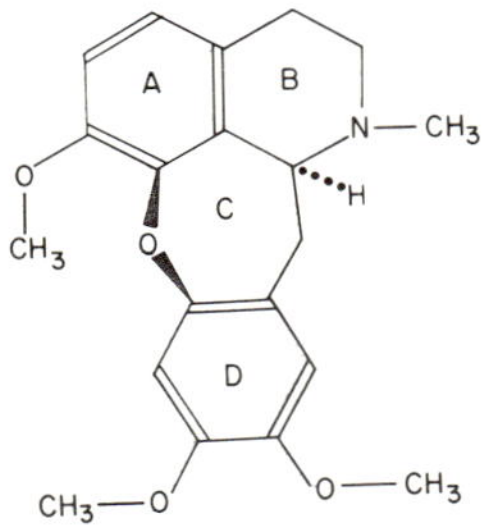


(55)

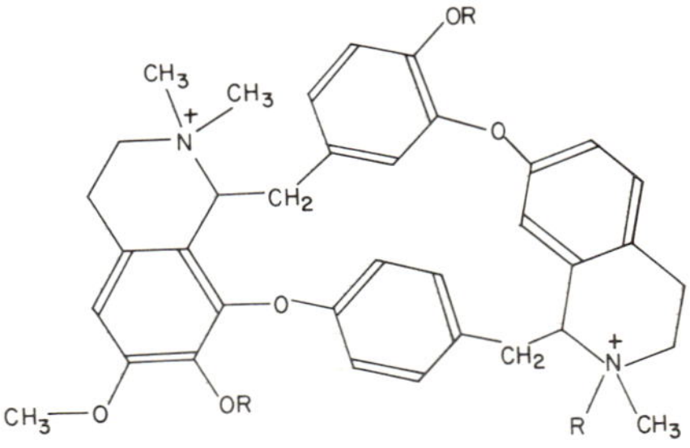


(56) (a) $R = H$ (b) $R = CH_3$

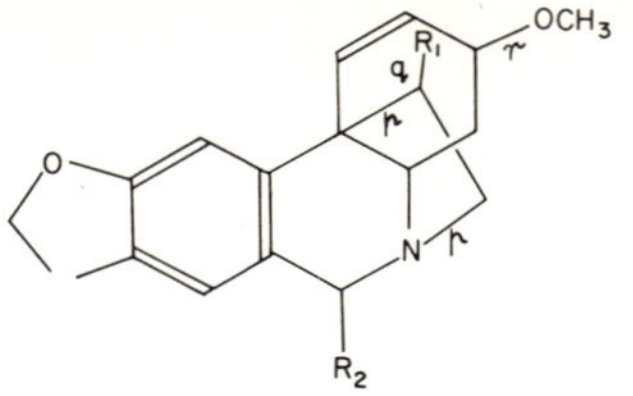


| | | |
|---|---|---|
| (57) | (a) $R_1 = OCOC_6H_4Br$, $R_2 = H$ | p = ···, q = ···, r = ▬ |
| | (b) $R_1 = H$, $R_2 = OH$ | p = ▬, r = ···, s = ··· |
| | (c) $R_1 = OH$, $R_2 = OH$ | p = ▬, q = ▬, r = ▬, s = ··· |

molecules in relation to the structural features depends largely on the particular model espoused.

Of the amaryllidaceae alkaloids, haemanthamine *p*-bromobenzoate (57a)[91], 6-hydroxybuphanidrine methiodide (57b)[91] and 6-hydroxycrinamine (57c)[92] have been analysed. To establish the configuration of the benzylic hydroxy group in this series, the structure of lycorenine methiodide (58)[93] was determined. Ring B is in a half-chair form, as also is ring C; ring D

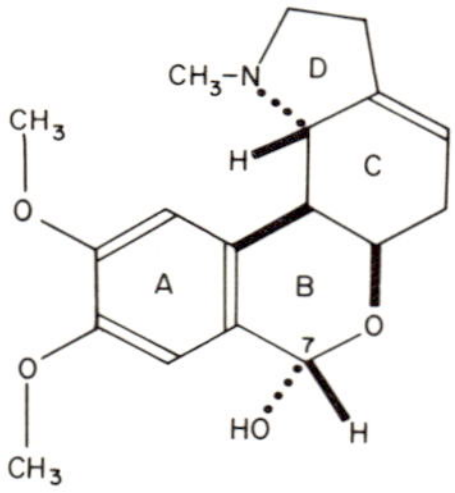


(58)

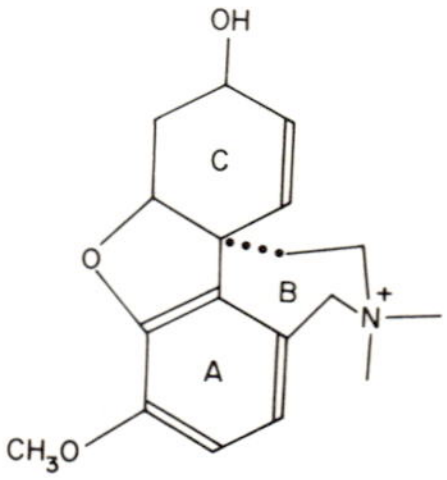


(59)

appears to be in an envelope form. Galanthamine methiodide (59)[94] indicated the existence of an internal hydrogen bond to the ether oxygen, with ring C in a half-chair form. In corynoline *p*-bromobenzoate (60)[95], the B/C ring function is *cis*-fused, the B and C rings having a half-chair and twist half-boat conformation, respectively; the dihedral angle between rings A and D is 73°.

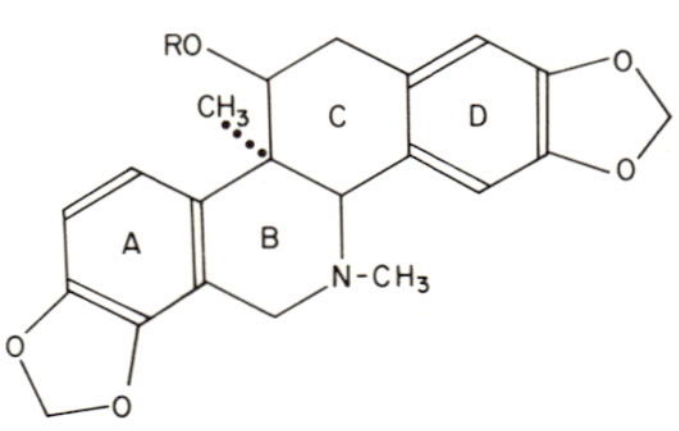


(60)

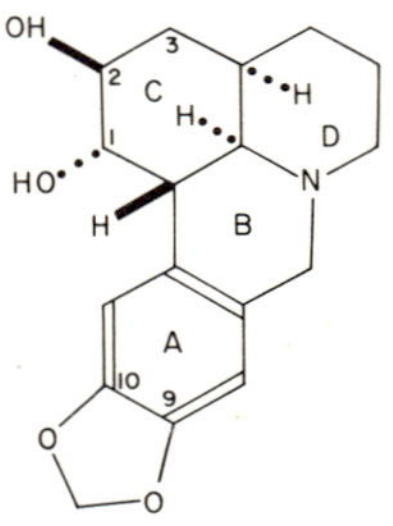


(61)

In dihydrolycorine hydrobromide (61)[96], ring B is in a distorted boat form and ring C a slightly distorted chair; ring A and its substituent atoms are essentially coplanar. Narcissidine, studied as the hydrobromide[97] and the methiodide (62)[98], has some similarities. In this molecule, ring D, as well as ring A, is planar while ring C is in the chair form and ring B is a boat form. In cryptostyline-I (63)[99], ring B appears to be in a sofa form with N out-of-plane.

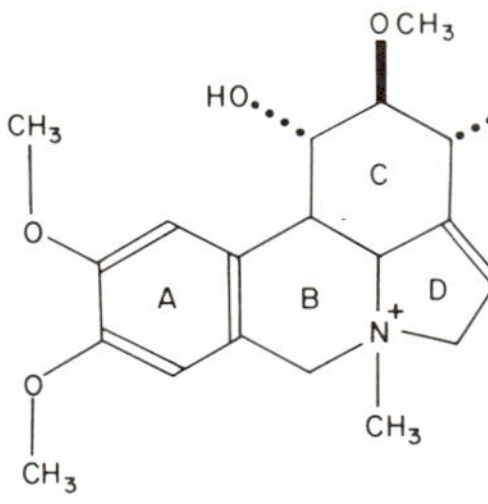


(62)

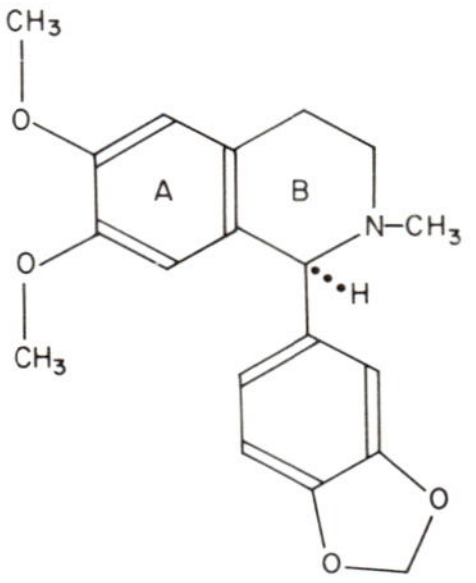


(63)

Other structures involving a methylenedioxy ring are *N*-methylrhogeagenine iodide (64)[100] in which ring B is in a chair form while ring C adopts a sofa form with C(5) 0.66 Å out of plane. The planes through rings A and D are almost at right angles. The structure of isolunine hydrobromide (65)[101] is

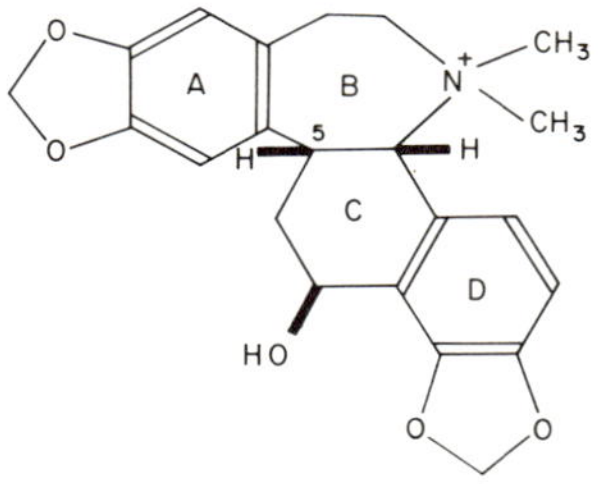


(64)

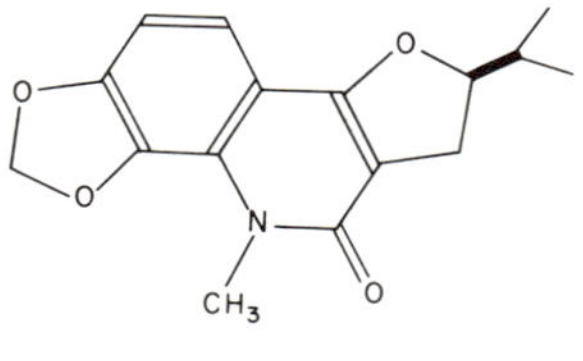


(65)

essentially planar, as also is that of bromodihydroacronycine (66)[102] although ring B is slightly boat shaped.

Of the hasubian alkaloids, 4-demethylhasubanonine *p*-bromobenzenesulphonate (67)[103] has been analysed. Ring B adopts a distorted half-chair conformation with C(9) and C(14) displaced +0.54 and −0.30 Å, respectively, relative to the plane through the other four atoms; ring C also has a distorted half-chair form. Ring D is close to an envelope form with C(14) out-of-plane. The dihedral angle of the ethylenic bond and C=O about (C6) —C(7) is 9°. In stephavanine hydrobromide (68)[104], the O-bridge between rings B and C causes considerable distortion in ring angles. Acutumine (69)[105] is related to hasubanonine but has a spiran junction. In this structure there

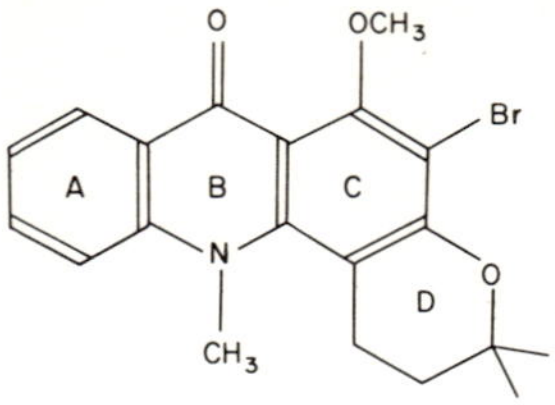


(66)

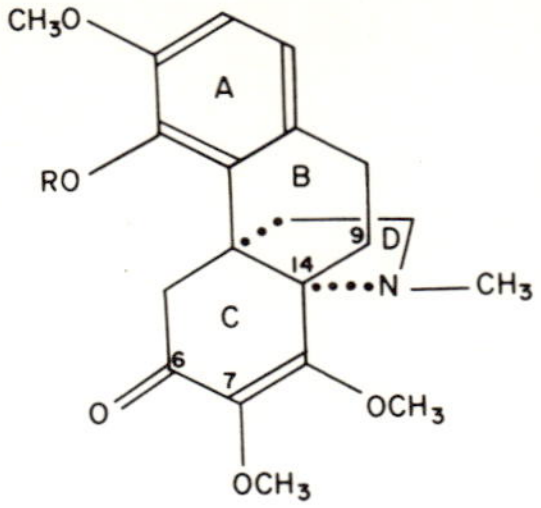


(67)

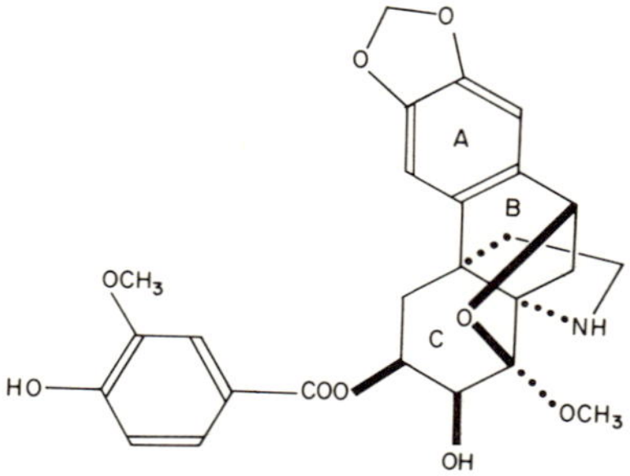


(68)

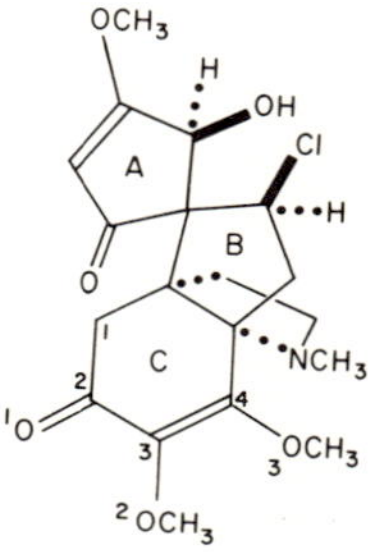


(69)

appears to be some ambiguity concerning O(1)—C(2) and C(3)—C(4) as being double bonds and O(3)—C(4) and C(2)—C(3) as single. Ring C is boat-shaped with C(1) and C(4), bow and stern, towards the heterocyclic ring. Acutumine acetate was also studied, the conformational details being closely similar to those of acutumine.

The following compounds have basic similarities to the codeine–morphine group. Codeine hydrobromide (P46)[106] has been further examined with reference to conformational details[107]. In the case of morphine (P46)[108], its absolute chirality was established later[109]. Because of a similarity in analgesic behaviour with that of morphine, the structure of pethidine hydrobromide (70)[110] has been examined. The hypothesised structural similarity was found

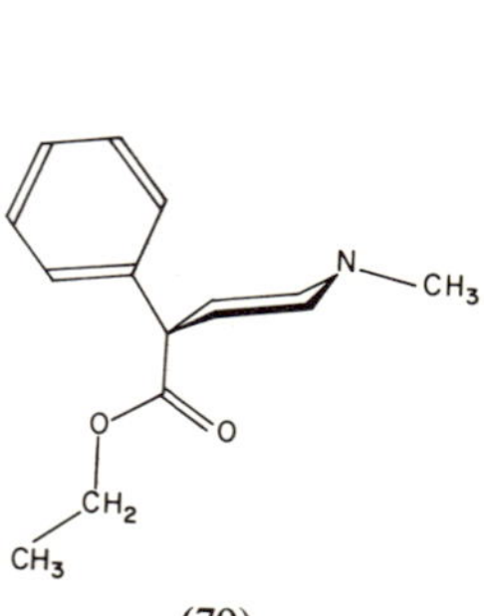


(70)

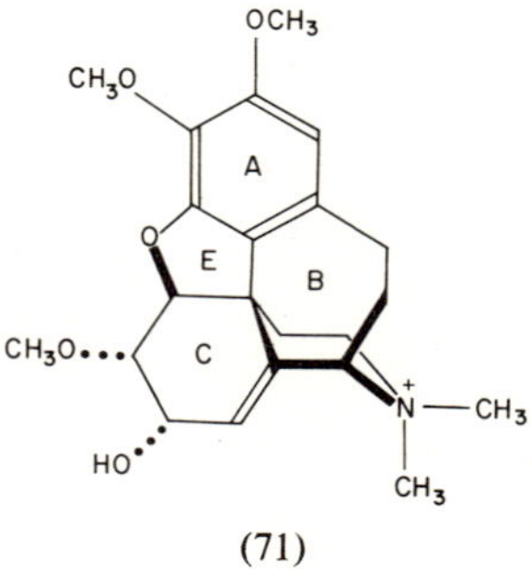


(71)

not to hold. In the homomorphine alkaloid kreysiginine (71)[109], increase of the size of ring B reduces the strain on ring E which exists in the morphine-type alkaloids. In the case of kreysiginine, acutumine and 4-demethylhasubaninone, their absolute configuration is established as opposite to that of the codeine group. In terms of their mode of synthesis, this difference does not have strong significance. A structure of similar type is the *O*-acetyl derivative of alkaloid CC-2 methiodide *O*-acetyl (72)[111]. Ring B adopts an envelope-like form with atoms C(4), C(5), C(8), C(11) and C(12) planar and this plane at 65° to the plane through atoms C(8)···C(11); ring D adopts a sofa form. A group of unusual dimeric benzylisoquinoline alkaloids is represented by dihydro-*O*-methylcancentrinemethine hydrobromide (73)[112]. This is composed of two fragments, one morphine-like and the other a modified cularine, joined by a spiro ring junction. The cularine fragment contains three heterocyclic ring systems. The seven-membered ring B is boat-like, ring D has a half-boat conformation while ring E is approximately planar.

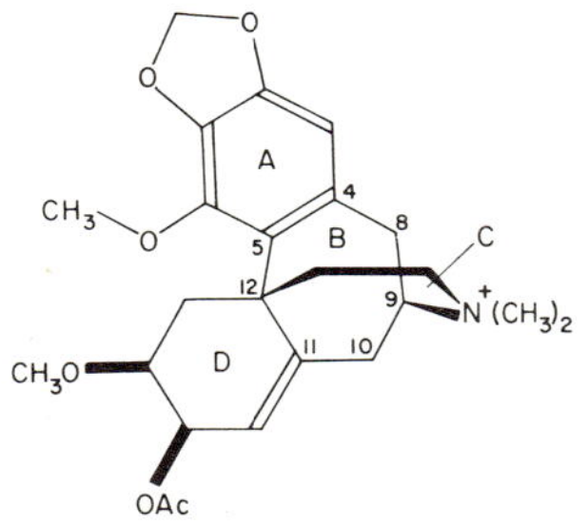


(72)

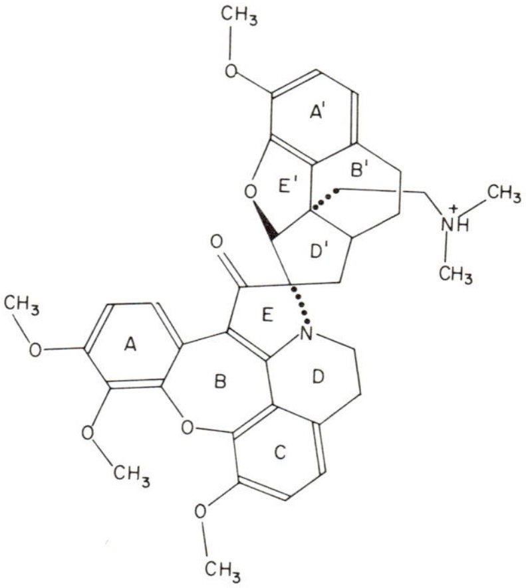


(73)

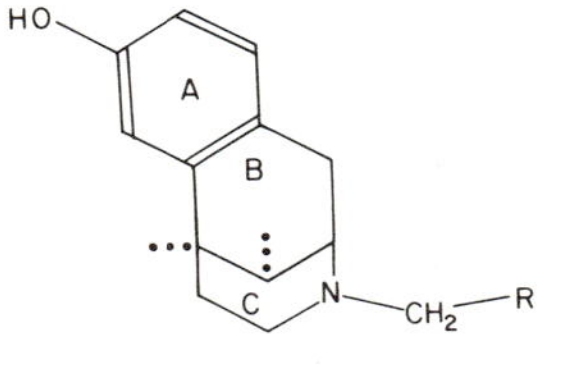


(74) (a) R = —CH——CH₂ / CH₂ (cyclopropyl: $-CH(CH_2)CH_2$)

(b) R = —CH=CH₂

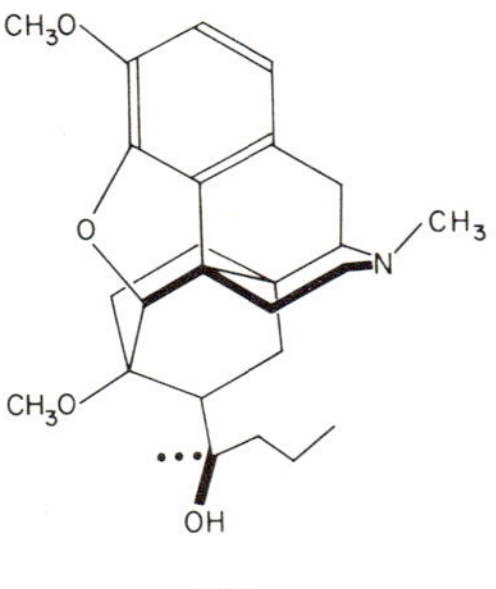


(75)

In the morphine-like moiety the stereochemical features are similar to the corresponding segments in morphine. The chirality is opposite to that of morphine (see kreysiginine, above). Rings D′ and E′, though puckered, are almost planar. Ring B′ has the form of a slightly-flattened boat. As part of studies to identify particular physiological activities of morphine, certain structural variants on fragments have been analysed. Thus 2-cyclopropyl-methyl-2′-hydroxy-5,9-dimethyl-6,7-benzomorphan (74a)[113] and 2-allyl-2′-hydroxy-5,9-dimethyl-6,7-benzomorphan hydrobromide (74b)[114] have both been studied. These molecules have the characteristic T-shape, ring B being in the half-chair form. In 19-propylthevinol hydrobromide (75)[118], the cage structure severely distorts the bonds to the benzene ring out of plane, although the benzene ring itself is planar.

## 6.5 QUINOLIZIDINE ALKALOIDS

Annotinine bromhydrin (P35)[116] was an early example of this group. It involved a complex grouping of ring structures, the four-membered ring being puckered while the six-membered rings were mainly in the chair form. Both *O*-methyl-lythrine hydrobromide (76a)[117] and lythridine methiodide (76b)[118] have been analysed. The biphenyl groups are skewed and linked to

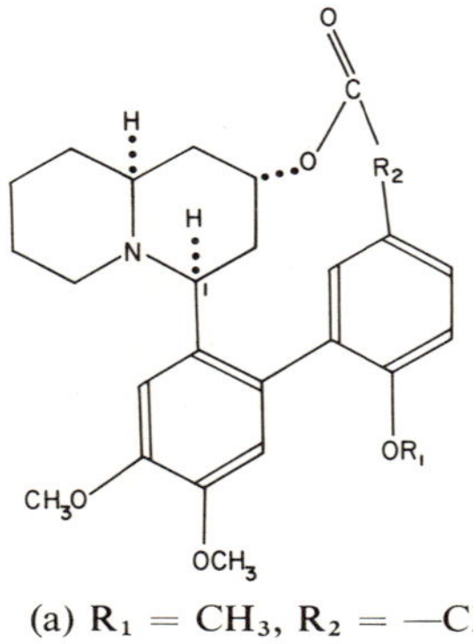


(76) (a) $R_1 = CH_3$, $R_2 =$ —CH=CH—

(b) $R_1 = H$, $R_2 =$ —CH—C(H)(OH)—

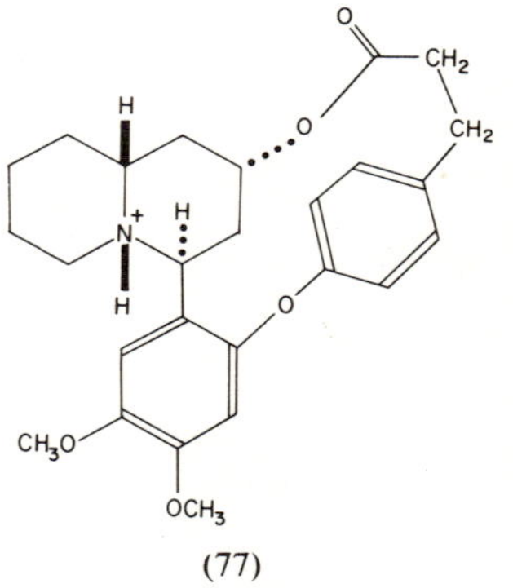


(77)

the quinolizidine ring system at two points, directly at C(1) and through an ester group. A related structure in which the biphenyl is changed to a biphenyl ether is vertaline hydrobromide (77)[119], where the ring fusion of the quinolizidine ring is *cis* and the lactone and biphenyl ether links are in the axial and equatorial positions, respectively. Another related biphenyl structure is that of bromolythranine hydrobromide (78)[120], where the torsion angle of the biphenyl is 75°. Other structures which may be placed in this group are cryptopleurine (P37)[121] neothiobinupharidine dihydrobromide (P38)[122],

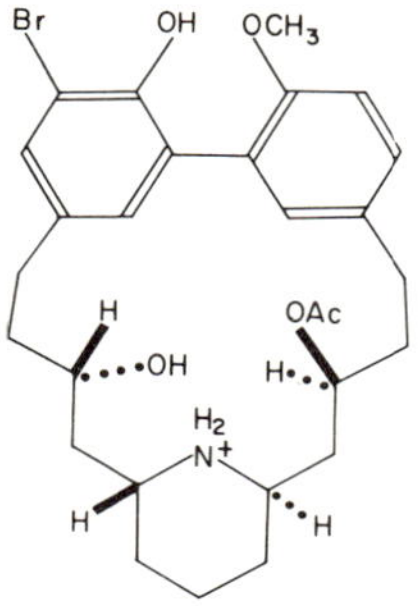


(78)

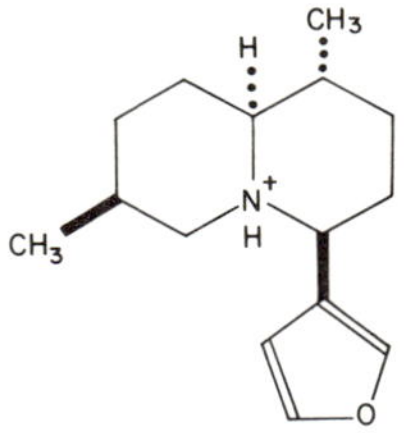


(79)

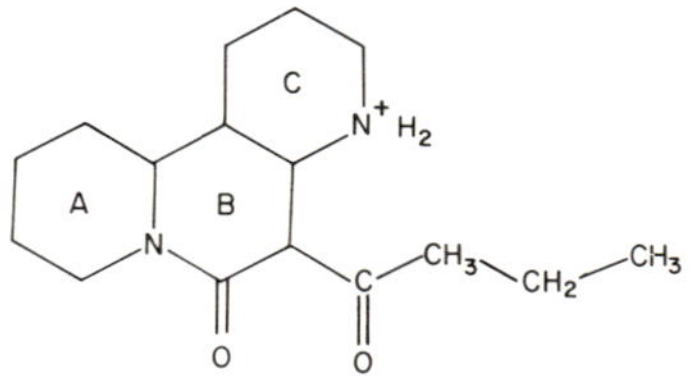


(80)

deoxynupharidine hydrobromide (79)[123] and haloxine hydrobromide (80)[124]. In the last, rings A and C are essentially in the chair form while ring B is in the half-chair form.

## 6.6 ERYTHRINA ALKALOIDS

Earlier studies in this group were of erythraline hydrobromide (P28)[125] and dihydro-β-erythroidine hydrobromide (P29)[126]. In addition to providing valuable dimensional details, the evidence on absolute chirality contradicted the prediction made from o.r.d. studies, which was based on the diene rule[127]. From well-established absolute structures based on x-ray studies, an explanation for the cause of the anomolies associated with the diene rule in such cases has been derived[128]. The A ring in both structures is in a half-chair form while ring D in the second compound is in a shallow boat form due to the association of the double bond and the lactone group. A compound closely related to erythraline is erythristemine, studied as the 2-bromo-4,6-dinitrophenolate (81)[129]. In tazettine methiodide (82)[130], ring A is planar but C(8) and C(12b) are slightly displaced (−0.07 and +0.04 Å, respectively). Rings B and C have a slightly distorted half-chair form. Ring D is a half-chair with atoms C(4a), N(5) and C(6) as the reference plane. Junction B/D is *cis* and C/D *trans*.

Two structures which involve an additional carbon in ring C are schelhammerine hydrobromide (83)[121] and cephalotaxine methiodide (84)[132], while

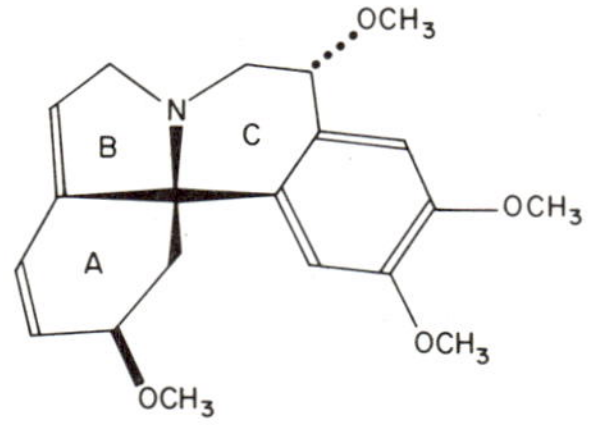


(81)

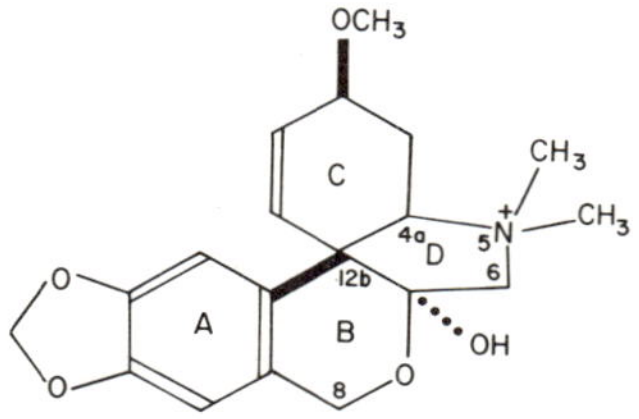


(82)

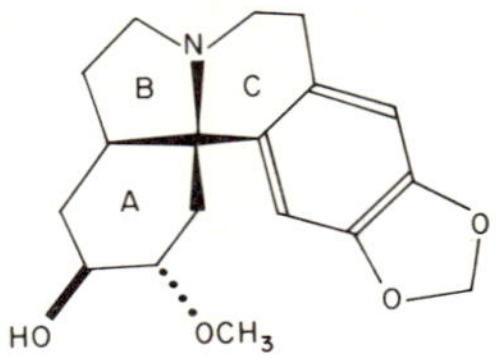


(83)

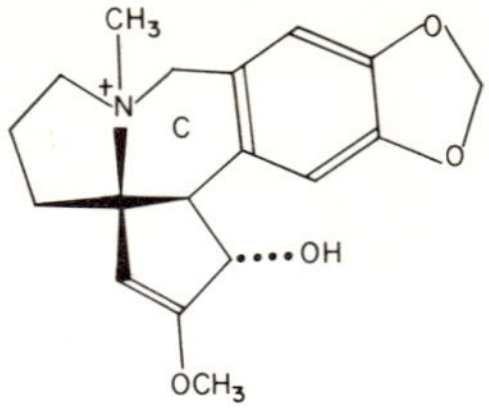


(84)

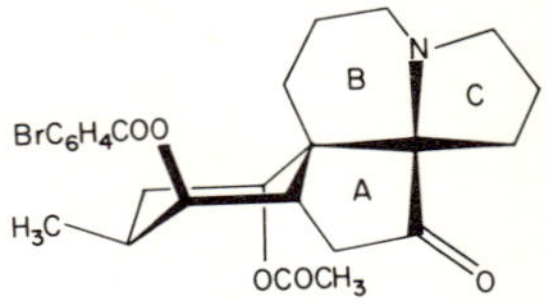


(85)

13-acetyl-8-*p*-bromobenzoylserratinine (85)[133] involves a rather different disposition of rings around the spiro junction. In this group, ochotensine (P30)[134] has been included on the basis of structural similarity.

## 6.7 PYRROLIZIDINE ALKALOIDS

In recent years, further studies have been added to the earlier work on jacobine bromhydrin (P39)[135], an alkaloid from *senecio kirkii* (P42i)[136] and of retusamine α′-bromo-D-camphor-*trans*-π-sulphonate (P41)[137]. In clivorine (86)[138], the configuration at C(7) is the same as for retusamine and jacobine, but the relationship at C(2) is opposite to that in jacobine and retusamine while at C(3) clivorine and jacobine are the same but differ from that in

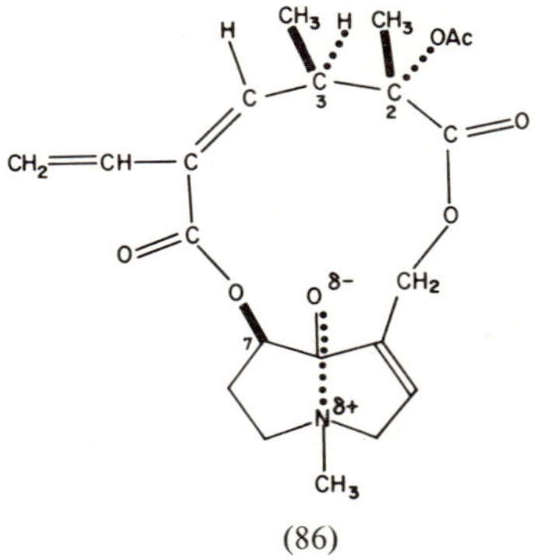


(86)

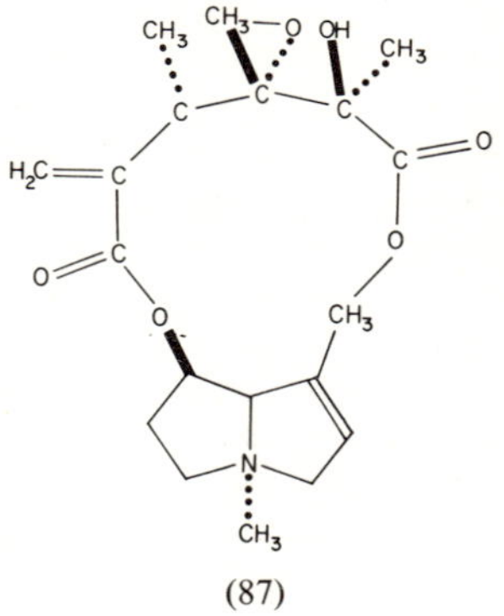


(87)

retusamine. Another structure closely related is that of swazine methiodide (87)[139]. The conformation of the 12-atom diester ring is similar to that found in jacobine and clivorine. Although at first sight potentially flexible, in fact the planar and preferred *trans* conformation of the ester groups greatly restricts its form. In swazine, the C═$CH_2$ and C═O groups are thus constrained to a *cis* disposition (dihedral angle 54°) and hence with limited conjugation, in accord with the u.v. spectrum. The ester groups (or mainly the C═O groups) point in opposite directions on either side of the mean plane of the 12-membered ring. Fulvine (88)[140] appears to have an 11-membered ring whereas

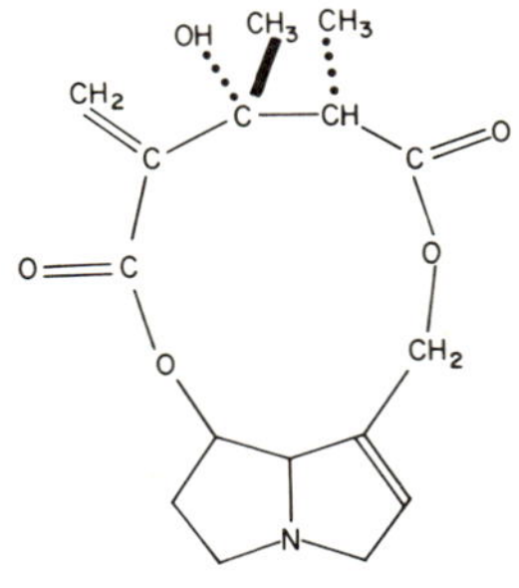

(88)

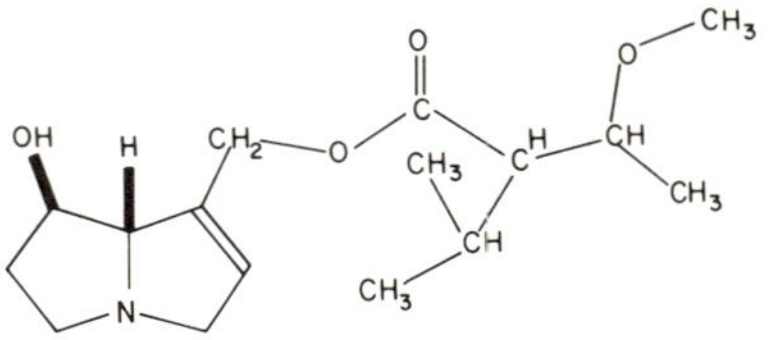

(89)

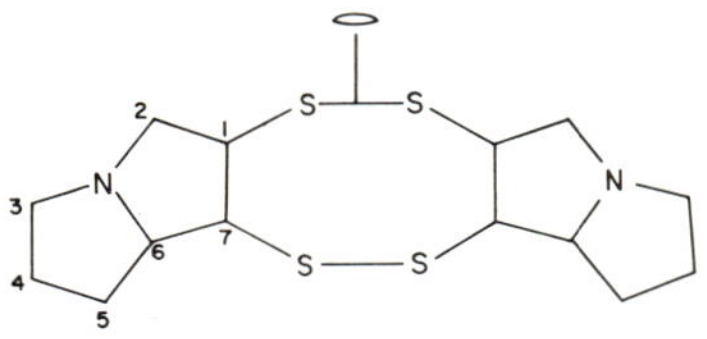

(90)

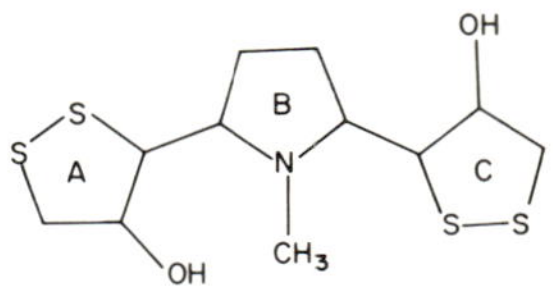

(91)

heliotrine (89)[141] involves only one ester linkage. The two five-membered rings are reported to have an *endo* buckling, in contrast to the situation in swazine in which the saturated five-membered ring is *exo* buckled. It has been observed in these structures, e.g. retusamine and clivorine, that the N—C(8) distance is ~2.0 Å and that the C(8)· · ·O distance at ~1.26 Å is longer than the normal value of 1.22 Å. This transannular effect in an 8-membered ring can be compared with similar effects noted in the 10-membered ring systems of protopine and cryptopine.

Thelepogine methiodide (P44)[142] may be grouped here for convenience. The molecule of cassipourine (90)[143] possesses a non-crystallographic two-fold axis which runs through the centres of the two disulphide bonds. Each pyrrolidine ring consists of four essentially planar atoms with the fifth displaced 0.59 Å out-of-plane, C(1) below and C(4) above the respective plane. A second alkaloid from the same source and also containing sulphur atoms is gerrardine (91)[144]. Ring systems B and C are nearly coplanar, to within 0.37 Å, while ring A is essentially at right angles to that plane. Strong intra-molecular hydrogen bonding between O(1) and N presumably determines this feature.

## 6.8 DAPHNIPHYLLINE ALKALOIDS

A number of these alkaloid structures have been analysed, all having a complex arrangement of fused rings. Although daphniphylline hydrobromide (P165)[145] was the first to be analysed, it is more appropriate to refer first to yuzurinine hydrobromide (92a)[146] and to macrodaphnine hydrobromide (92b)[147], which have very similar ring systems differing only in that the former

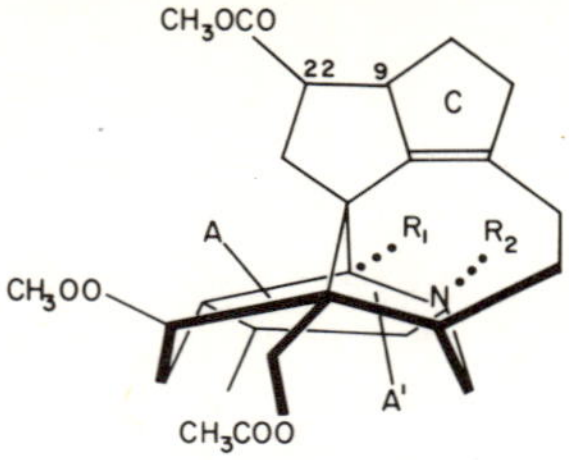


(92) (a) $R_1$ = OH; $R_2$ = H (with $N^+$) (b) $R_1$ = H; $R_2$ = O

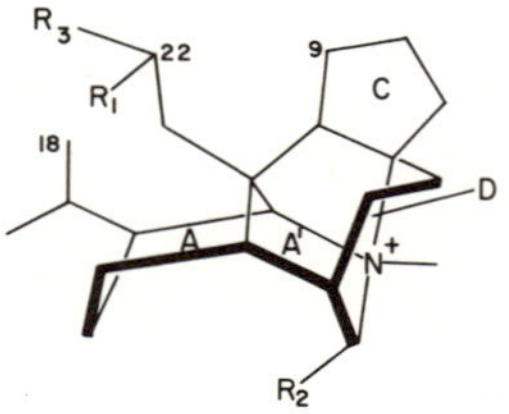


(93)

(a) $R_1$ = —$OCOCH_3$ $R_2$ = H (b) $R_1$ = H $R_2$ = H (c) $R_1$ = H $R_2$ = OH

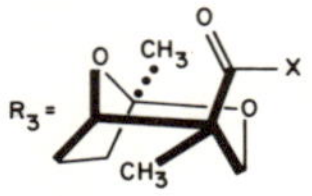


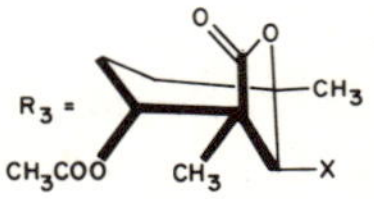


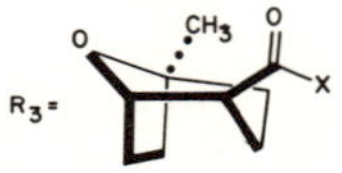


has a hydroxy group at position 2 and the latter is an *N*-oxide. This ring structure has common features with the remaining structures studied, rings A, A′ and C being retained. Thus the structure of daphniphylline hydrobromide (93a)[145], which is the mirror structure of (P165), can be related to that of yuzurnine (and macrodaphnine) by bond formation C(18)→N(1) and C(22)→C(9) and bond fission N(1)↛C(12). Closely related structures are those of daphmacrine methiodide (93b)[148] and alkaloid II methiodide (93c)[149]. All three compounds consist of two cage structures linked by a flexible chain, the attached ring systems being detailed as $R_3$ in (93). The main ring structure is closely similar with ring D adopting a boat form.

From a structural view the daphniphyllum alkaloids were earlier classified into two groups—yuzurimine and daphniphylline. Further study suggested the existence of a third group, represented by secodaphniphylline and analysed as methyl *N*-bromoacetylhomosecodaphniphyllate (94)[150]. With continued investigation a further two groups have been proposed, represented by daphnilactone A (95)[151] and daphnilactone B (96)[152]. All involve a

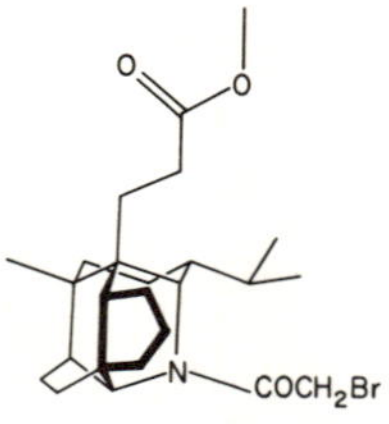


(94)

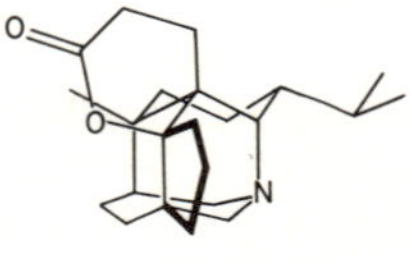


(95)

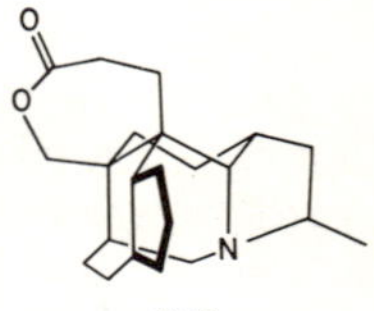


(96)

2-azabicyclo[3,3,1]nonane ring in their system. These structures consist of complex cross-linked ring systems and are therefore mainly rigid in form and the shapes of the individual rings are therefore largely determined by these linkages.

## 6.9 STEROIDAL ALKALOIDS

In buxenine-G dihydroiodide (97)[153], atoms 1, 5, 10, 19, 9, 8, 11 and 12 are virtually coplanar, ring A is in a chair form and ring C in a slightly distorted half-chair form. Ring B may be described as having an envelope form with

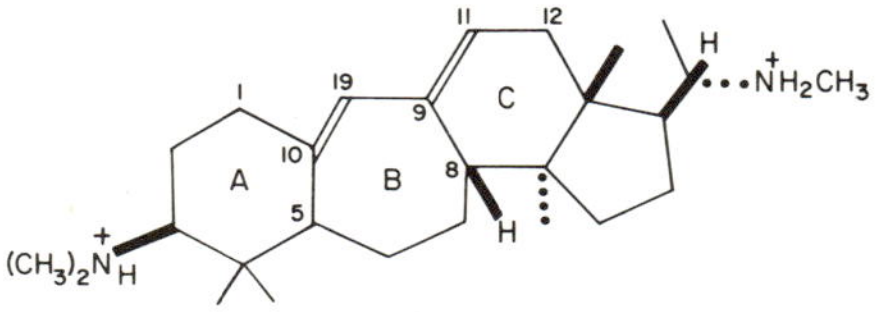

(97)

C(7) out-of-plane. The analysis of veralkamine hydroiodide (98)[154] was not sufficiently definitive to yield dimensional details. The subsequent analysis of

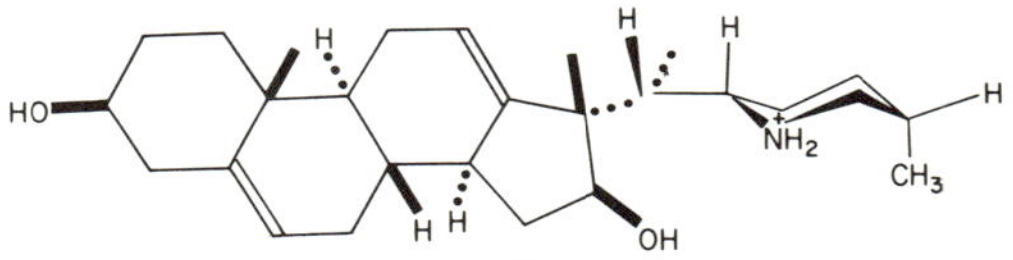

(98)

tetrahydroveralkamin-16-one hydroiodide (99)[155] indicates the simultaneous existence of both chair and boat conformations (2:1) of ring C. This apparently

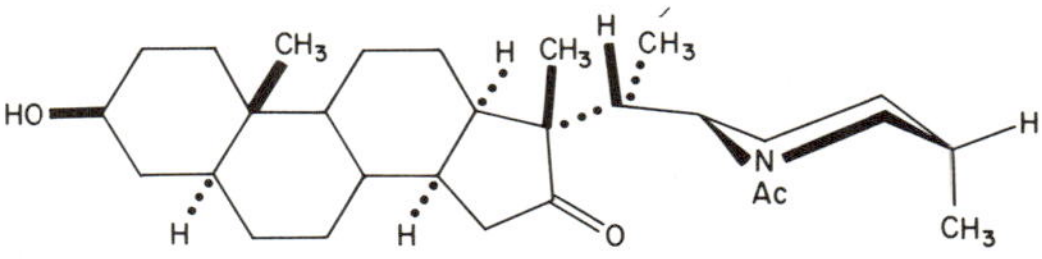

(99)

involves only a displacement of one atom, C(12). In solaphyllidine (100)[156], rings A, B, C and E are in the chair form with ring D a 13-envelope. The mean planes of the *O*-acetyl and 3-hydroxy-5-methyl piperidyl groups lie near perpendicular to the average plane of the steroid. With tomatidine hydrobromide (101)[157], ring closure yields ring D as a 14-envelope and E a 22-envelope, the DE ring system having a butterfly shape. Demissidine hydroiodide (102)[158] also involves a butterfly shape for the DE ring system. Vèratrobasine (103)[159] has a skeleton like that of jervine, except for a hydroxy

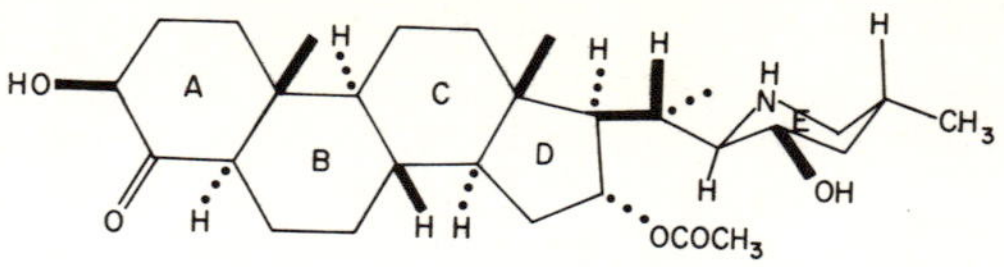

(100)

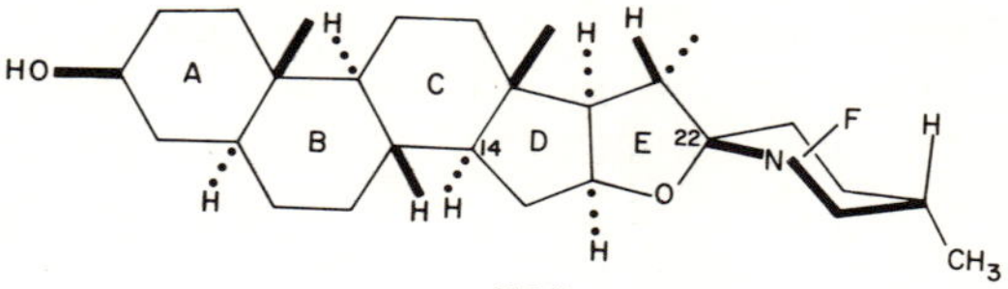

(101)

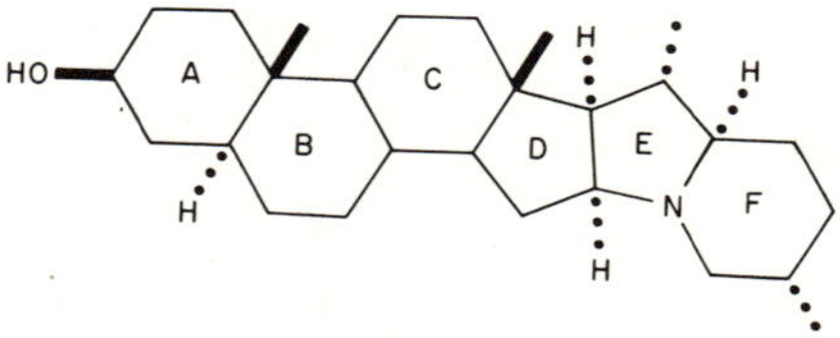

(102)

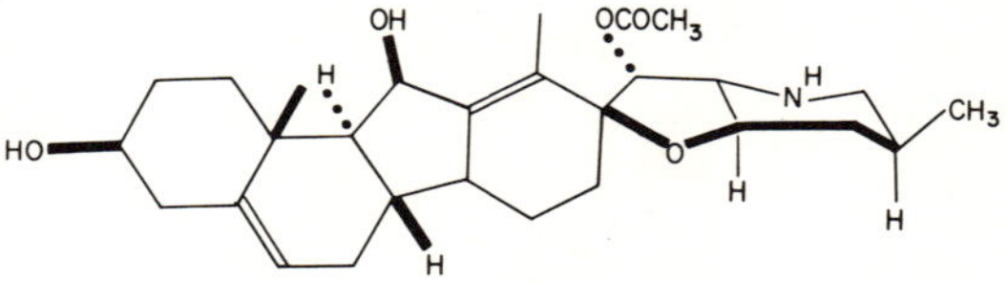

(103)

group at C(11) rather than a keto group. Another veratrum alkaloid derivative is *N*-ethylbaikeine (104)[160]. Also derived from *Veratrum grandiflorum* is zygacine, studied[161] as the acetonide hydroiodide (105). The two seven-membered rings formed by the hemiacetal function and rings A and B are in conformations between boat and twist-boat forms. Verticinone (106), which

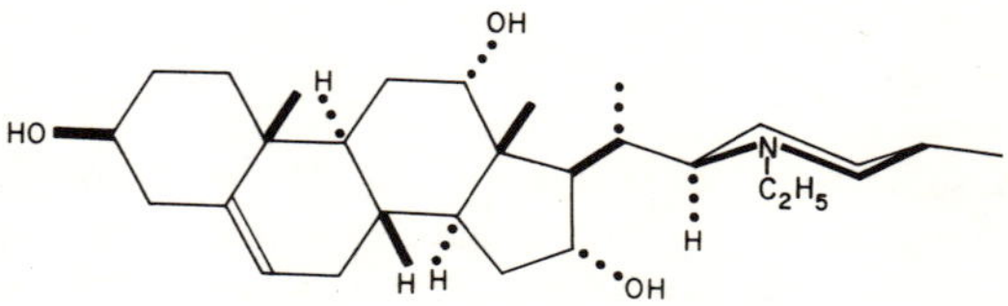

(104)

has been studied[162] in two crystalline products, involves two *cis* fusions, one between rings C/D and the other between rings E/F. In the latter case it is

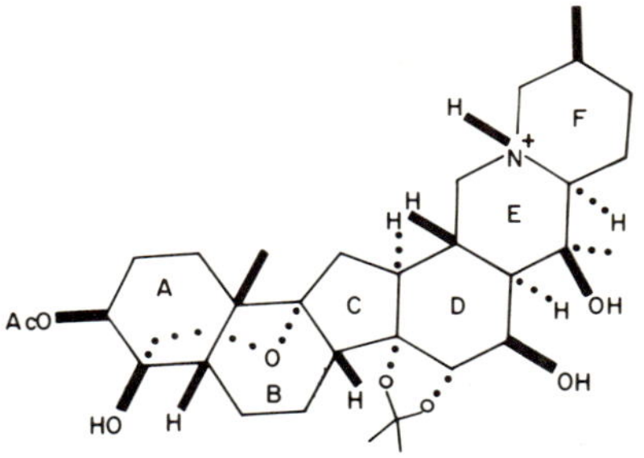


(105)

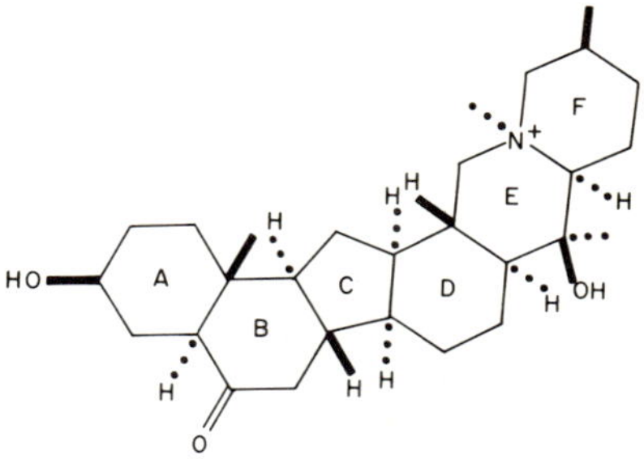


(106)

presumed to occur on formation of the quaternary ion. All rings are in the chair form. Only a brief report[163] is available on cevine (P50).

## 6.10 TOXINS

In addition to batrachotoxin A, which has been dealt with in the earlier series[1], other alkaloids of animal origin have been studied by x-rays. Samandarine hydrobromide (P51)[164] was from the skin gland of a salamander species while pumiliotoxin C hydrochloride (107)[165], from the skin of a

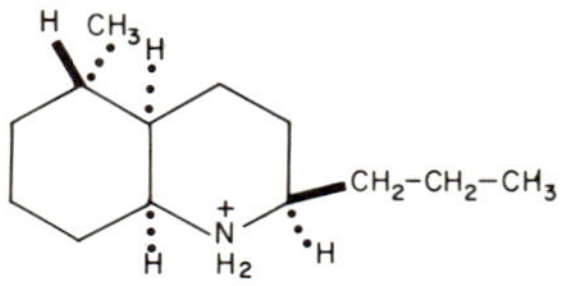


(107)

Panamanian frog, is structurally related to coniine. Both rings are in the chair form, *cis* fused. In the case of tetrodotoxin, from the Japanese puffer fish, x-ray techniques were fully deployed to elucidate the structure of this relatively small but unexpectedly complex molecule. Four complete analyses were carried out, successive structures involving less modification relative to the original structure itself. These were anhydrobromotetrodoic lactone hydrobromide (P47a)[166], tetrodoic acid hydrobromide (P47b)[167], *O′*,*O″*-isopropylidene-*O*-methyltetrodotoxin hydrochloride (P47c)[168] and 6,11-diacetylanhydrotetrodotoxin hydroiodide (108)[169], all of which have individually interesting conformational features, the most unusual being the hemiacetal group in tetrodotoxin itself (P47d). An analysis of tetrodotoxin hydrobromide (109) has also appeared[170]. In this, the apical C—OH bond is apparently shortened to 1.37 Å, the effect being ascribed to resonance. There are small conformational differences between the two molecules of the asymmetric unit. In one, C(8) is *trans* to C(6) and in the other *trans* to C(4). Also, in one, the guanidinium group is akin to a shallow boat form while in the other

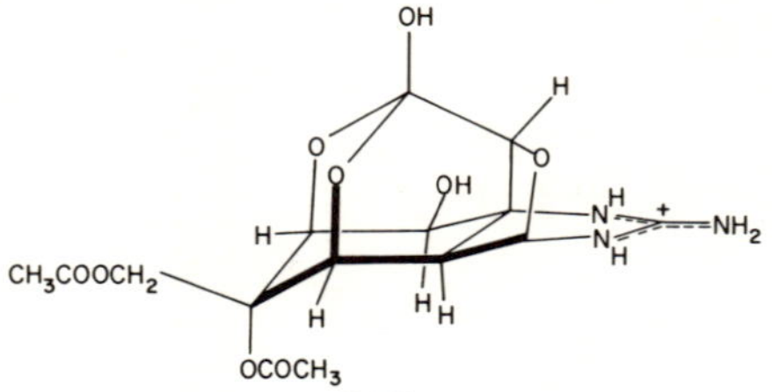


(108)

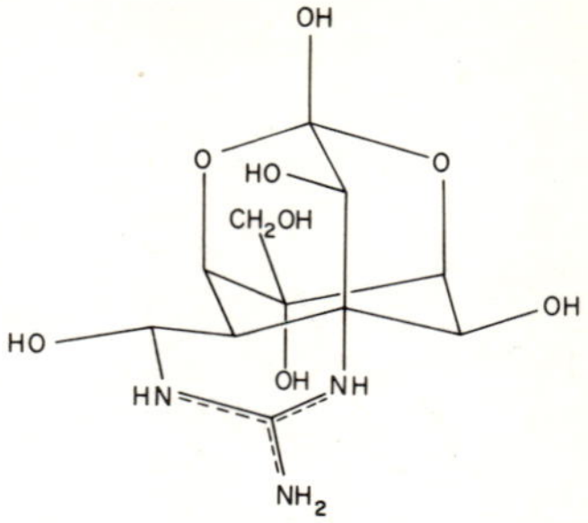


(109)

it tends towards a shallow chair form. Another alkaloid of marine origin is dibromophakellin, analysed[171] as the monoacetyl derivative (110). This also

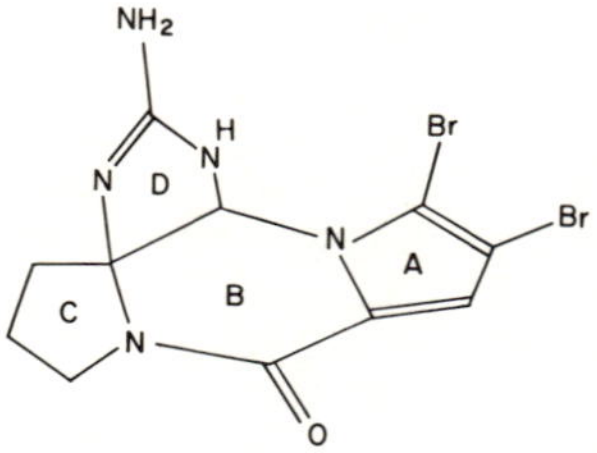


(110)

has, in its structure, a guanidine unit which lies almost at right angles to the plane of the other three rings.

## 6.11 OTHER ALKALOIDS

Where the number of examples studied is insufficient to warrant separate group treatment, they have been treated in this section. In most of the following, only isolated cases are available and are presented in respect of increasing structural elaboration around the nitrogen atom(s).

Open-chain examples are ephidrine hydrochloride (111)[172] and D-methadone hydrobromide (112)[173]. In the former the molecule is in the most

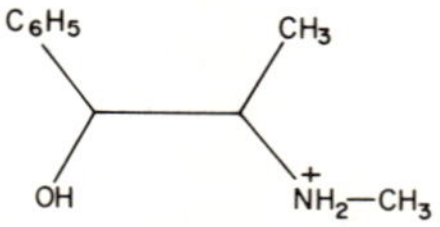


(111)

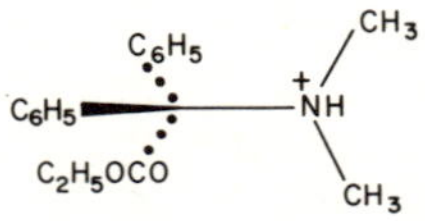


(112)

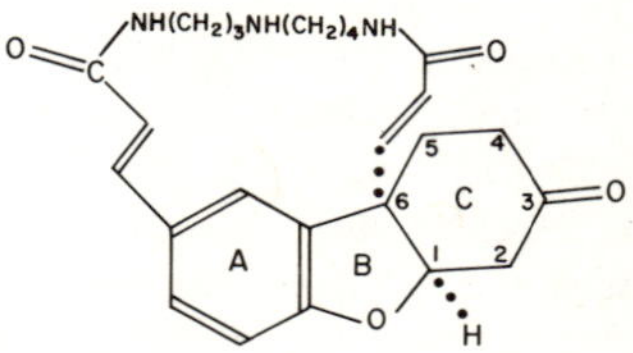


(113)

extended conformation, as in the case of noradrenaline[174] and dopamine[175]. In nobiline methiodide (P53)[176], ring B has a sofa form with C(1) out of plane while ring C in lunarine hydroidide[177a] and hydrobromide (113)[177b] has a skew boat form with atoms 1, 2, 3 and 6 deviating by ~0.1 Å from the mean plane and atoms 2 and 5 ~0.5 Å on the same side of that plane. This variation from the chair form is attributed to the *cis* fusion of rings B and C. A simple base involving the piperidine ring is *ψ*-conhydrine hydrobromide (114)[178] while himbacine (P33) has a *trans*-decalin ring system attached[179]. A further elaboration of this structure by fusion of the decalin and piperidine

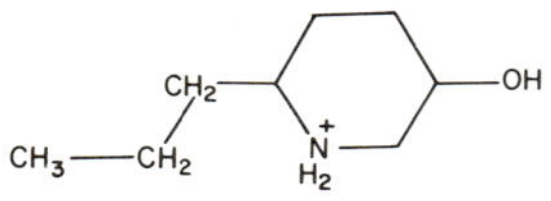

(114)

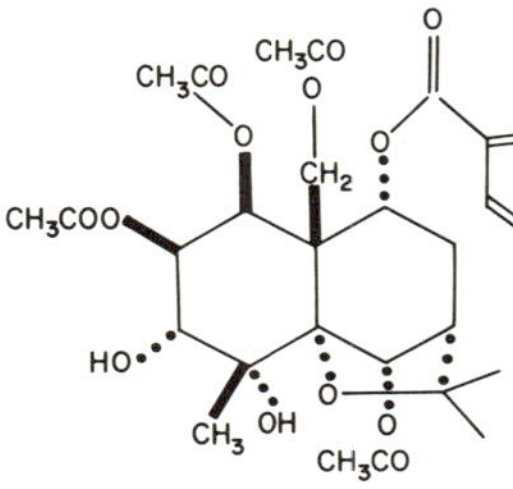

(115)

rings is seen in himbosine (P44)[180]. With this group involving a nitrogen atom on an isolated ring system we may include maytoline methiodide (115)[181]. This is characterised by the presence of a nicotinyl ester substituted on a highly oxygenated sesquiterpene nucleus where rings A and B are both chair form with the latter distorted by the oxido bridge. Where no obtruding factors exist the ester groups adopt the conformation preferred in such circumstances. Derived from the same family is bromoacetylneoevonine (116)[182]. The sesquiterpene nucleus has a closely similar substituent pattern, the pyridine

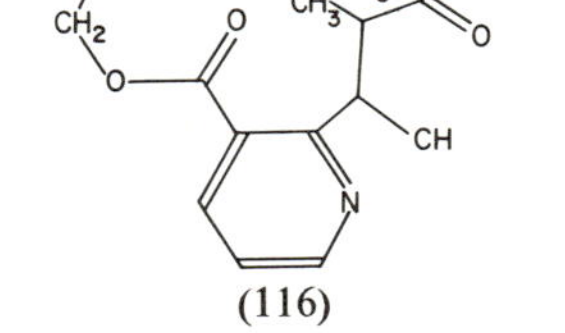

(116)

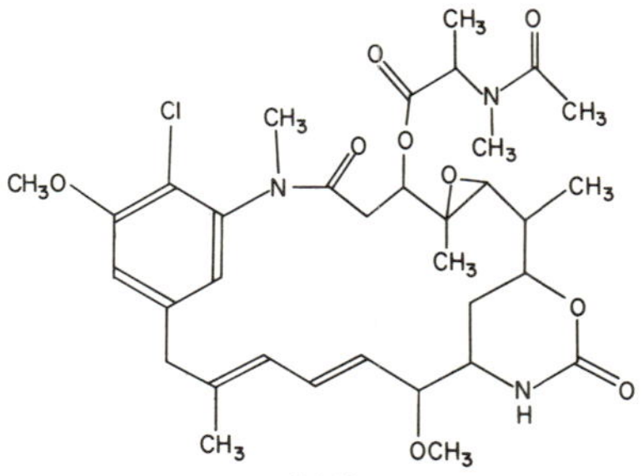

(117)

ring being linked at two points. From the same plant as maytoline is maytansine, studied[183] as its 3-bromopropyl ether (117). The ring structure has two longer sides which are roughly parallel and separated by ~5.4 Å so that there is a hole in the centre of the ring. The two faces of the ring differ in character, one surface being predominantly hydrophilic while the other is more hydrophobic.

An imidazole ring occurs in casimidine dihydrochloride (P49)[184] and also in pilocarpine trichlorogermanate (118)[185], where the lactone group is planar with C(6) 0.52 Å out of plane. In the latter the imidazole ring is planar and

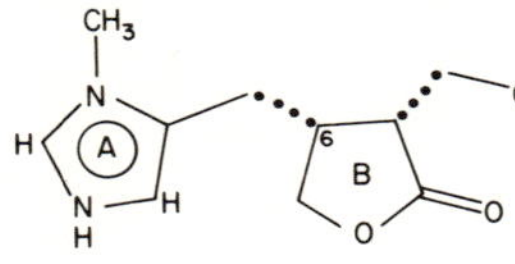


(118)

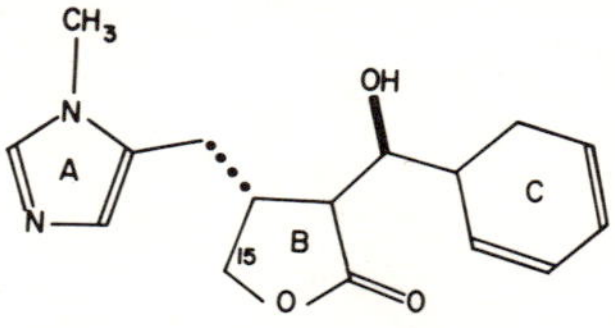


(119)

lies approximately at right angles to ring B. A related structure is that of isopilosine (119)[186]. In this structure ring B is reported as almost perfectly planar. In an orchidacene alkaloid (120)[187], ring B is planar and rings A and C have a half-chair form.

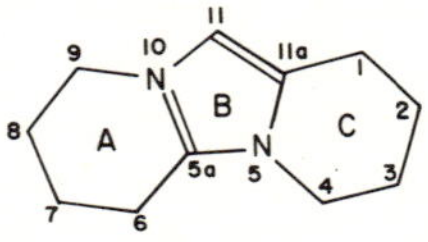


(120)

Tropine (121a)[188] and pseudo-tropine (121b)[189] differ only in the disposition of the hydroxy group at position 1, the disposition of the methyl at N being similar. In bellendine (122)[190] ring B is essentially planar. Its presence produces

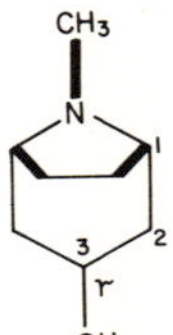


(121) (a) r = ▬
(b) r = ····

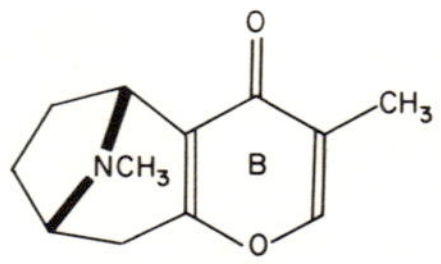


(122)

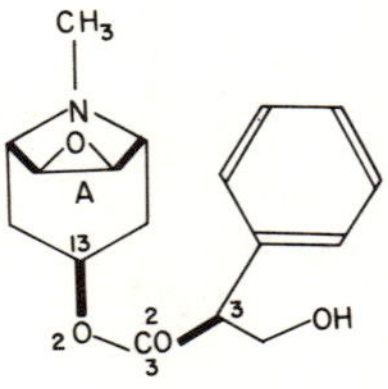


(123)

minor distortions from symmetry in the tropane ring system. In (—)-hyoscine hydrobromide (123)[191] the six-membered ring is again in the chair form and the plane of the ester group C(13)O(2)C(2)O(3)C(3) is 44.5° to the reflection plane of the A ring system. Similar nitrogen–oxygen distances are observed in 1-hyoscyamine hydrobromide (124)[192]. The two ester groups in (—)-cocaine

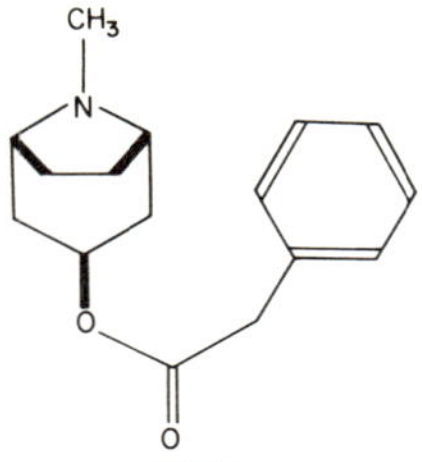


(124)

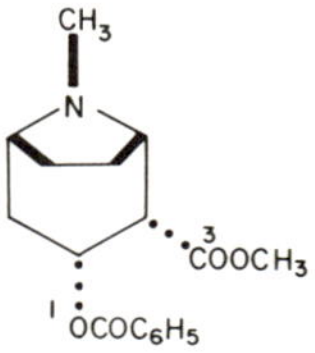


(125)

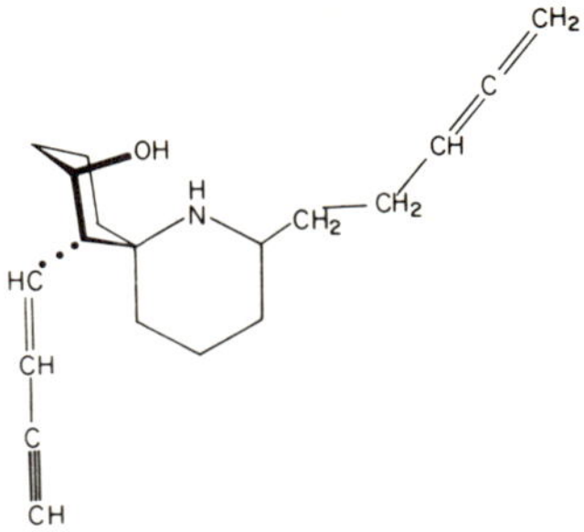


(126)

(125) are disposed so that the separation of O(3) to 0(1) is only 2.79 Å, an unexpected feature[193]. The anticholinergic agent dihydroisohistrionicotoxin (126), as the hydrochloride, has a relatively simple ring system[194] with an N· · ·O distance of 2.71 Å.

The early work[195] on α-isosparteine (127a) confirmed the *cis-cis* configuration while that on 7-hydroxy-β-isosparteine perchlorate (127b) showed the *trans-trans* configuration[196]. In this latter compound, N(16) is protonated while N(1) is not and this difference is reflected in the mean bond length values of N—C being ~0.03 Å smaller than those for $N^+$—C. Both α- and β-sparteine have real or effective two-fold symmetry in respect of the ring skeleton. The third member of this group is sparteine *N*-oxide sesquiperchlorate (127c)[197]. All the rings in these compounds have a chair form. In

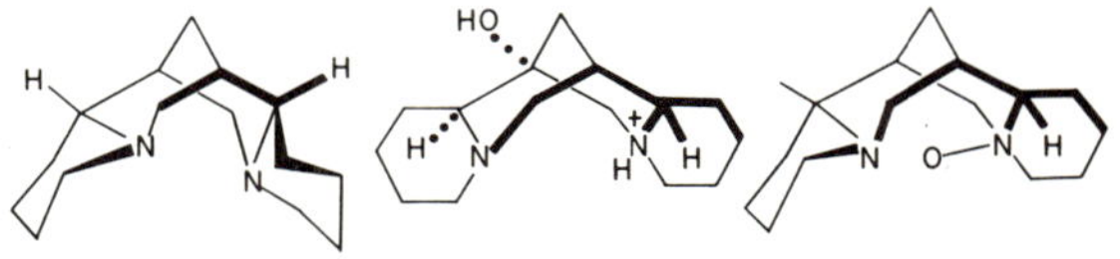


(127) (a) (b) (c)

lupinane *N*-oxide perchlorate (128)[198] ring A′ probably has a half-chair conformation due to the lactone group while ring C has a boat form. Rings B and D are chair in conformation.

Elaeocarpine hydrobromide (129)[199] represents a new class of indolizidine alkaloids. The structure is roughly planar. Securinine hydrobromide (130a)[200]

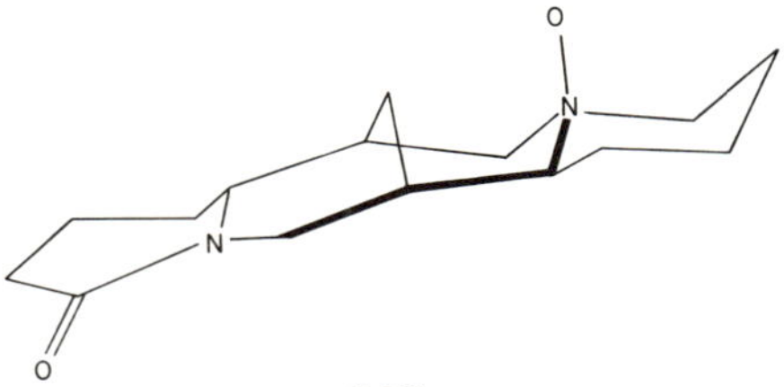


(128)

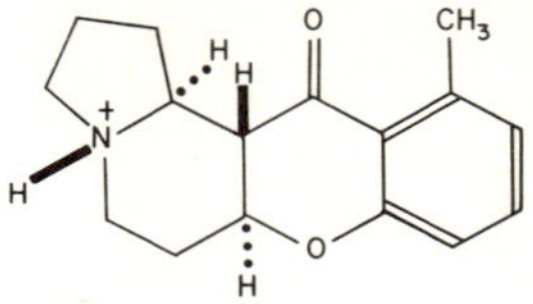


(129)

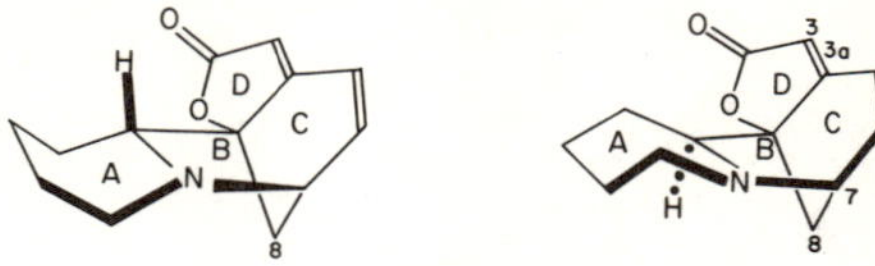


(a) (130) (b)

and phyllochrysine methiodide (130b)[201] are diastereoisomers. In both, ring A is in a boat form, slightly skewed, while ring C is a sofa form with C(8) out of plane. Ring D is nearly planar but is 10° and 13°, respectively, relative to the plane through 3a, 4, 5, and 7.

A group of stemona alkaloids have been analysed. Stemonine hydrobromide (131)[202] has ring B in a slightly distorted chair form while rings A

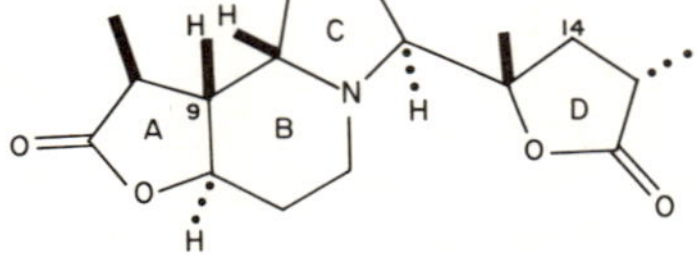


(131)

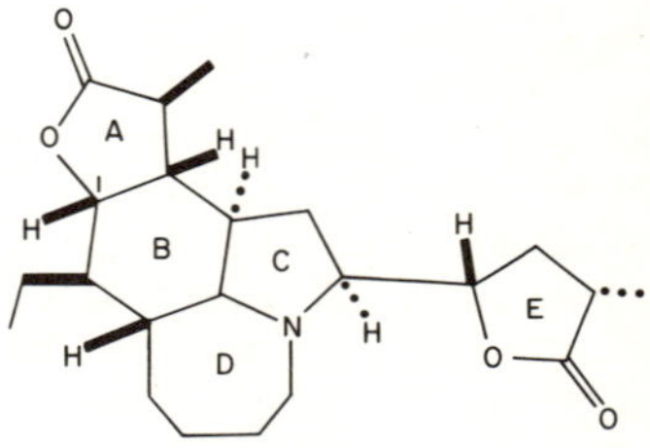


(132)

and D have 9- and 14-envelope forms and C a half-chair form. In tuberostemonine methobromide (132)[203], ring A appears to be a 1-envelope and ring B in a chair form. Oxotuberostemonine [204] has the structure shown in (133), with ring A in the half-chair form and fused *cis* to ring B. Ring D is in the chair

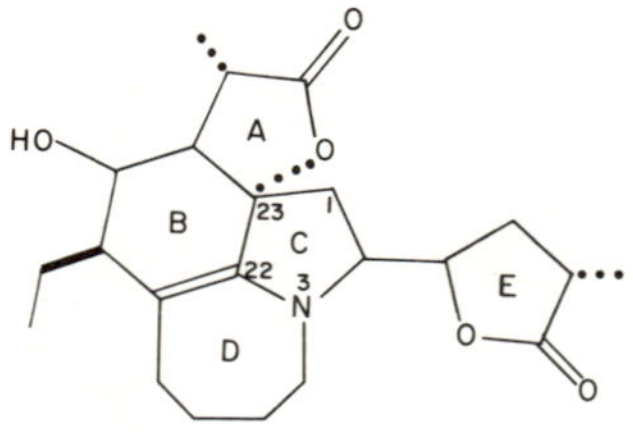


(133)

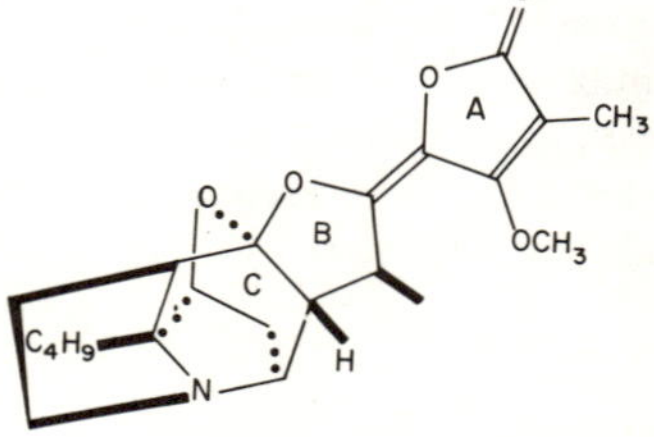


(134)

form and ring C is puckered with C(1) deviating most (~0.8 Å) from the plane through N(3), C(22) and C(23). Stemofoline hydrobromide[205] has the structure (134), with ring C in a chair form. A related compound, of a more open structure, is protostemonine (135)[206]. In the case of stemonamine hydrochloride (136)[207] the plant product is racemic. Ring B has a chair form.

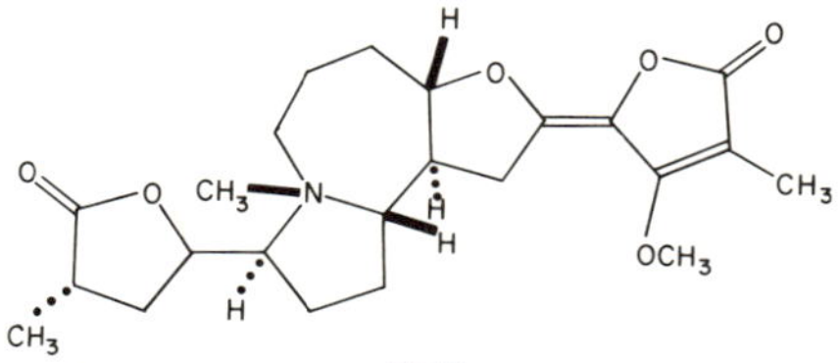


(135)

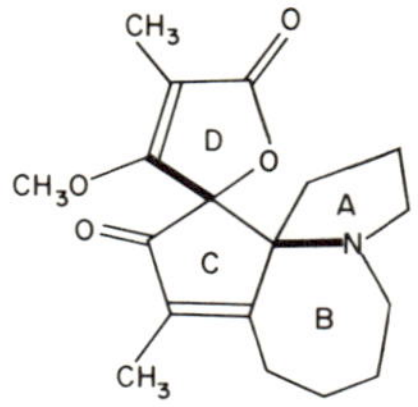


(136)

Both alopecurine methobromide (137) and debenzoylalopecurine hydrobromide[208] have been analysed. Ring C is in a twist form while ring B is apparently distorted to almost a half-chair form. A closely related compound, lycopecurine hydrobromide (138), has also been analysed[209].

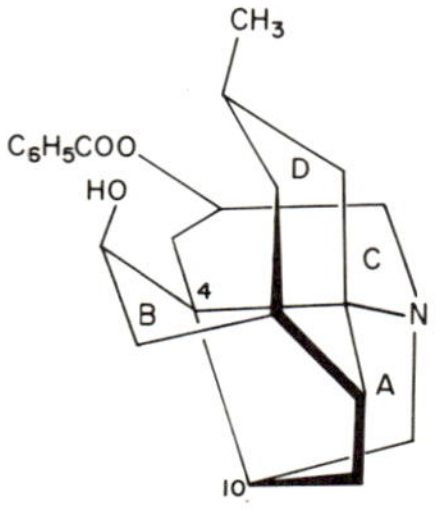


(137)

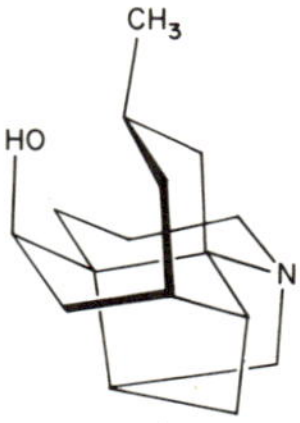


(138)

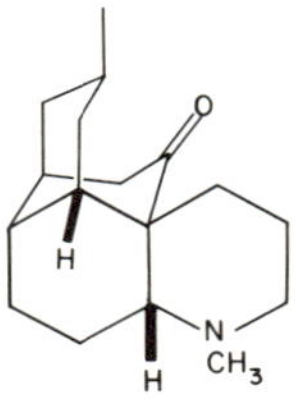


(139)

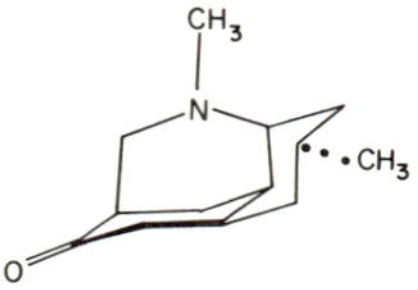


(140)

The Hoffmann base from lycopodine methiodide studied as the hydrobromide[210] has been shown to be (139). A structure from a similar source is that of *p*-bromobenzoyldihydroluciduline (140); rings A and C are both chair in form[211]. In mesembranol (141a) ring B adopts a chair conformation[212] whereas, in 6-epimesembranol methiodide (141b)[213], the presence of the quaternary nitrogen forces the adoption of a boat form with slight distortions and in the case of 'sceletum alkaloid $A_4$' (141c) the fusion of a pyridine ring results in a half-chair conformation[214]. To clarify a reversible isomerisation

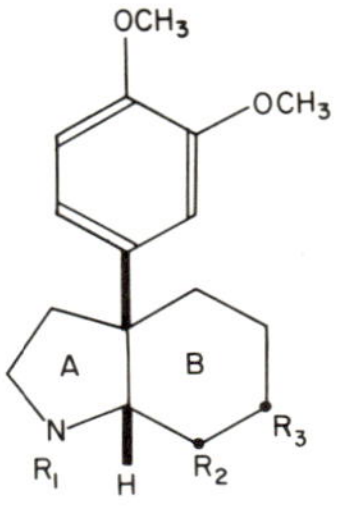


(141) (a) $R_1 = -CH_3$
$R_2 = H$
$R_3 =$ ▬OH

(b) $R_1 = (CH_3)_2$
$R_2 = H$
$R_3 = \cdots OH$

(c) $R_1 = -CH_3$
$R_2 = R_3 =$ 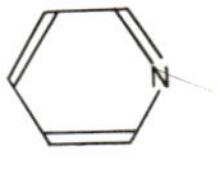

of an ergot alkaloid, the *aci-p*-iodobenzoylaminocyclol derivative (142) was analysed[215] and, *inter alia*, showed an interesting close approach of ~3.07 Å between a benzene ring and hydroxy group. Ring B is more correctly described as a distorted boat form with N(4) and C(12) displaced −0.29 and −0.58 Å from the plane through atoms 5, 6, 7, 8, 11 and 15. Rings A and C are 2- and 10-envelopes.

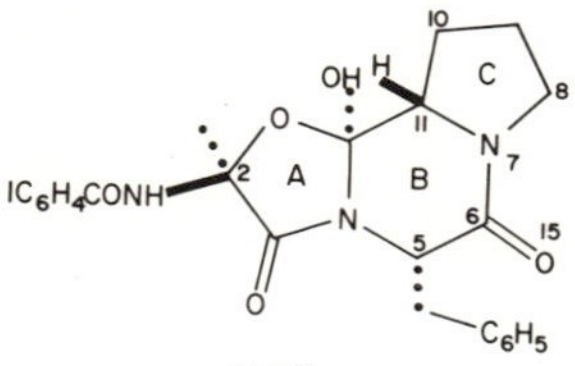


(142)

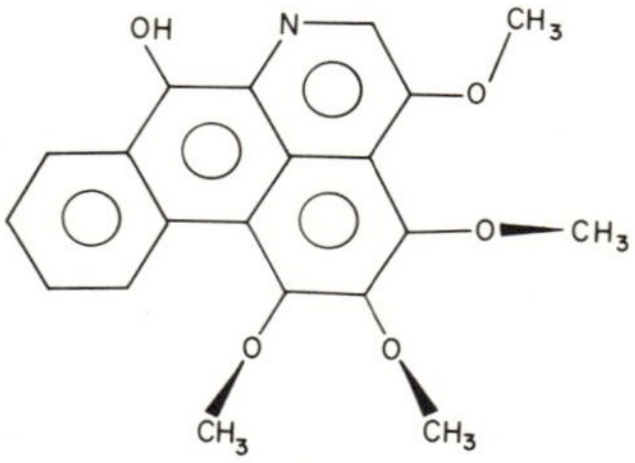


(143)

The structure of imenine (143) is essentiallly planar[216], as is the case for perloline mercurichloride (P45)[217]. Also isolated from *Lolium perenne* is perlolyrine hydrobromide (144)[218]. The ormosia alkaloids jamine (P31)[219] and panamine diperchlorate (P32)[220] involve six-membered rings mainly in the chair form except for ring C in jamine which has a boat form, the C/D ring

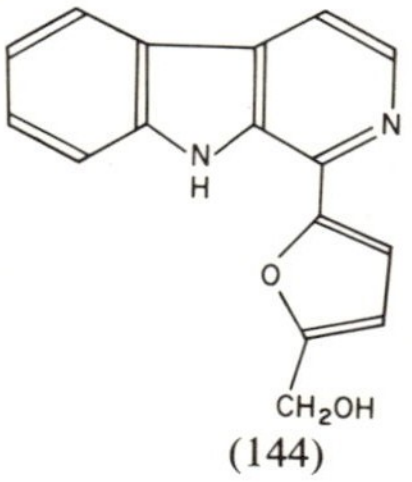


(144)

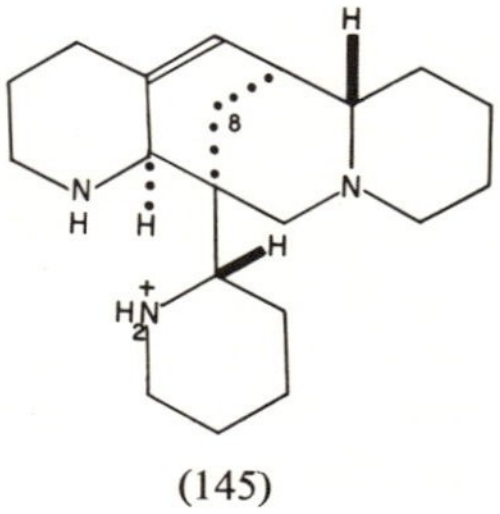


(145)

junction being *trans*. A related structure is that of podopetaline hydrobromide (145)[221]. In this rings A, C, D and E are in the chair form while ring B is in a sofa form with C(8) out-of-plane; the C/D ring junction is *cis*.

In nicotine hydriodide (146) the two rings lie approximately at right angles[222]. Harmidine (harmaline) (147) is essentially a planar molecule; the form studied[223] proved to be the enamine tautomer. In alchorneine, studied[224]

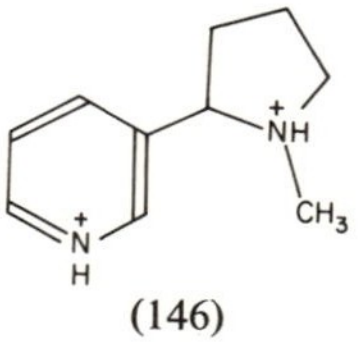


(146)

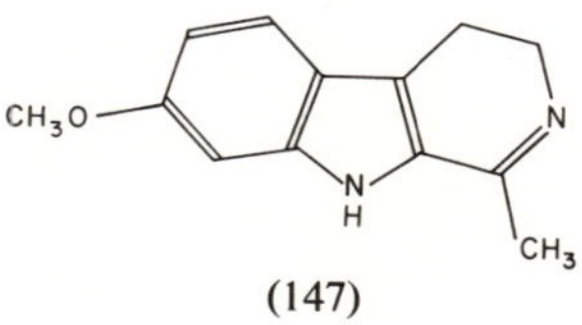


(147)

as the bromomethylate (148), ring A is in a shallow boat form with a dihedral angle of 17°; N(15) and N(16) appear to be shallowly pyramidal. Anthramycin methyl ether (149) is said to have a strained conformation due to the

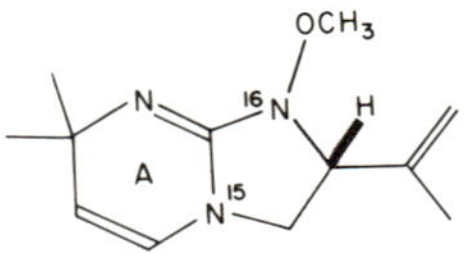


(148)

fusion of the seven-membered ring to the benzene ring[225]. Another fungal product is 5-bromo-12*S*-tetrahydroaustamide (150)[226] in which the diketopiperazine ring is folded[226]; the seven-membered ring C has an approximate chair form.

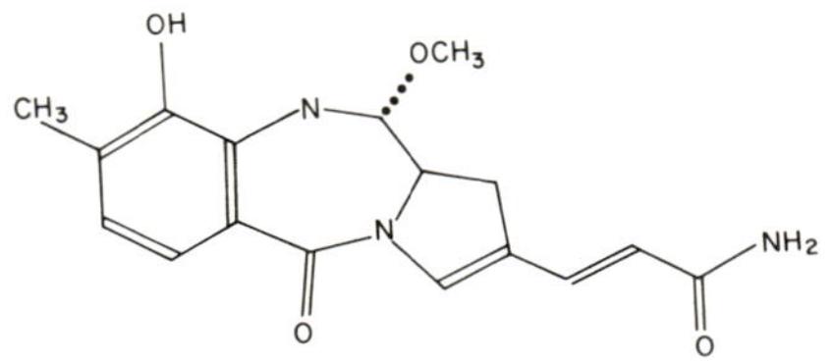


(149)

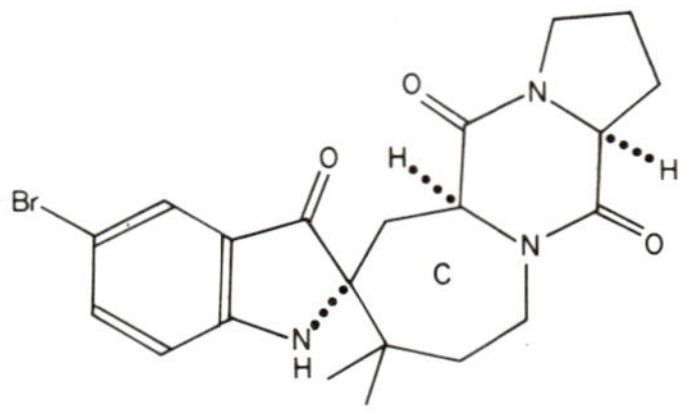


(150)

Coccinellin hemihydrochloride (151)[227] is an alkaloid derived from the ladybird insect. Its mode of synthesis is probably similar to that of the poranthera alkaloids, porantherine (152)[228], poranthericine (153)[229], porantheridine

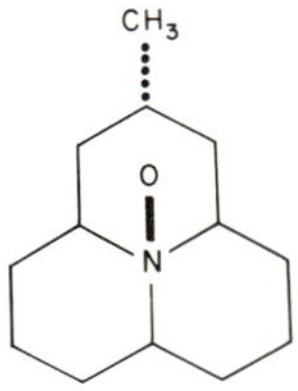


(151)

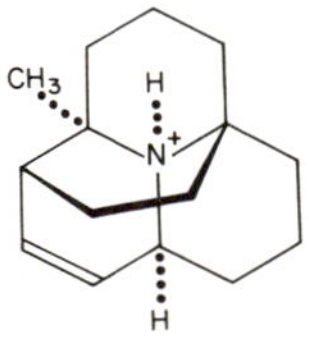


(152)

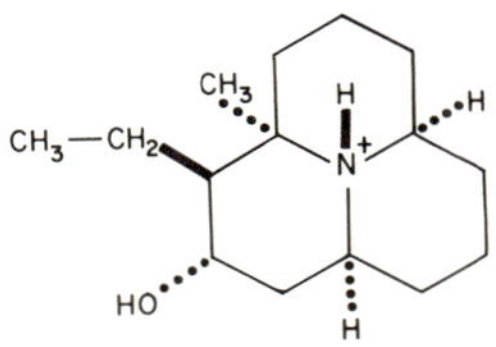


(153)

(154)[229a, 230] and 4-methyl-6-(2′benzoyloxypentyl)quinolizidine (155)[231]. These various structures appear to represent stages in the postulated biosynthetic sequence.

In crepidine, an orchid alkaloid, studied[232] as the methiodide (156), rings A and B are in the chair form, while ring C is in the envelope form with N out of plane. Narciclasine tetra-acetate (157)[233] has ring B apparently in the sofa form while ring C may also approximate to the sofa form. Papaverine

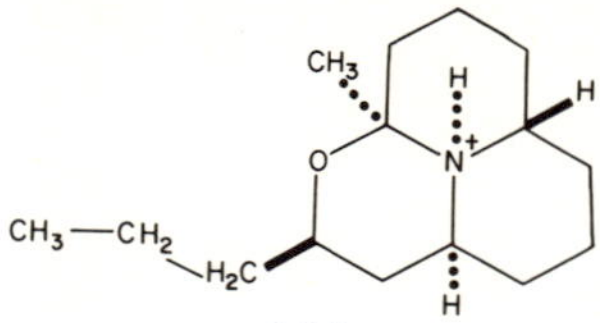

(154)

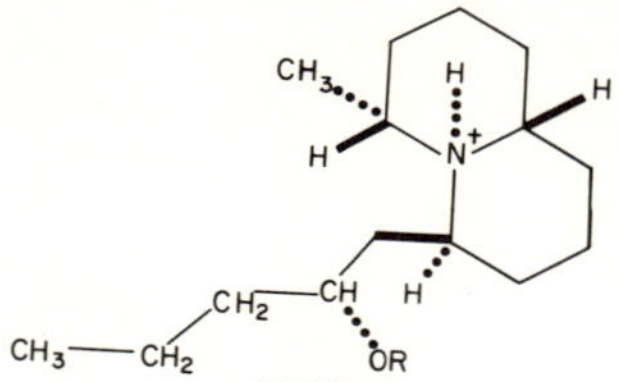

(155)

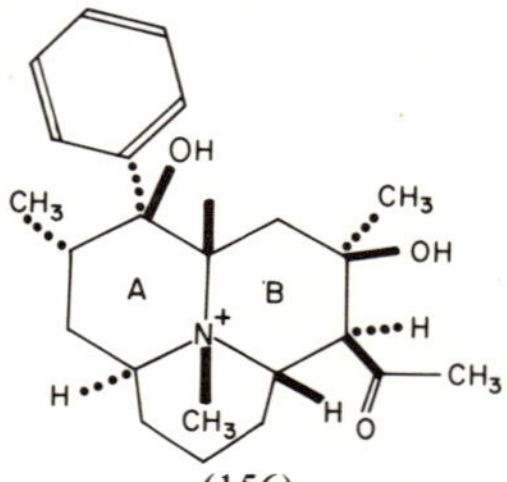

(156)

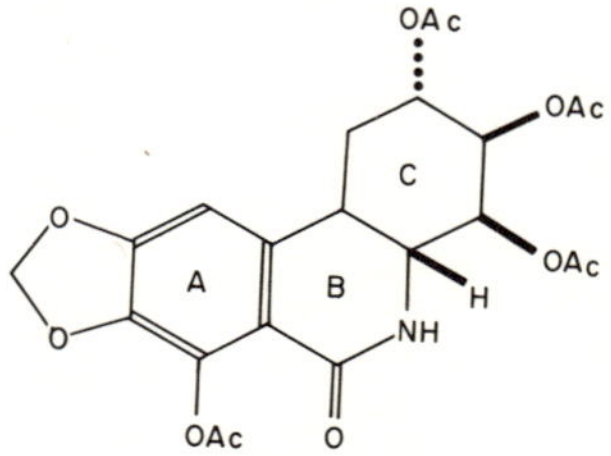

(157)

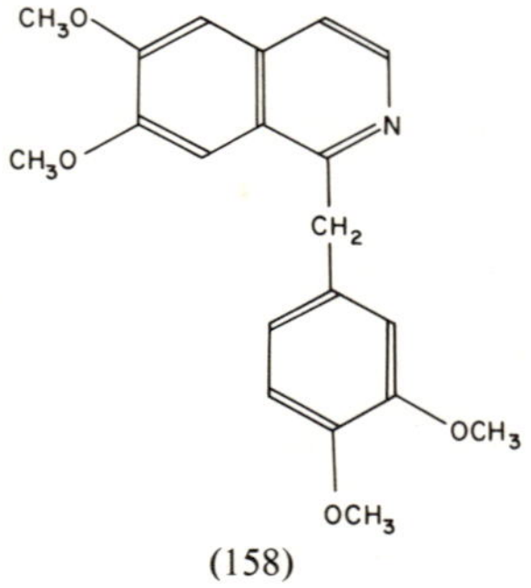

(158)

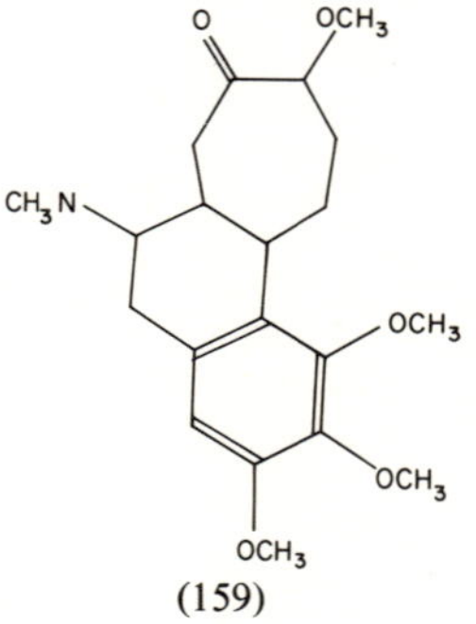

(159)

(158)[234] illustrates the tendency of methoxy groups to lie near the plane of the associated benzene ring, unless extraneous factors override this. In *N*-deacetyl-*N*-methylcolchicine (159)[235], two of the methoxy groups are moved

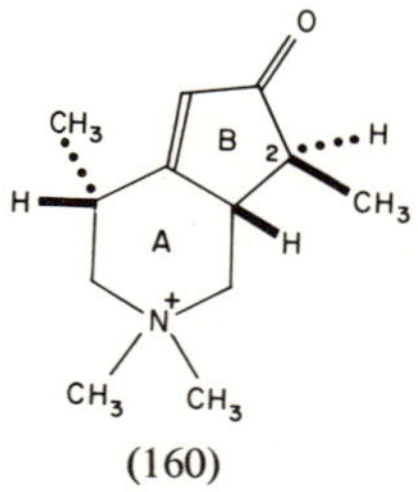

(160)

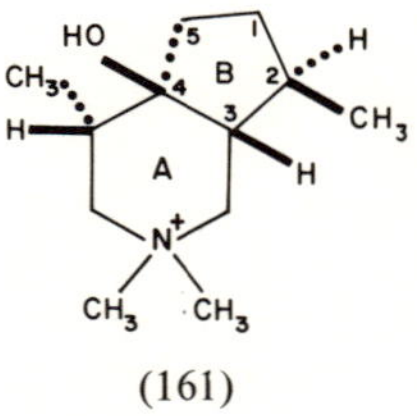

(161)

out-of-plane by intramolecular factors. The tropolonoid ring is planar but shows an alternation of bond lengths.

Two monoterpene alkaloids which have been analysed are tecomanine methoperchlorate (160)[236] and alkaloid C methiodide (161)[236]. In both, ring A is in the chair form. In the former, ring B is in the envelope form with C(2) out-of-plane, while in the latter it is a distorted envelope with C(1) 0.65Å and C(5) 0.11 Å out of the plane of C(2), C(3) and C(4).

## References


1. Mathieson, A. McL. (1972). In *MTP International Review of Science, Physical Chemistry, Series One*, Vol. 11, 311 (J. M. Robertson, editor) (London: Butterworths)
2. Mathieson, A. McL. (1967). *Perspectives in Structural Chemistry*, Vol. 1, 41 (New York: Wiley)
3. (1970). *Molecular Structures and Dimensions* (O. Kennard and D. G. Watson, editors) Vols. 1 and 2 (Utrecht: Oosthoetz)
4. Petcher, T. J. and Pauling, P. (1973). *Nature* (*London*), **241,** 277
5. James, M. N. G. and Williams, G. J. B. (1972). *Can. J. Chem.*, **50,** 2407

6a. Scott, C. C., Sim, G. A. and Robertson, J. M. (1962). *Proc. Chem. Soc.*, 355

6b. Asher, J. D. M., Robertson, J. M. and Sim, G. A. (1965). *J. Chem. Soc.*, 6355

7. Zacharias, D. E., Rosenstein, R. D. and Jeffrey, G. A. (1965). *Acta Crystallogr.*, **18,** 1039
8. Cesario, M. and Pascard-Billy, C. (1973). *Acta Crystallogr.*, **B29,** 529
9. Camerman, N. and Trotter, J. (1964). *Acta Crystallogr.*, **17,** 384
10. Arai, G., Coppola, J. and Jeffrey, G. A. (1960). *Acta Crystallogr.*, **13,** 553
11. Silvers, S. J. and Tulinsky, A. (1963). *Acta Crystallogr.*, **16,** 579
12. McPhail, A. T., Robertson, J. M. and Sim, G. A. (1963). *J. Chem. Soc.*, 1832
13. Wulf, H. and Niggli, A. (1967). *Helv. Chim. Acta*, **50,** 1011
14. Burnell, R. H., Chapelle, A., Khalil, M. F. and Bird, P. H. (1970). *Chem. Commun.*, 772
15. Guilhem, J. (1970). *Acta Crystallogr.*, **B26,** 2029

16a. Hamilton, J. A., Hamor, T. A., Robertson, J. M. and Sim, G. A. (1962). *J. Chem. Soc*, 5061

16b. Manohar, H. and Ramaseshan, S. (1962). *Z. Kristallogr.*, **117,** 273

17. Bevan, C. W. L., Patel, M. B., Rees, A. H., Harris, D. R., Marshak, M. L. and Mills, H. H. (1965). *Chem. and Ind.*, 603
18. Jaggi, H. and Renner, U. (1964). *Chimia* (*Switz.*), **19,** 173

19a. Mills, J. F. D. and Nyburg, S. C. (1960). *J. Chem. Soc.*, 1458

19b. Craven, B. M. and Zacharias, D. E. (1968). *Experientia*, **24,** 770

20. Camerman, A., Camerman, N. and Trotter, J. (1965). *Acta Crystallogr.*, **19,** 314
21. Ling, N. C. and Djerassi, C. (1970). *Tetrahedron Lett.*, 3015
22. Sakabe, N., Sendo, Y., Iljima, I. and Ban, Y. (1969). *Tetrahedron Lett.*, 2527
23. Kennard, O., Kerr, K. A., Watson, D. G., Fawcett, J. K. and Riva Di Sanseverino, L. (1970). *J. Chem. Soc. A*, 1779
24. Craven, B. M. (1969). *Acta. Crystallogr.*, **B25,** 2131
25. Kahrl, J., Gebreyesus, T. and Djerassi, C. (1971). *Tetrahedron Lett.*, 2527
26. Brown, S. H., Djerassi, C. and Simpson, P. G. (1968). *J. Amer. Chem. Soc.*, **90,** 2445
27. Oberhansli, W. E. (1969). *Helv. Chim. Acta*, **52,** 1905

28a. Pascard-Billy, C. (1969). *Acta Crystallogr.*, **B25,** 166

28b. Pascard-Billy, C. (1968). *Bull. Soc. Chim. Fr.*, 3289

29. Lovell, F. M., Pepinsky, R. and Wilson, A. J. C. (1959). *Tetrahedron Lett.*, 1
30. Przybylska, M. (1962). *Acta Crystallogr.*, **15,** 301
31. Nordman, C. E. and Nakatsu, K. (1964). *J. Amer. Chem. Soc.*, **85,** 353

32a. Robertson, J. H. and Beevers, C. A. (1951). *Acta Crystallogr.*, **4,** 270

32b. Bokhoven, C., Schoone, J. C. and Bijvoet, J. M. (1951). *Acta Crystallogr.*, **4,** 275

33. Peerdeman, A. F. (1956). *Acta Crystallogr.*, **9,** 824

34. Karle, J. M. and Le Quesne, P. W. (1972). *J. Chem. Soc. Chem. Commun.*, 416
35. Jones, N. D. and Nowacki, W. (1972). *J. Chem. Soc. Chem. Commun.*, 805
36a. Chothia, C. H. and Pauling, P. (1970). *Proc. Nat. Acad. Sci. USA*, **65,** 477
36b. Baker, R. W., Chothia, C. H., Pauling, P. and Petcher, T. J. (1971). *Nature (London)*, **230,** 439
37. McPhail, A. T. and Sim, G. A. (1965). *J. Chem. Soc.*, 1663
38. Fehlmann, M., Koyama, H. and Niggli, A. (1965). *Helv. Chim. Acta*, **48,** 303
39. Gemmell, K. W., Robertson, J. M., Sim, G. A., Bernauer, K., Guggisberg, A., Hesse, M., Schmid, H. and Karrer, P. (1969). *Helv. Chim. Acta*, **52,** 689
40. Zacharias, D. E. (1970). *Acta Crystallogr.*, **B26,** 1455
41. Moncrief, J. W. and Lipscomb, W. M. (1966). *Acta Crystallogr.*, **21,** 322
42. Nordman, C. E. and Kumra, S. K. (1965). *J. Amer. Chem. Soc.*, **87,** 2059
43. Guilhem, J. (1972). *Acta Crystallogr.*, **A28,** 519, and personal communication
44. Lefebvre-Soubeyran, O. (1972). *Acta Crystallogr.*, **A28,** 512, and personal communication
45a. Hamor, T. A. and Robertson, J. M. (1962). *J. Chem. Soc.*, 194
45b. Beecham, A. F., Hurley, A. C., Lamberton, J. A. and Mathieson, A. McL. (1973). *Nature (London) Phys. Sci.*, **244,** 30
46. Grant, I. J., Hamor, T. A., Robertson, J. M. and Sim, G. A. (1965). *J. Chem. Soc.*, 5678
47. Fridrichsons, J., Mackay, M. F. and Mathieson, A. McL. (1967). *Tetrahedron Lett.*, 3521
48. Ling, N. C., Djerassi, C. and Simpson, P. G. (1970). *J. Amer. Chem. Soc.*, **92,** 222
49. Fennessey, J. P. and Nowacki, W. (1970). *Z. Kristallogr.*, **131,** 342
50. Ambody, G. and Kartha, G. (1972). *Amer. Cryst. Assn., Winter Meeting, Abstr.*, 75
51. Finch, N., Gemenden, C. W., Hsu, I. H. C., Kerr, A., Sim, G. A. and Taylor, W. I. (1965). *J. Amer. Chem. Soc.*, **87,** 2229
52. Karle, I. L. and Karle, J. (1968). *Acta Crystallogr.*, **B24,** 81
53. Irie, H., Ishizuka, K., Kawashima, S., Masaki, N., Osaki, K., Shingu, T. and Uyeo, S. (1972). *J. Chem. Soc. Chem. Commun.*, 871
54. Baker, R. W., Chothia, C., Pauling, P. and Weber, H. P. (1972). *Science*, **178,** 614
55. McPhail, A. T. and Sim, G. A. (1968). *J. Chem. Soc. B*, 923
56. Yoshino, A. and Iitaka, Y. (1966). *Acta Crystallogr.*, **21,** 57
57a. Brisse, F. (1971). *Acta Crystallogr.*, **B27,** 1047
57b. Wright, L. H., Newton, M. G., Pelletier, S. W. and Singh, H. (1970). *Chem. Commun.*, 359
58. Denne, W. A., Johns, S. R., Lamberton, J. A., Mathieson, A. McL. and Suares, H. (1972). *Tetrahedron Lett.*, 2727
59. Birnbaum, K. B. (1972). *Acta Crystallogr.*, **B28,** 1551
60. Birnbaum, G. I. (1970). *Acta Crystallogr.*, **B26,** 755
61. Przybylska, M. (1965). *Acta Crystallogr.*, **18,** 536
62. Mair, G. A. and Przbylska, M. Unpublished work
63. Przybylska, M. (1961). *Acta Crystallogr.*, **14,** 424
64. Przybylska, M. (1961). *Acta Crystallogr.*, **14,** 429
65. Przybylska, M. (1963). *Acta Crystallogr.*, **16,** 871
66. Wright, H. E., Newton, G. and Pelletier, S. W. (1969). *Chem. Commun.*, 507
67. Pelletier, S. W., Wright, L. H., Newton, M. G. and Wright, H. (1970). *Chem. Commun.* 98
68a. Pelletier, S. W., Page, S. W. and Newton, M. G. (1970). *Tetrahedron Lett.*, 4825
68b. Page, S. W., Pelletier, S. W. and Newton, M. G. (1971). *Amer. Crystallogr. Assoc., Summer Meeting, Abstr.*, 34
69a. Goto, G., Sasaki, S., Sakabe, N. and Hirata, Y. (1968). *Tetrahedron Lett.*, 1369
69b. Sasaki, K., Sakabe, N. and Hitata, Y. (1971). *J. Chem. Soc. B*, 354
70. Srimanouchi, H., Sasata, Y. and Takeda, T. (1971). *Bull. Chem. Soc. Jap.*, **44,** 3020
71. Pelletier, S. W., Iyer, K. N., Bhalla, V. K., Newton, M. G. and Aneja, R. (1970). *Chem. Commun.*, 393
72a. Birnbaum, K. B. (1969). *Tetrahedron Lett.*, 5245
72b. Birnbaum, K. B. (1971). *Acta Crystallogr.*, **B27,** 1169
73. Pelletier, S. W., Iyer, K. N., Bright, L. H., Newton, M. G. and Singh, N. (1971). *J. Amer. Chem. Soc.*, **93,** 5942

74. Pelletier, S. W., Katiadi, A. H., Wright, L. H., Page, S. W. and Newton, M. G. (1972). *J. Amer. Chem. Soc.*, **94,** 1754
75. Hall, S. R. and Ahmed, F. R. (1968). *Acta Crystallogr.*, **B24,** 346
76. Hall, S. R. and Ahmed, F. R. (1968). *Acta Crystallogr.*, **B24,** 337
77. Stermitz, F. R., Coomes, R. M. and Harris, D. R. (1968). *Tetrahedron Lett.*, 3915
78. Shimanouchi, H., Sasada, Y., Ihara, M. and Kametani, T. (1969). *Acta Crystallogr.*, **B25,** 1310
79. Kametani, T., Ihara, N., Kitahara, Y., Kabuto, C., Shimanouchi, H. and Sasada, Y. (1970). *Chem. Commun.*, 1241
80. Shimanouchi, H., Sasada, Y., Wakisaki, K., Kametani, Tl. and Ihara, M. (1970). *Acta Crystallogr.*, **B26,** 607
81. Brossi, A., Blount, J. F., O'Brien, J. and Teitel, S. (1971). *J. Amer. Chem. Soc.*, **93,** 6248
82. Fridrichsons, J. and Mathieson, A. McL. (1968). *Tetrahedron*, **24,** 5785
83. Ashida, T., Pepinsky, R. and Okaya, Y. (1963). *Acta Crystallogr.*, **16,** A48
84. Hall, L. H. and Brown, G. M. (1968). *Amer. Crystallogr. Assoc., Winter Meeting, Abstr.*, 14
85. Brown, G. M. (1968). *Amer. Crystallogr. Assoc., Summer Meeting, Abstr.*, 58
86. Oberhansli, W. E. (1971). *Helv. Chim. Acta*, **54,** 1389
87. Shamma, M., Moniot, J. L., Yao, S. Y. and Stanko, J. A. (1972). *J. Chem. Soc. Chem. Commun.*, 508
88. Kametani, T., Honda, T., Shimanouchi, H. and Sasada, Y. (1972). *J. Chem. Soc. Chem. Commun.*, 1072
89. Codding, P. W. and James, M. N. G. (1972). *J. Chem. Soc. Chem. Commun.*, 1174
90. Sobell, H. M., Sakore, T. D., Tavale, S. S., Canepa, F. G., Dowling, P. and Petcher, T. J. (1972). *Proc. Nat. Acad. Sci. USA*, **69,** 2212
91. Clardy, J., Hauser, F. M., Dahm, D., Jacobson, R. A. and Wildman, W. C. (1970). *J. Amer. Chem. Soc.*, **92,** 6337
92. Karle, J., Estlin, J. A. and Karle, I. L. (1967). *J. Amer. Chem. Soc.*, **89,** 6510
93. Clardy, J., Chan, J. A. and Wildman, W. C. (1972). *J. Org. Chem.*, **37,** 49
94. Williams, D. J. and Rogers, D. (1964). *Proc. Chem. Soc.*, 357
95. Kametani, T., Honda, T., Ihara, M., Shimanouchi, H. and Sasada, Y. (1972). *Tetrahedron Lett.*, 3729
96. Shiro, M., Sato, T. and Koyama, H. (1968). *J. Chem. Soc. B*, 1544
97. Clardy, J. C., Wildman, .W C. and Hauser, F. M. (1970). *J. Amer. Chem. Soc.*, **92,** 1781
98. Immirzi, A. and Guganti, C. (1971). *J. Chem. Soc. B*, 1218
99. Westin, L. (1972). *Acta Chem. Scand.*, **26,** 2305
100a. Huber, C. S. (1970). *Acta Crystallogr.*, **B26,** 373
100b. Huber, C. S. (1972). *Acta Crystallogr.*, **B28,** 982
101. Mackay, M. F. and Mathieson, A. McL. (1969). *Acta Crystallogr.*, **B25,** 1925
102. Gougoutas, J. Z. and Kaski, B. A. (1970). *Acta Crystallogr.*, **B26,** 853
103a. Kupchan, S. M., Suffness, M. I., White, D. N. J., McPhail, A. T. and Sim, G. A. (1968). *J. Org. Chem.*, **23,** 4529
103b. White, D. N. J., McPhail, A. T. and Sim, G. A. (1972). *J. Chem. Soc. Perkin Trans. II*, 1280
104. Kupchan, S. M., Suffness, M. I., McLure, R. J. and Sim, G. A. (1970). *J. Amer. Chem. Soc.*, **92,** 5756
105. Nishikawa, M., Kamiya, K., Tomita, M., Okamoto, Y., Kikuchi, T., Osaki, K., Tomiie, Y., Nitta, I. and Goto, K. (1968). *J. Chem. Soc. B*, 652
106. Kartha, G., Ahmed, F. R. and Barnes, W. H. (1962). *Acta Crystallogr.*, **15,** 326
107. Donahue, J., Sharma, B. D. and Marsh, R. E. (1964). *Acta Crystallogr.*, **17,** 249
108. Mackay, M. F. and Hodgkin, D. C. (1955). *J. Chem. Soc.*, 3261
109. Fridrichsons, J., Mackay, M. F. and Mathieson, A. McL. (1968). *Tetrahedron Lett.*, 2887
110. Van Koningsveld, H. (1970). *Rec. Trav. Chim. Pays-Bas*, **89,** 375
111. Cameron, A. F. and Hannaway, C. (1973). *J. Chem. Soc. Perkin Trans. II*, 1002
112. Clark, G. R. and Palenik, G. R. (1972). *J. Chem. Soc. Perkin Trans. II*, 1219
113. Karle, I. L., Gilardi, R. D., Fratini, A. V. and Karle, J. (1969). *Acta Crystallogr.*, **B25,** 1469
114. Fedeli, W., Giacomello, G., Cerrini, S. and Vaciago, A. (1970). *J. Chem. Soc. B*, 1190
115. van den Hende, J. H. and Nelson, N. R. (1967). *J. Amer. Chem. Soc.*, **89,** 2901

116. Przybylska, M. and Ahmed, F. R. (1958). *Acta Crystallogr.*, **11,** 718
117. Zacharias, D. E., Jeffrey, G. A., Douglas, B., Weisbach, J. A., Kirkpatrick, J. L., Ferris, J. P., Boyce, C. B. and Briner, R. C. (1965). *Experientia*, **21,** 247
118. Chu, S. C., Jeffrey, G. A., Douglas, B., Kirkpatrick, J. L. and Weisbach, J. A. (1966). *Chem. and Ind.*, 1795
119. Hamilton, J. A. and Steinrauf, L. K. (1971). *J. Amer. Chem. Soc.*, **93,** 2939
120. McClure, R. J. and Sim, G. A. (1972). *J. Chem. Soc. Perkin Trans. II*, 2073
121. Fridrichsons, J. and Mathieson, A. McL. (1955). *Acta Crystallogr.*, **8,** 761
122. Birnbaum, G. I. (1967). *Acta Crystallogr.*, **23,** 526
123. Oda, K. and Koyama, H. (1970). *J. Chem. Soc. B*, 1450
124. Nilsson, B. (1968). *Acta Crystallogr.*, **B24,** 252
125. Nowacki, W. and Bonsma, G. F. (1958). *Z. Kristallogr.*, **110,** 89
126. Hanson, A. W. (1963). *Acta Crystallogr.*, **15,** 939
127. Weiss, U. and Ziffer, H. (1963). *Experientia*, **19,** 108
128. Beecham, A. F., Mathieson, A. McL., Johns, S. R., Lamberton, J. A., Sioumis, A. A., Batterham, T. J. and Young, I. G. (1971). *Tetrahedron*, **27,** 3725
129. Barton, D. H. R., Jenkins, P. N., Letcher, R., Widdowson, D. A., Hough, E. and Rogers, D. (1970). *Chem. Commun.*, 391
130. Sato, T. and Koyama, H. (1971). *J. Chem. Soc. B*, 1070
131. Kowala, C. and Wunderlich, J. A. (1969). *Z. Kristallogr.*, **130,** 121
132. Abraham, D. J., Rosenstein, R. D. and McGandy, E. L. (1969). *Tetrahedron Lett.*, 4085
133. Nishio, K., Fujiwara, T., Tomita, K., Ishii, H., Inubushi, Y. and Harayama, T. (1969). *Tetrahedron Lett.*, 861
134. MacDonald, A. C. and Trotter, J. (1966). *J. Chem. Soc. B*, 929
135. Fridrichsons, J., Mathieson, A. McL. and Sutor, D. J. (1963). *Acta Crystallogr.*, **16,** 1075
136. Dodson, G. G. and Hall, D. (1966). *Acta Crystallogr.*, **20,** 42
137. Wunderlich, J. A. (1967). *Acta Crystallogr.*, **23,** 846
138. Birnbaum, K. B. (1972). *Acta Crystallogr.*, **B28,** 2825
139. Laing, M. and Sommerville, P. (1973). *Tetrahedron Lett.*, 5183
140. Wodak, S., Sussman, J. L. and Levinthal, C. (1971). *Amer. Crystallogr. Assoc., Summer Meeting, Abstr.*, 34
141. Wodak, S. (1972). *Amer. Crystallogr. Assoc., Winter Meeting, Abstr.*, 25
142. Fridrichsons, J. and Mathieson, A. McL. (1963). *Acta Crystallogr.*, **16,** 206
143. Gafner, G. and Admiraal, L. J. (1969). *Acta Crystallogr.*, **B25,** 2114
144. Gafner, G. and Admiraal, L. J. (1971). *Acta Crystallogr.*, **B27,** 565
145. Sakabe, N. and Hirata, Y. (1966). *Tetrahedron Lett.*, 965
146. Sakurai, H., Sakabe, N. and Hirata, Y. (1966). *Tetrahedron Lett.*, 6309
147. Nakano, T. and Nilsson, B. (1969). *Tetrahedron Lett.*, 2883
148. Gibbons, C. S. and Trotter, J. (1969). *J. Chem. Soc. B*, 840
149. Kamijo, N., Nakano, T., Terao, Y. and Osaki, K. (1966). *Tetrahedron Lett.*, 2889
150. Sasaki, K. and Hirata, Y. (1971). *J. Chem. Soc. B*, 1565
151. Sasaki, K. and Hirata, Y. (1972). *J. Chem. Soc. Perkin Trans. II*, 1411
152. Sasaki, K. and Hirata, Y. (1972). *Acta Crystallogr.*, **A28,** S19
153. Puckett, R. T., Sim, G. A. and Waite, M. G. (1971). *J. Chem. Soc. B*, 935
154. Hohne, E., Adam, G., Schreiber, K. and Tomko, J. (1968). *Tetrahedron*, **24,** 4875
155. Hohne, E., Seidel, I., Adam, G., Schreiber, K. and Tomko, J. (1972). *Tetrahedron*, **28,** 4019
156a. Karle, I. L. (1970). *Acta Crystallogr.*, **B26,** 1639
156b. Usubillaga, A., Seelkopf, C., Karle, I. L., Daly, J. W. and Witkop, B. (1970). *J. Amer. Chem. Soc.*, **92,** 700
157a. Kennard, O., Riva di Sanseverino, L. and Rollett, J. S. (1967). *J. Chem. Soc. C*, 956
157b. Hohne, E., Ripperger, H. and Schreiber, K. (1967). *Tetrahedron*, **23,** 3705
158. Hohne, E. (1972). *J. Prakt. Chem.*, **314,** 371
159. Keeke, G. N. Jr., Vincent, R. L. and Lipscomb, W. N. (1968). *J. Amer. Chem. Soc.* **90,** 1663
160. Ito, S., Miyashita, M., Fukazawa, Y., Mon, A., Iwai, I. and Yoshimura, M. (1972) *Tetrahedron. Lett.*, 2962

161. Bryan, R. F., Restivo, R. J. and Kupchan, S. M. (1973). *J. Chem. Soc. Perkin Trans. II*, 386
162a. Brisse, F. (1970). *Acta Crystallogr.*, **B26**, 171
162b. Ito, S., Fukazawa, Y. and Okuda, T. (1968). *Tetrahedron Lett.*, 5373
163. Eeles, W. T. (1960). *Tetrahedron Lett.*, 24
164. Weitz, G. and Wolfel, E. (1972). *Acta Crystallogr.*, **15**, 484
165. Daly, J. W., Tokuyama, T., Havermehl, G., Karle, I. L. and Witkop, B. (1968). *Ann. Chem.*, **729**, 198
166. Tomiie, Y., Furusaki, A., Kasami, K., Yasuoka, N., Miyake, K., Haisa, M. and Nitta, I. (1963). *Tetrahedron Lett.*, 2101
167a. Tsudo, K., Tamura, C., Tachikawa, R., Sakai, K., Amaksu, O., Kawamura, M. and Ikuma, S. (1963). *Chem. Pharm. Bull.*, **11**, 1473
167b. Tamura, C., Amakasu, O., Sasada, Y. and Tsuda, K. (1968). *Acta Crystallogr.*, **21**, 219
168a. Woodward, R. B. (1964). *Pure Appl. Chem.*, **9**, 49
168b. Woodward, R. B. and Gougoutas, J. Z. (1964). *J. Amer. Chem. Soc.*, **86**, 5030
169. Tamura, C., Amakasu, O., Sasada, Y. and Tsuda, K. (1966). *Acta Crystallogr.*, **21**, 226
170. Furusaki, A., Tomiie, Y. and Nitta, I. (1970). *Bull. Chem. Soc. Jap.*, **43**, 3332
171. Sharma, G. M. and Burkholder, P. R. (1971). *J. Chem. Soc. Chem. Commun.*, 151
172a. Phillips, D. C. (1954). *Acta Crystallogr.*, **7**, 159
172b. Bergin, R. (1971). *Acta Crystallogr.*, **B27**, 381
173. Hanson, A. W. and Ahmed, F. R. (1958). *Acta Crystallogr.*, **11**, 724
174. Bergin, R. and Carlström, D. (1967). *Acta Crystallogr.*, **23**, 313
175. Bergin, R. and Carlström, D. (1968). *Acta Crystallogr.*, **B24**, 1506
176. Tomiie, Y., Furusaki, A., Sakabe, N., Nishikawa, M., Kamiya, K., Matsuda, H. and Nitta, I. (1966). *Acta Crystallogr.*, **21**, A131
177a. Jeffreys, J. A. D. and Ferguson, G. (1970). *J. Chem. Soc. B*, 826
177b. Tamura, C. and Sim, G. A. (1970). *J. Chem. Soc. B*, 991
178. Yanai, H. S. and Lipscomb, W. N. (1959). *Tetrahedron*, **6**, 103
179. Fridrichsons, J. and Mathieson, A. McL. (1962). *Acta Crystallogr.*, **15**, 119
180. Lovell, F. M. (1964). *Proc. Chem. Soc.*, 58
181. Bryan, R. F. and Smith, R. M. (1971). *J. Chem. Soc. B*, 2159
182. Sasaki, Y. and Hirata, Y. (1972). *J. Chem. Soc. Perkin Trans. II*, 1268
183. Bryan, R. F., Gilmore, C. J. and Haltiwanger, R. C. (1973). *J. Chem. Soc. Perkin Trans. II*, 897
184. Raman, S., Reddy, J. M. and Lipscomb, W. N. (1963). *Acta Crystallogr.*, **16**, 364
185. Fregerslev, S. and Rasmussen, S. E. (1968). *Acta Chem. Scand.*, **22**, 2541
186. Oberhäusli, W. E. (1972). *Cryst. Struct. Commun.*, **1**, 203
187. Soderberg, E. and Kierkegaard, P. (1970). *Acta Chem. Scand.*, **24**, 397
188. Visser, J. W., Manassen, J. and De Vries, J. L. (1954). *Acta Crystallogr.*, **7**, 288
189. Schenk, H., Macgillavry, C. H., Skolnik, S. and Laan, J. (1967). *Acta Crystallogr.*, **23**, 423
190. Motherwell, W. D. S., Isaacs, N. W., Kennard, O., Bick, I. R. C., Bremner, J. B. and Gillard, J. (1971). *J. Chem. Soc. Chem. Commun.*, 133
191. Pauling, P. and Petcher, T. J. (1969). *Chem. Commun.*, 1001
192. Kussäther, E. and Haase, J. (1972). *Acta Crystallogr.*, **B28**, 2896
193. Gabe, E. J. and Barnes, W. H. (1963). *Acta Crystallogr.*, **16**, 796
194. Karle, I. L. (1973). *Amer. Crystallogr. Assoc., Winter Meeting, Abstr.*, 22
195. Przybylska, M. and Barnes, W. H. (1953). *Acta Crystallogr.*, **6**, 377
196. Pinkerton, J. M. H. and Steinrauf, L. K. (1967). *J. Org. Chem.*, **32**, 1828
197. Srivastava, S. N. and Przybylska, M. (1969). *Acta Crystallogr.*, **B25**, 1651
198. Kaluski, Z., Gusiev, A. I., Sruchkov, Y. T., Skolik, J., Baranowski, P. and Wiewiorowski, M. (1972). *Bull. Acad. Polon. Sci., Sci. Chim.*, **20**, 1
199. Wunderlich, J. A. (1969). *Acta Crystallogr.*, **B25**, 1436
200. Imado, S., Shiro, M. and Horii, Z. (1965). *Chem. Pharm. Bull. Jap.*, **13**, 643
201. Pascard-Billy, C. (1966). *Bull. Soc. Chim. Fr.*, 369
202. Koyama, H. and Oda, K. (1970). *J. Chem. Soc. B*, 1330
203. Harada, H., Irie, H., Masaki, N., Osaki, K. and Uyeo, S. (1967). *Chem. Commun.*, 460
204. Huber, C. P., Hall, S. R. and Maslen, E. N. (1968). *Tetrahedron Lett.*, 4081

205. Irie, H., Masaki, N., Ohno, K., Osaki, K., Taga, T. and Uyeo, S. (1970). *Chem. Commun.*, 1066
206. Ishizuka, K., Ohno, K., Taga, T. and Masaki, N. (1972). *Acta Crystallogr.*, **A28,** S20
207. Iizuka, H., Irie, H., Masaki, N., Osaki, K. and Uyeo, S. (1973). *J. Chem. Soc. Chem. Commun.*, 125
208. Ayer, W. A., Altenkirk, B., Masaki, N. and Valverde-Lopez, S. (1969). *Can. J. Chem.*, **47,** 2449
209. Ayer, W. A. and Masaki, N. (1971). *Can. J. Chem.*, **49,** 524
210. Hall, D. and Masaki, N. (1973). *Cryst. Struct. Commun.*, **2,** 271
211. Chin-You, N., MacLean, D. B., Prakash, A. and Calvo, C. (1971). *Can. J. Chem.*, **49,** 3240
212. Luhan, P. A. and McPhail, A. T. (1973). *J. Chem. Soc. Perkin Trans. II*, 51
213. Coggon, P., Farrier, D. S., Jeffs, P. W. and McPhail, A. T. (1970). *J. Chem. Soc. B*, 1267
214. Luhan, P. A. and McPhail, A. T. (1972). *J. Chem. Soc. Perkin Trans. II*, 2006
215. McPhail, A. T., Sim, G. A., Frey, A. J. and Ott, H. (1966). *J. Chem. Soc. B*, 377
216. Cook, R. E. and Glick, M. D. (1970). *Acta Crystallogr.*, **B26,** 2102
217. Ferguson, G., Jeffreys, J. A. D. and Sim, G. A. (1966). *J. Chem. Soc. B*, 454
218. Jeffreys, J. A. D. (1970). *J. Chem. Soc. C*, 1091
219. Karle, I. L. and Karle, J. (1964). *Acta Crystallogr.*, **17,** 1356
220. Karle, I. L. and Karle, J. (1966). *Acta Crystallogr.*, **21,** 860
221. Hart, N. K., Johns, S. R., Lamberton, J. A., Mackay, M. F., Mathieson, A. McL. and Satzke, L. (1972). *Tetrahedron Lett.*, 5333
222. Koo, C. H. and Kim, H. S. (1965). *Daehan Hwahak Hwoejee*, **9,** 134
223. Rakman, A.-U., Serantoni, E. F. and Riva di Sanseverino, L. (1971). *Pak. J. Sci. Ind. Res.*, **14,** 480
224. Cesario, M. and Guilhem, J. (1972). *Acta Crystallogr.*, **B28,** 151
225. Sherfinski, J. S., Marsh, R. E., Sweet, R. M. and Dahl, L. F. (1973). *Amer. Crystallogr. Assoc., Winter Meeting, Abstr.*, 26
226. Coetzer, J. and Steyn, P. S. (1973). *Acta Crystallogr.*, **B29,** 685
227. Karlsson, R. and Losman, D. (1972). *J. Chem. Soc. Chem. Commun.*, 626
228a. Denne, W. A., Johns, S. R., Lamberton, J. A. and Mathieson, A. McL. (1971). *Tetrahedron Lett.*, 3107
228b. Denne, W. A. and Mathieson, A. McL. (1973). *J. Cryst. Mol. Struct.*, **3,** 79
229a. Denne, W. A., Johns, S. R., Lamberton, J. A., Mathieson, A. McL. and Suares, H. (1972). *Tetrahedron Lett.*, 1767
229b. Denne, W. A. and Mathieson, A. McL. (1973). *J. Cryst. Mol. Struct.*, **3,** 87
230. Denne, W. A. and Mathieson, A. McL. (1973). *J. Cryst. Mol. Struct.*, **3,** 139
231. Denne, W. A. (1973). *J. Cryst. Mol. Struct.*, **3,** 367
232. Pilotti, A. M. (1971). *Acta Crystallogr.*, **B27,** 877
233. Immirzi, A. (1973). *Cryst. Struct. Commun.*, **2,** 359
234. Baggio, R. F. and Baggio, S. (1973). *Cryst. Struct. Commun.*, **2,** 251
235. Margulis, T. N. (1973). *Amer. Crystallogr. Assoc., Winter Meeting, Abstr.*, 95
236. Jones, G., Ferguson, G. and Marsh, W. C. (1971). *Chem. Commun.*, 994

# Index